DIGITAL FILTERS AND SIGNAL PROCESSING

DIGITAL FILTERS AND SIGNAL PROCESSING

LELAND B. JACKSON
University of Rhode Island

KLUWER ACADEMIC PUBLISHERS
A MEMBER OF THE KLUWER ACADEMIC PUBLISHERS GROUP
Boston Dordrecht Lancaster

DISTRIBUTORS:

For the United States and Canada: Kluwer Academic Publishers, 101 Philip Drive, Assinippi Park, Norwell, MA, 02061, USA

For the UK and Ireland: Kluwer Academic Publishers, MTP Press Limited, Falcon House, Queen Square, Lancaster LA1 1RN, UK

For all other countries: Kluwer Academic Publishers Group, Distribution Centre, P.O. Box 322, 3300 AH Dordrecht, The Netherlands

Jonathan Allen, Consulting Editor

Library of Congress Cataloging in Publication Data

Jackson, Leland B.
Digital filters and signal processing.

Bibliography: p.
Includes index.
1. Electric filters, Digital. 2. Signal processing—Digital techniques. I. Title.
TK7872.F5J33 1986. 621.38'043 85-9812
ISBN 0-89838-174-6

To my wife Diana
and daughter Anita

CONTENTS

PREFACE

This book provides a broad introduction to the field of digital signal processing and contains sufficient material for a two-semester sequence in this multifaceted subject. It is also written with the practicing engineer or scientist in mind, having many observations and examples of practical significance drawn from the author's industrial experience.

The first semester, at the senior undergraduate or first-year graduate level, could cover chapters 1 through 7 with topics perhaps from chapters 8 and 9, depending upon the background of the students. The only requisite background is linear systems theory for continuous-time systems, including Fourier and Laplace transforms. Many students will also have had some previous exposure to discrete-time systems; in which case, chapters 2 through 4 may serve to review and expand that preparation. Note, in particular, that knowledge of probability theory and random processes is not required until chapters 10 and 11, except for section 7.5 on the periodogram. A second, advanced course could utilize material from chapters 8 through 13. A comprehensive one-semester course for suitably prepared graduate students might cover chapters 4 through 9 and additional topics from chapters 10 through 13.

Notable features of the book include the following:

1. Numerous useful filter examples early in the text, in chapters 4 and 5.
2. State-space representation and structures in chapters 4 and 11.
3. Practical considerations in A/D and D/A conversion, as well as the fundamentals of sampling, in chapter 6.

4. Concise, but insightful coverage of the DFT, FFT, and periodogram in chapter 7, including zero padding and windows.
5. Filter design and modern spectrum analysis by AR, MA, and ARMA modeling in chapter 10, synthesis of lattice structures.
6. Design of digital filter structures for low coefficient sensitivity and roundoff noise, with an analysis of limit cycles and overflow oscillations, in chapter 11.
7. Hardware implementation of digital filters in chapters 12 with VLSI considerations. Also block FIR and IIR structures.
8. Significant filter and system examples, including interpolation and decimation, in chapter 13.

Chapter 10 is somewhat different from the rest of the book in that it addresses an area of particularly wide scope, and hence the pace is accelerated. However, a unified approach is employed based upon the least-squares solution of overdetermined sets of simultaneous equations via pseudoinverse matrices. If the students have not previously studied this technique in a linear algebra course, the instructor can quickly provide a sufficient introduction. For an overview of modeling, it may be appropriate to cover only section 10.1 on AR filter design and perhaps section 10.4 on lattice structures. Likewise, section 10.5 on spectrum analysis by modeling can be covered without the material on MA and ARMA models in sections 10.2 and 10.3.

Problems are included at the end of each chapter. Some allow the student to complete straightforward derivations that were omitted in the text; while the majority provide important practice with the concepts and techniques presented. An answer book is available to instructors through the publisher.

I am indebted to my colleagues over 20 years who have instructed me, stimulated my research, and encouraged my teaching in digital signal processing. These colleagues include Drs. James Kaiser, Henry McDonald, and Irwin Sandberg at Bell Laboratories, Dr. John Bertrand, Rodger Hosking, and S. N. Thanos at Rockland Systems Corporation, and Drs. Steven Kay, Ramdas Kumaresan, Allen Lindgren, Louis Scharf, John Spence, and Donald Tufts at the University of Rhode Island. I also appreciate the support and encouragement of Carl Harris, publisher, of Kluwer Academic Publishers. I am grateful to Howard Lieberman and Eric Rosenfeld of the Interval Corporation for their support of the INTERVAL signal processing language which was so helpful in preparing the examples and graphics. Donald Borsay and David Dugal of the URI Engineering Computer Laboratory have been very helpful in the preparation of the text. The technical drawings were ably done by Diana Norton Jackson. Finally, I wish to thank the students of ELE606 over the years for their enthusiasm and feedback.

Leland B. Jackson

DIGITAL FILTERS AND SIGNAL PROCESSING

1. INTRODUCTION: TERMINOLOGY AND MOTIVATION

This book is concerned with the processing of *discrete-time signals* or data sequences. Such signals arise in two distinct ways: They may be inherently discrete in time, or they may be sampled versions of signals that are continuous in time. Examples of data sequences that are inherently discrete in time abound in our daily lives; for example, our daily or monthly checking account balances, the daily high/low temperatures and other weather data, monthly or quarterly sales figures and inventory levels, the annual GNP and other economic data, and so forth. Meteorologists, economists, and business people often process such data to determine cyclic patterns, averages, or norms, and long-term trends. In so doing, they usually employ filters to smooth out noisy data or to enhance certain patterns of interest, whether or not they call them by that name.

The second source of discrete-time signals—the sampling of continuous-time or *analog* signals—is more common in engineering and the physical sciences. Electrical engineers sample speech and other communication signals for high-quality and/or secure communications, and attempt to filter out interference and noise. They also analyze the frequency spectra of sonar, radar, biomedical, and instrumentation signals with filter algorithms in computers, following the conversion of these signals to discrete-time form. After all, computers can only process sequences of numbers or symbols. Likewise, mechanical engineers, seismologists, astronomers, acousticians, etc., analyze computer-generated spectra of their respective signals—all in discrete-time form.

If the book is about discrete-time filters and signal processing, why is its title "*digital* filters and signal processing"? Because, to be processed by computers, the signals must be discrete not only in time, but in amplitude as well. That is, the amplitude of each sample is *quantized* to a given accuracy and represented by a corresponding number of binary digits or *bits*. The resulting digital sequence is then input to the computer or other digital circuitry for processing and analysis. Interestingly, it is not until chapters 11 and 12 that we actually deal with the digital nature of the signals and systems, i.e., with their quantization. This is because, fortunately, the nonlinear effects produced by quantization are typically small and can be modeled as either small linear perturbations or additive noise. Hence, the beauty and power of linear systems theory can be brought to bear fully on the analysis and design of these systems.

Digital signal processing (DSP) is a technology-driven field which dates its growth as a separate discipline from the mid-1960s when computers and other digital circuitry became fast enough to process large amounts of data efficiently. The implementation of DSP systems falls into two general categories—hardware and software. On the one hand, many systems can be implemented completely by software in general-purpose computers, especially for relatively low sampling rates or nonreal-time operation. Widespread application of software signal processing has resulted from the ever-increasing size and speed of computers, the introduction of fast array-processor peripherals, and the explosive growth of the ubiquitous microprocessor. On the other hand, many applications with high sampling rates, high production volumes, and/or low cost and space requirements can only be satisfied by special-purpose digital hardware. Implementation of these systems had been less widespread until the advent of custom VLSI (very-large-scale integration) circuits made them attractive alternatives to existing technologies. Indeed, digital filters and other DSP systems are well suited to VLSI implementation since they have highly parallel algorithmic structure, local connectivity, and circuit modularity. Hence, they can be realized by *systolic* and *completely pipelined* VLSI circuits, as discussed in chapter 12.

Two broad areas of interest can be identified within the field of digital signal processing —digital filters and spectrum analysis—with many points of overlap between them. As our title suggests, there is somewhat more emphasis on digital filters in this book although a substantial amount of material on spectrum analysis is also included. In particular, the discrete Fourier transform (DFT) and fast Fourier transform (FFT) algorithms are presented in chapter 7; while spectrum analysis by signal modeling is covered in chapter 10.

2. DISCRETE-TIME SIGNALS AND SYSTEMS

2.0 INTRODUCTION

The basic concepts and relationships of the theory of discrete-time signals and systems are analogous to those for continuous-time signals and systems. In some respects, however, they are more simply derived and perhaps easier to visualize in the discrete-time case. In this chapter, we will introduce these basic concepts and relationships, developing them further in subsequent chapters.

2.1 DISCRETE-TIME SIGNALS, OR SEQUENCES

A discrete-time signal x consists of a sequence of numbers denoted alternatively by x_n, $x(n)$, or $x(nT)$, with n being an integer index. The latter notation implies that the sequence is derived from or related to a continuous-time signal $x(t)$ by, for example, sampling $x(t)$ at the time instants $t = nT$. However, since the independent variable can be other than time, or the sequence may be produced directly with no underlying continuous-time signal, we will generally use the notation $x(n)$ unless it is necessary to relate the sequence directly to a continuous-time signal.

Strictly speaking, $x(n)$ is a single number from the sequence for a given index value n; while $\{x(n)\}$ denotes the entire finite, or countably infinite set comprising the sequence. However, we will follow the general practice of using $x(n)$ to represent the entire sequence as well as a number from the sequence, depending on whether n is assumed to be arbitrary or fixed. The set notation will be used only when necessary to avoid confusion, as in the following definitions of the sum of two sequences and the product of a constant and a sequence:

$$\{x(n)\} + \{y(n)\} = \{x(n) + y(n)\} \tag{2.1.1}$$

$$a\{x(n)\} = \{ax(n)\}. \tag{2.1.2}$$

Several sequences of particular importance have been given special names and notation. The unit-sample or *impulse* sequence is defined for all n by

$$\delta(n) = \begin{cases} 1, & n = 0 \\ 0, & n \neq 0. \end{cases} \tag{2.1.3}$$

As implied by its name and notation, the impulse sequence $\delta(n)$ is similar to the impulse or Dirac delta function utilized in the theory of continuous-time systems; and, as we shall see, it plays the same central role in discrete-time systems theory. Happily, however, as seen from equation 2.1.3, there is no analogous mathematical problem with the definition of $\delta(n)$.

The *unit-step* sequence $u(n)$ is defined by

$$u(n) = \begin{cases} 1, & n \geq 0 \\ 0, & n < 0. \end{cases} \tag{2.1.4}$$

In addition to its direct usage, the unit step is often employed to describe other sequences such as the *exponential* sequence

$$a^n u(n) = \begin{cases} a^n, & n \geq 0 \\ 0, & n < 0. \end{cases} \tag{2.1.5}$$

The impulse, unit-step, and exponential sequences are illustrated in figure 2.1.

An important operation on a sequence $x(n)$ is its delay by n_d to produce another sequence $y(n)$, i.e.,

$$y(n) = x(n - n_d).$$

The term *delay* reflects our assumption that the index n corresponds to discrete-time values, and the term *shift* may be more appropriate in other contexts. If the value of n_d is actually negative, we may use the term *advance* to describe this operation, and with $n_a = -n_d$, we will then write

$$y(n) = x(n + n_a).$$

A sequence $x(n)$ is said to be *periodic* if, and only if,

$$x(n) = x(n \pm n_p) \tag{2.1.6}$$

for some integer n_p and all n. The smallest nonzero value of n_p for which (2.1.6) holds is the *period* of $x(n)$. Note that the sampling of a periodic continuous-signal to produce $x(n)$

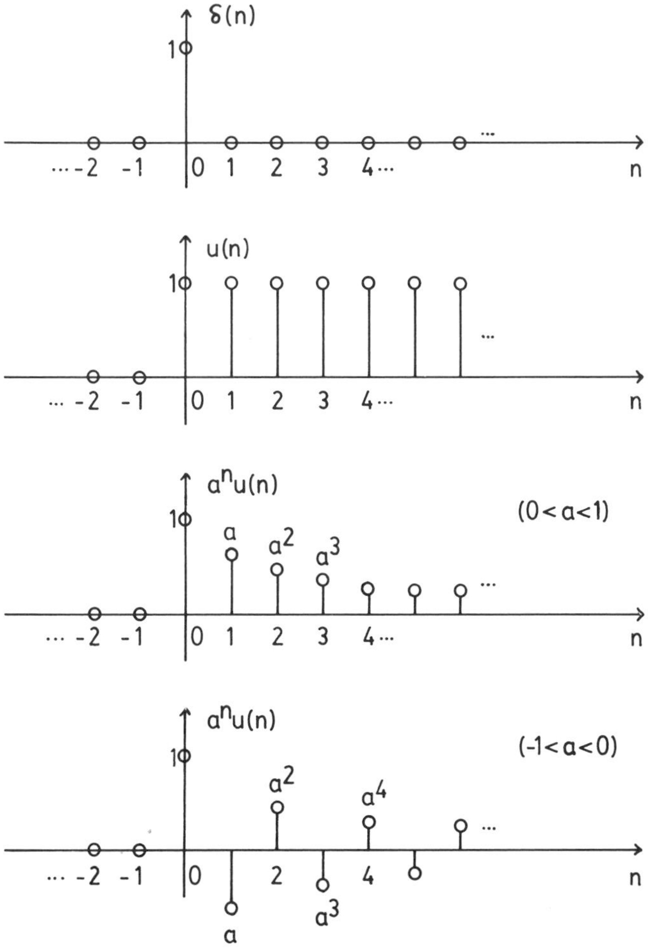

Figure 2.1. The impulse, unit-step, and exponential sequences.

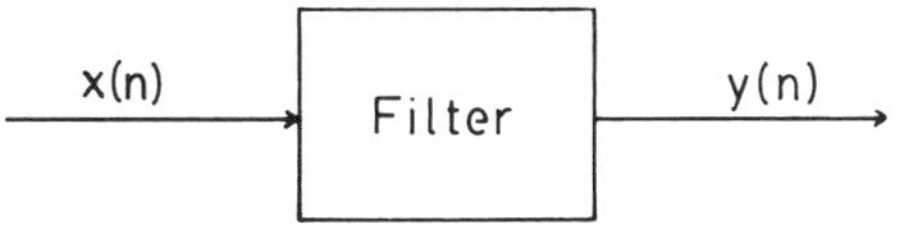

Figure 2.2. A discrete-time filter with input $x(n)$ and output $y(n)$.

does not ensure that $x(n)$ is a periodic sequence unless $n_p T$ is, in fact, an integral multiple of the period of the continuous-time signal, where T is the sampling interval.

2.2. DISCRETE-TIME SYSTEMS AND FILTERS

If a sequence $x(n)$ is operated upon to produce another sequence $y(n)$, we may think of these sequences as the input and output, respectively, of a discrete-time system or *filter*, as depicted in figure 2.2. If $x(n)$ and $y(n)$ can assume only a finite number of possible

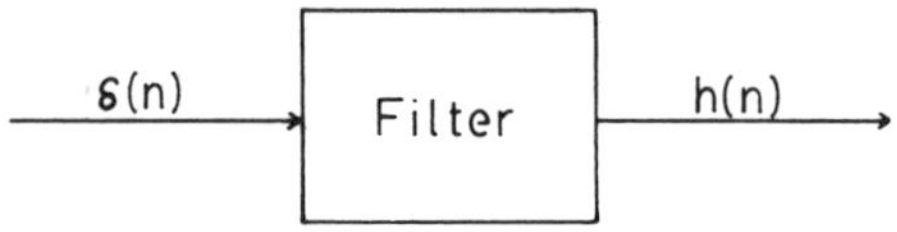

Figure 2.3. A discrete-time filter with impulse input $\delta(n)$ and response $h(n)$.

amplitude values, we will call this instead a *digital filter*, as discussed previously in chapter 1. For now, however, we assume that $x(n)$ and $y(n)$ can take on any (real) values and hence that our filter is simply discrete-time.

A discrete-time filter is said to be *linear* if, for any two input sequences $x_1(n)$ and $x_2(n)$ that produce, respectively, the output sequences $y_1(n)$ and $y_2(n)$, the input sequence

$$x(n) = ax_1(n) + bx_2(n)$$

produces the output sequence

$$y(n) = ay_1(n) + by_2(n)$$

for all a and b. A *time-invariant* (or *shift-invariant*) filter, on the other hand, implies that if $x(n)$ produces $y(n)$, then $x(n - n_d)$ produces $y(n - n_d)$ for all n and any value of n_d.

If the input is the impulse sequence $\delta(n)$, the resulting output is called the *impulse response* (or *unit-sample response*) of the filter and is denoted by $h(n)$. (See figure 2.3.) The input and output of a linear time-invariant discrete-time filter may be easily related via the impulse response of the filter as follows: Any input $x(n)$ can be thought of as the sum of an infinite number of delayed and weighted impulse sequences, with the kth impulse $\delta(n - k)$ weighted by $x(k)$, as illustrated in figure 2.4. Mathematically, we may thus write

$$x(n) = \sum_{k=-\infty}^{\infty} x(k)\delta(n - k). \tag{2.2.1}$$

But by time-invariance, the input $\delta(n - k)$ will produce the output $h(n - k)$; and by linearity, the output corresponding to the weighted sum in (2.2.1) is thus

$$y(n) = \sum_{k=-\infty}^{\infty} x(k)h(n - k). \tag{2.2.2}$$

This is the *convolution sum* relating the input and output of a discrete-time filter. It is left for the reader to show (by change of variables) that the convolution sum may also be written as

$$y(n) = \sum_{k=-\infty}^{\infty} x(n - k)h(k). \tag{2.2.3}$$

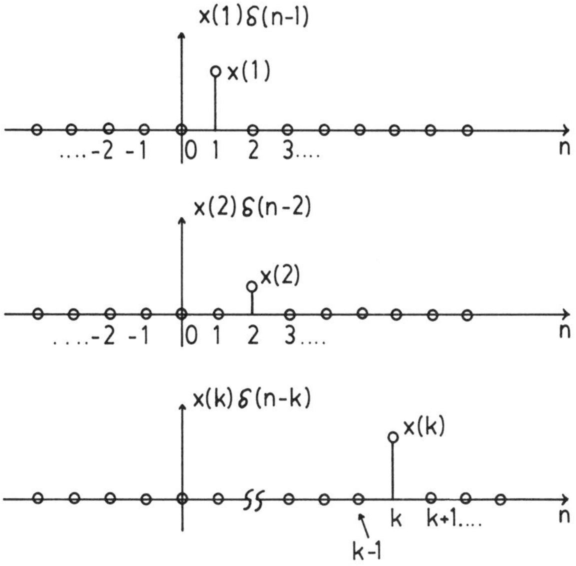

Figure 2.4. Individual impulse components comprising the sequence $x(n)$.

We will use the symbol $*$ to denote convolution, and hence

$$y(n) = x(n) * h(n). \tag{2.2.4}$$

implies either (2.2.2) or (2.2.3).

EXAMPLE. As an example of convolution, we will find the output of a linear time-invariant filter with an exponential impulse response to a unit-step input, i.e.,

$$x(n) = u(n),$$

$$h(n) = a^n u(n),$$

$$y(n) = u(n) * a^n u(n).$$

Using the form of the convolution sum in (2.2.2), we must sum over k the elements of the sequence $x(k)h(n - k)$ for a given value of n to produce $y(n)$, as illustrated in figure 2.5. For n negative, the non-zero elements of the sequences $x(k)$ and $h(n - k)$ do not overlap, and thus $y(n) = 0$ for all $n < 0$. For $n = 0$, only the elements at $k = 0$ overlap, producing $y(0) = 1$. For $n = 1$, the elements at $k = 0$ and 1 overlap to produce $y(1) = 1 + a$. Continuing in this manner, we see that

$$y(n) = \begin{cases} 0, & n < 0 \\ 1 + a + a^2 + \cdots + a^n, & n \geq 0 \end{cases}$$

or

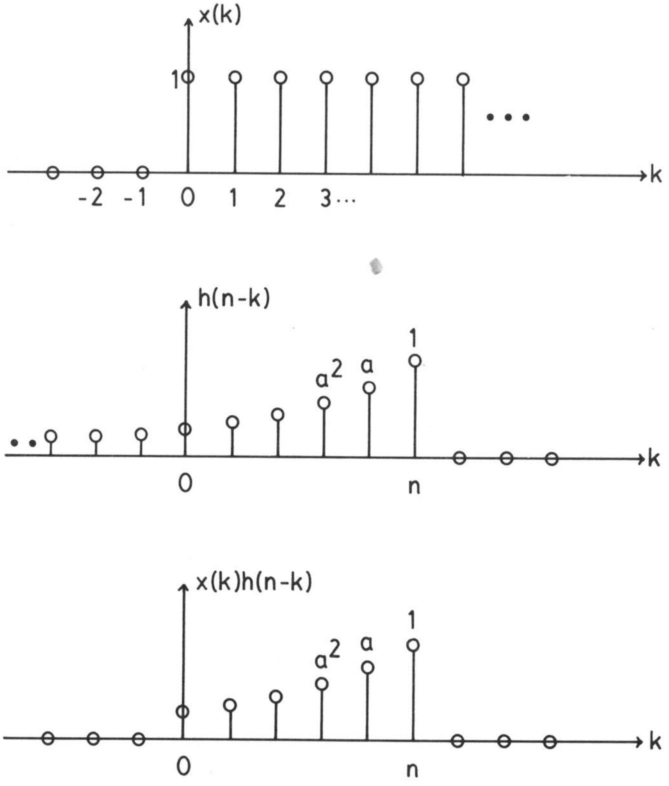

Figure 2.5. Visualizing the sequence $x(k)h(n - k)$ in the convolution sum.

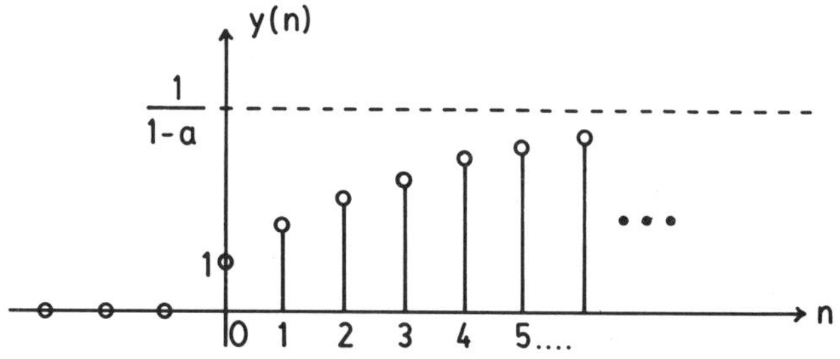

Figure 2.6. The output $y(n) = u(n) * a^n u(n)$ for $0 < a < 1$.

$$y(n) = \left(\frac{1 - a^{n+1}}{1 - a}\right) u(n).$$

The response $y(n)$ is shown in figure 2.6 for $0 < a < 1$.

Two filters are said to be in *cascade* if the output from the first is the input to the second, as depicted in figure 2.7. For linear time-invariant filters, we must have

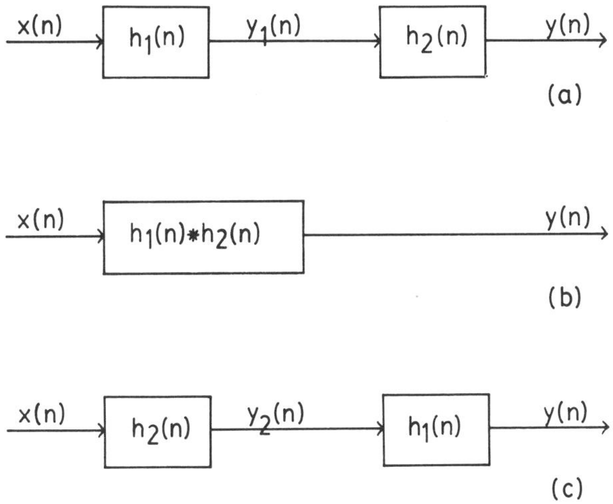

Figure 2.7. Equivalent representations of the cascade interconnection of $h_1(n)$ and $h_2(n)$.

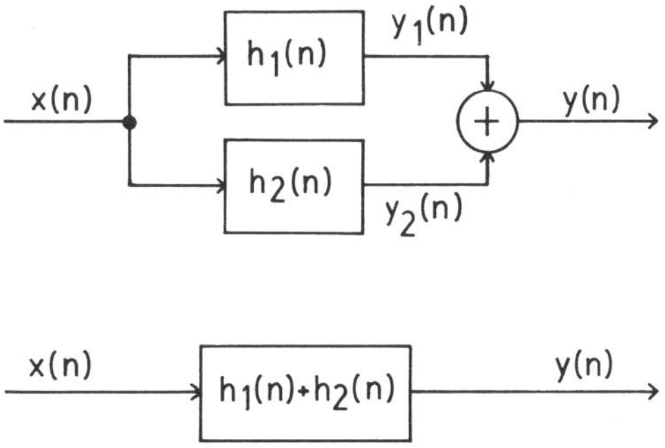

Figure 2.8. Equivalent representations of the parallel interconnection of $h_1(n)$ and $h_2(n)$.

$$
\begin{aligned}
y_1(n) &= x(n) * h_1(n), \\
y(n) &= y_1(n) * h_2(n) \\
&= [x(n) * h_1(n)] * h_2(n) \\
&= x(n) * [h_1(n) * h_2(n)].
\end{aligned}
\tag{2.2.5}
$$

The overall impulse response of the cascaded filters is thus

$$h(n) = h_1(n) * h_2(n).$$

This is illustrated in figure 2.7(b); and reversing the order of the filters as in figure 2.7(c) does not, therefore, change the output $y(n)$.

Two filters are said to be in *parallel* if their inputs are the same and their outputs are summed, as shown in figure 2.8. Hence,

$$\begin{aligned} y(n) &= y_1(n) + y_2(n) \\ &= [x(n) * h_1(n)] + [x(n) * h_2(n)] \\ &= x(n) * [h_1(n) + h_2(n)], \end{aligned} \tag{2.2.6}$$

and the overall impulse response of the paralleled filters is the same as a single filter with impulse response

$$h(n) = h_1(n) + h_2(n).$$

2.3 STABILITY AND CAUSALITY

A discrete-time filter is *stable* if a bounded input sequence produces a bounded output sequence, i.e., if

$$|x(n)| \leq M_1$$

implies that

$$|y(n)| \leq M_2$$

for some finite M_1 and M_2. For a linear time-invariant filter, we will show that stability holds if, and only if,

$$\sum_{k=-\infty}^{\infty} |h(k)| < \infty. \tag{2.3.1}$$

To prove that (2.3.1) is sufficient, we use the convolution sum in (2.2.3) as follows:

$$\begin{aligned} |y(n)| &= \left| \sum_{k=-\infty}^{\infty} x(n-k)h(k) \right| \\ &\leq \sum_{k=-\infty}^{\infty} |x(n-k)| \cdot |h(k)| \\ &\leq M_1 \sum_{k=-\infty}^{\infty} |h(k)| = M_2 < \infty \end{aligned}$$

since by assumption the sum of the absolute values of $h(n)$ is finite. To prove the necessity of (2.3.1), we assume that $x(n) = \pm M_1$ for all n with a sign sequence such that for $n = n_0$

$$\operatorname{sgn}[x(n_0 - k)] = \operatorname{sgn}[h(k)], \qquad \text{for all } k,$$

and thus the output for real-valued $h(k)$ is given by

$$y(n_0) = \sum_{k=-\infty}^{\infty} M_1 |h(k)|$$

$$= M_1 \sum_{k=-\infty}^{\infty} |h(k)|.$$

Hence, if this summation is not finite, then neither is $y(n_0)$, and the output is not bounded.

EXAMPLE. As an example of stability, consider the exponential impulse response

$$h(n) = a^n u(n).$$

The corresponding filter is stable if, and only if,

$$\sum_{k=-\infty}^{\infty} |h(k)| = \sum_{k=0}^{\infty} |a|^k < \infty.$$

But the indicated summation converges only for $|a| < 1$, in which case it equals $(1 - |a|)^{-1}$, and thus the filter is stable only for $-1 < a < 1$.

The final important concept to be introduced in this chapter is that of *causality*. A discrete-time filter is *causal* if for any two input sequences $x_1(n)$ and $x_2(n)$ that are equal for all $n \leq n_0$, the corresponding output sequences $y_1(n)$ and $y_2(n)$ are also equal for $n \leq n_0$. For a linear, time-invariant filter, this is equivalent (problem 2.1) to the condition that

$$h(n) = 0, \qquad n < 0. \tag{2.3.2}$$

In words, causality simply means that the filter does not respond to an input before that input is, in fact, applied. Although any physically implemented filter is always causal, it will be convenient at times to assume noncausal versions of the filter for analysis purposes; and if the filter is implemented in computer software on a data array in memory, for example, it may indeed be noncausal. Any sequence $x(n)$ will be said to be causal if $x(n) = 0$, $n < 0$, or in other words, if the sequence *could* be the impulse response of a causal filter.

PROBLEMS

2.1 Show that the definition of causality implies that $h(n) = 0$, $n < 0$, for a causal linear time-invariant system.

2.2 Find and sketch the output $y(n)$ for a filter with the pulse input $x(n) = u(n) - u(n - 6)$ and the impulse response $h(n) = 0.8^n u(n)$. Repeat for $h(n) = (-0.8)^n u(n)$.

2.3 Let $h(n) = a^n u(n)$ and $x(n) = b^n u(n)$. Find and sketch the output $y(n) = x(n) * h(n)$ for $a \neq b$. Repeat for $a = b$.

2.4 Show that for $h(n) = a^n u(n)$, $1/2 < a < 1$, the time constant n_T of the filter (i.e., the number of samples such that $h(n_T) \approx 1/e$) is approximately given by $1/(1 - a)$.

2.5 Let $h_1(n) = u(n) - u(n - 6)$ and $h_2(n) = u(n + 4) - u(n)$.
   a. Find and sketch $h(n)$ for the parallel combination of $h_1(n)$ and $h_2(n)$.
   b. Find and sketch $h(n)$ for the cascade combination of $h_1(n)$ and $h_2(n)$.

2.6 Let $h_1(n) = (ja)^n u(n)$ and $h_2(n) = (-ja)^n u(n)$.
   a. Find and sketch $h(n)$ for the parallel combination of $h_1(n)$ and $h_2(n)$.
   b. Find and sketch $h(n)$ for the cascade combination of $h_1(n)$ and $h_2(n)$.

2.7 The step response $s(n)$ of a system is defined to be the output for a unit-step input, i.e.,

$$s(n) = u(n) * h(n).$$

   a. Express $s(n)$ for a causal system as a summation of $h(n)$.
   b. Give the inverse relationship for $h(n)$ in terms of $s(n)$.
   c. Find and sketch $s(n)$ when $h(n) = u(n) - u(n - 6)$.
   d. Find and sketch $s(n)$ when $h(n) = (-\frac{1}{2})^n u(n)$. What is the asymptotic value of $s(n)$ as $n \to \infty$?

2.8 The autocorrelation function $r(m)$ of a real-valued deterministic signal $x(n)$ is defined by

$$r(m) = \sum_{n=-\infty}^{\infty} x(n)x(n + m).$$

Express $r(m)$ as a convolution, i.e.,

$$r(m) = x(m) * v(m)$$

where $v(m)$ is some version of $x(m)$.

2.9 Determine the conditions on the parameters of the following systems for stability:
   a. $h(n) = a^n u(-n)$.
   b. $h(n) = a^n [u(n) - u(n - 100)]$.
   c. $h(n) = r^n \sin(n\omega_0 T) u(n)$.
   d. $h(n) = a^{|n|}$.
   e. $h(n) = K(-1)^n u(n)$.

2.10 Can the cascade of a causal system and a noncausal system be causal? If so, give an example.

3. THE z TRANSFORM

3.0 INTRODUCTION

The z transform is to discrete-time systems what the Laplace transform is to continuous-time systems. For instance, the relationship between the input and output of a discrete-time system involves multiplication of the appropriate z transforms, rather than convolution as for the signals themselves. Poles and zeros can be defined from the z transform and have the same useful role and intuitive appeal as for continuous-time systems. And finally, the frequency response of the system is readily derived from the z transform and can be related to an appropriately defined Fourier transform.

3.1 DEFINITION OF THE z TRANSFORM

The z transform of the sequence $x(n)$ is defined by

$$X(z) = \sum_{n=-\infty}^{\infty} x(n) z^{-n}, \tag{3.1.1}$$

where z is a complex variable. This is sometimes referred to as the *two-sided z transform*, with the *one-sided z transform* being the same except for a summation from $n = 0$ to ∞. The primary use of the one-sided transform, however, is for causal sequences, in which case the two transforms are the same anyway. We will not, therefore, make this distinction and will refer to (3.1.1) as simply the z transform of $x(n)$.

Note that the function $X(z)$ is, in fact, a Laurent series in the complex variable z, and thus all of the properties and theorems for such series from the theory of complex variables

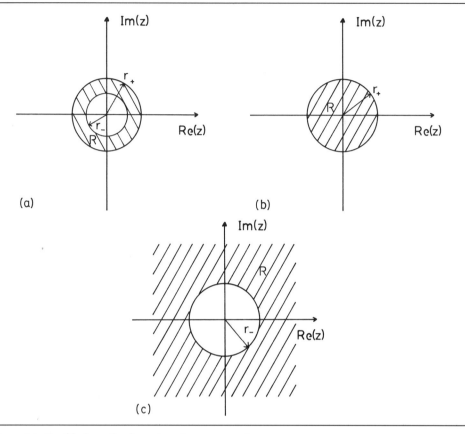

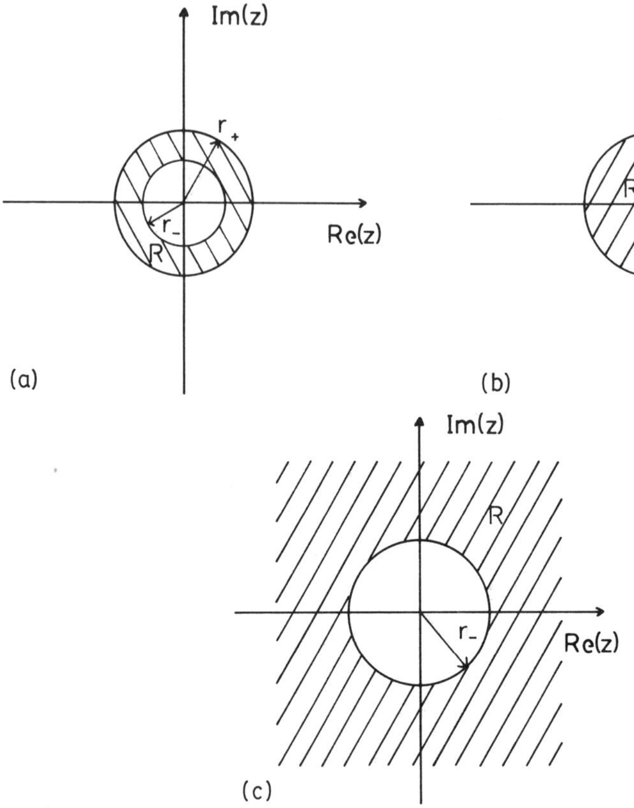

Figure 3.1. Possible forms for the region of convergence R.

apply to the z transform. In particular, the region of convergence R for $X(z)$ is an annular ring in the z plane of the form

$$r_- < |z| < r_+; \tag{3.1.2}$$

and *unless otherwise obvious, r_- and r_+ must be included in the specification of $X(z)$ in order for the z transform to be complete.* The three possible forms for R are depicted in figure 3.1.

Note that in figure 3.1(b), r_- equals zero; while in figure 3.1(c), r_+ equals infinity. In these cases, R may or may not actually contain $z = 0$ or $z = \infty$, respectively. That is, we can have in case (b) either

$$0 < |z| < r_+ \quad \text{or} \quad 0 \leq |z| < r_+,$$

while in case (c) we can have either

$$r_- < |z| < \infty \quad \text{or} \quad r_- < |z| \leq \infty.$$

All three cases become the same if $r_- = 0$ and $r_+ = \infty$, and $X(z)$ then converges everywhere except, possibly, at $z = 0$ and/or $z = \infty$.

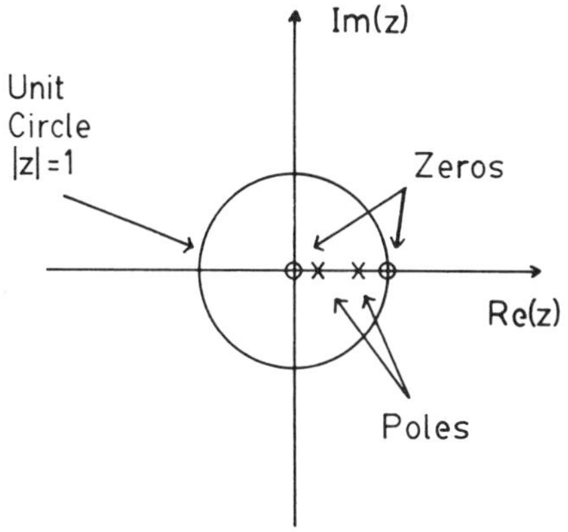

Figure 3.2. A sample pole/zero diagram.

An important class of z transforms with which we will be concerned almost exclusively is that where $X(z)$ is a rational function, i.e., a ratio of polynomials in z. The roots of the numerator polynomial are called the *zeros* of $X(z)$ since for these values of z, $X(z)$ equals zero. The roots of the denominator polynomial are called the *poles* of $X(z)$, and since $X(z)$ is infinite at these values of z, the poles lie outside the region of convergence by definition. In fact, R is bounded by poles or by infinity. The zeros may, of course, fall anywhere in the z plane. A typical pole/zero diagram is shown in figure 3.2. The *unit circle* $|z| = 1$ has special significance, as will later show.

The above definition and discussion are illustrated in the following examples:

Impulse: For $x(n) = \delta(n)$, we have simply

$$X(z) = 1, \qquad 0 \leq |z| \leq \infty$$

and thus $X(z)$ converges everywhere.

Delayed Impulse: For $x(n) = \delta(n - n_d)$ with $n_d > 0$,

$$X(z) = z^{-n_d}, \qquad 0 < |z| \leq \infty;$$

while for $x(n) = \delta(n + n_a)$ with $n_a > 0$,

$$X(z) = z^{n_a}, \qquad 0 \leq |z| < \infty.$$

Unit Step: For $x(n) = u(n)$, we have

$$X(z) = \sum_{n=0}^{\infty} z^{-n} = \frac{1}{1 - z^{-1}}, \qquad |z| > 1$$

and thus $X(z)$ has a single pole at $z = 1$. By multiplying the numerator and denominator by z, we may also write $X(z)$ as

$$X(z) = \frac{z}{z-1}, \qquad |z| > 1$$

and we then see that $X(z)$ has a zero at $z = 0$, as well.

Exponential Sequence: For $x(n) = a^n u(n)$,

$$X(z) = \sum_{n=0}^{\infty} a^n z^{-n} = \sum_{n=0}^{\infty} (az^{-1})^n$$

$$= \frac{1}{1 - az^{-1}} = \frac{z}{z-a}, \qquad |z| > |a|.$$

Hence, $X(z)$ has a pole at $z = a$ and a zero at $z = 0$. On the other hand, if $x(n) = -a^n$, $n < 0$, and $x(n) = 0$, $n \geq 0$,

$$X(z) = -\sum_{n=-\infty}^{-1} a^n z^{-n} = -\sum_{n=1}^{\infty} \left(\frac{z}{a}\right)^n$$

$$= -\frac{z}{a} \sum_{n=0}^{\infty} \left(\frac{z}{a}\right)^n = \frac{-z/a}{1 - z/a}$$

$$= \frac{z}{z-a}, \qquad |z| < |a|.$$

We thus see the necessity for including the region of convergence in $X(z)$, for otherwise, the z transforms for these two different sequences would be exactly the same!

Finite-Length Sequences: For $x(n) = 0$, $n < n_1$ and $n > n_2$, it is apparent from the definition of the z transform that it converges everywhere except, possibly, at $z = 0$ and/or $z = \infty$. In particular, if $n_2 \leq 0$, then $x(n) = 0$ for all $n > 0$, at least, and $X(z)$ converges for $z = 0$. If, on the other hand, $n_1 \geq 0$, then $x(n)$ is a causal sequence, and $X(z)$ converges for $z = \infty$. The latter case is especially useful since, *if R contains $z = \infty$, we know immediately that the sequence is causal.*

3.2 INVERSE z TRANSFORM

Very often, we will be able to analyze or design discrete-time signals and systems using their z transforms without having to convert the transforms back to the corresponding sequences. However, such conversion is sometimes desired or necessary and is accomplished via the *inverse z transform.* The formal definition of the inverse z transform is simple in concept, but somewhat cumbersome to use; and for causal sequences, in particular, we will obtain simpler methods to invert the z transform.

The basis of the inverse z transform is the Cauchy Integral Theorem from the theory of complex variables, which states that

$$\frac{1}{2\pi j} \oint_{\Gamma} z^{k-1}\, dz = \begin{cases} 1, & k = 0 \\ 0, & k \neq 0 \end{cases}, \tag{3.2.1}$$

where Γ is a counterclockwise contour of integration enclosing the origin. Therefore, to find $x(n)$ from $X(z)$, we multiply both sides of (3.1.1) by $z^{k-1}/2\pi j$, and integrate along a suitable Γ in R to obtain

$$\frac{1}{2\pi j}\oint_\Gamma X(z)z^{k-1}\,dz = \frac{1}{2\pi j}\oint_\Gamma \sum_{n=-\infty}^{\infty} x(n)z^{-n+k-1}\,dz$$

$$= \sum_{n=-\infty}^{\infty} x(n)\frac{1}{2\pi j}\oint_\Gamma z^{-n+k-1}\,dz = x(k).$$

Thus, the inverse z transform is given by

$$x(n) = \frac{1}{2\pi j}\oint_\Gamma X(z)z^{n-1}\,dz, \tag{3.2.2}$$

where Γ is a counterclockwise contour in the region of convergence of $X(z)$ enclosing the origin. We know that a suitable Γ enclosing the origin can always be found since R is an annular ring centered on the origin.

In the usual case where $X(z)$ is a rational function of z, the Cauchy Residue Theorem states that (3.2.2) can be evaluated by

$$x(n) = \sum_i \varrho_i, \tag{3.2.3}$$

where the ϱ_i are the residues of $X(z)z^{n-1}$ at the poles inside Γ. To show the k poles at $z = p_i$ explicitly, we write

$$X(z)z^{n-1} = \frac{\Phi_i(z)}{(z - p_i)^k} \tag{3.2.4}$$

and the residue at p_i is then given by

$$\varrho_i = \frac{1}{(k-1)!}\left.\frac{d^{k-1}\Phi_i(z)}{dz^{k-1}}\right|_{z=p_i} \tag{3.2.5}$$

Very often, $k = 1$, in which case (3.2.5) becomes simply

$$\varrho_i = \Phi_i(p_i). \tag{3.2.6}$$

EXAMPLE. Consider the z transform

$$X(z) = \frac{z}{z - a}, \qquad |z| > |a|.$$

The function $X(z)z^{n-1} = z^n/(z - a)$ has poles at $z = a$ and, for $n < 0$, at $z = 0$. Any Γ in the region of convergence $|z| > |a|$ will enclose all of these poles. Thus, for $n \geq 0$, we have only the residue

$$\varrho_1 = z^n\big|_{z=a} = a^n, \qquad n \geq 0.$$

For $n = -1$, there are residues at both $z = a$ and $z = 0$ given by

$$\varrho_1 = z^{-1}\big|_{z=a} = a^{-1} \quad \text{and} \quad \varrho_2 = \left.\frac{1}{z - a}\right|_{z=0} = -a^{-1}$$

and, therefore,

$$x(-1) = \varrho_1 + \varrho_2 = 0.$$

For all $n < -1$, we must use the general form of (3.2.5) to obtain the residues, and the reader can verify that $x(n) = 0,\ n \leq -1$. Thus, we have determined that

$$x(n) = a^n u(n),$$

which checks with our previous derivation of this particular $X(z)$.

3.3 INVERSE z TRANSFORM FOR CAUSAL SEQUENCES

If the region of convergence includes $z = \infty$, i.e., if R is of the form $|z| > r$, then we know from section 3.1 that the sequence is causal. If, in addition, $X(z)$ is a rational function of z, then $x(n)$ may be obtained much more simply than by the direct use of (3.2.2) or (3.2.3). Specifically, $X(z)$ can then be expressed as the ratio of two polynomials of the form

$$X(z) = \frac{N(z)}{D(z)} = \frac{\sum_{m=0}^{M} b_m z^{-m}}{\sum_{k=0}^{N} a_k z^{-k}}, \qquad |z| > r \tag{3.3.1}$$

and either of the following two techniques can be used to invert the z transform.

LONG DIVISION. Starting with the *lowest* powers of z^{-1}, we divide $N(z)$ by $D(z)$ to expand $X(z)$ in the original power series of (3.1.1), to wit:

$$a_0 + a_1 z^{-1} + \ldots + a_N z^{-N} \overline{\big)\, b_0 + b_1 z^{-1} + \cdots + b_M z^{-M}}^{\,x(0) + x(1)z^{-1} + x(2)z^{-2} + \ldots}. \tag{3.3.2}$$

The $x(n)$ are thus obtained directly as the coefficients of the resulting power series for $X(z)$.

PARTIAL-FRACTION EXPANSION. If $M < N$ and $X(z)$ has no multiple poles, it may be expanded in a partial-fraction expansion of the form

$$X(z) = \sum_{k=1}^{N} \frac{A_k}{1 - p_k z^{-1}}, \qquad |z| > r, \tag{3.3.3}$$

with the p_k being the poles of $X(z)$. But each term in (3.3.3) is just the z transform of an exponential sequence, and thus the inverse z transform for $X(z)$ is given by

$$x(n) = \sum_{k=1}^{N} A_k p_k^n u(n). \tag{3.3.4}$$

If $M \geq N$, we divide $N(z)$ by $D(z)$ starting with the *highest* powers of z^{-1} to produce

$$a_N z^{-N} + \cdots + a_1 z^{-1} + a_0 \overline{\big) \, b_M z^{-M} + \cdots + b_1 z^{-1} + b_0} \quad \begin{matrix} C_{M-N} z^{-M+N} + \cdots + C_1 z^{-1} + C_0 \end{matrix} + \frac{R(z)}{D(z)} \tag{3.3.5}$$

where the remainder polynomial $R(z)$ is of order $M' = N - 1$, or less. Then, $R(z)/D(z)$ can be expanded in a partial-fraction expansion as before and $x(n)$ is given by

$$x(n) = \sum_{i=0}^{M-N} C_i \delta(n - i) + \sum_{k=1}^{N} A'_k p_k^n u(n). \tag{3.3.6}$$

EXAMPLE. Assume that

$$X(z) = \frac{z^{-2} + 2z^{-1} + 2}{z^{-1} + 1}, \qquad |z| > 1.$$

By long division as in (3.3.2), we obtain

$$X(z) = 2 + z^{-2} - z^{-3} + z^{-4} - z^{-5} + \cdots$$

and thus

$$x(n) = \begin{cases} 0, & n < 0 \\ 2, & n = 0 \\ 0, & n = 1 \\ (-1)^n, & n \geq 2. \end{cases}$$

By the partial-fraction expansion method

$$X(z) = z^{-1} + 1 + \frac{1}{z^{-1} + 1}, \qquad |z| > 1$$

and thus

$$x(n) = \delta(n - 1) + \delta(n) + (-1)^n u(n),$$

which checks with our previous result.

The above techniques can also be employed even if $x(n)$ is not causal, with suitable modification. Common z transform pairs are given in table 3.1.

3.4 PROPERTIES OF THE z TRANSFORM

The following important properties of the z transform following readily from its definition.

Table 3.1 Common z Transform Pairs

Sequence	z Transform	ROC
$\delta(n)$	1	all z
$\delta(n-m), \quad m>0$	z^{-m}	$\lvert z\rvert > 0$
$\delta(n+m), \quad m>0$	z^{m}	$\lvert z\rvert < \infty$
$u(n)$	$\dfrac{1}{1-z^{-1}}$	$\lvert z\rvert > 1$
$-u(-n-1)$	$\dfrac{1}{1-z^{-1}}$	$\lvert z\rvert < 1$
$a^n u(n)$	$\dfrac{1}{1-az^{-1}}$	$\lvert z\rvert > \lvert a\rvert$
$-a^n u(-n-1)$	$\dfrac{1}{1-az^{-1}}$	$\lvert z\rvert < \lvert a\rvert$
$na^n u(n)$	$\dfrac{az^{-1}}{(1-az^{-1})^2}$	$\lvert z\rvert > \lvert a\rvert$
$[\cos n\omega_0 T]u(n)$	$\dfrac{1-(\cos\omega_0 T)\,z^{-1}}{1-2(\cos\omega_0 T)z^{-1}+z^{-2}}$	$\lvert z\rvert > 1$
$[\sin n\omega_0 T]u(n)$	$\dfrac{(\sin\omega_0 T)z^{-1}}{1-2(\cos\omega_0 T)z^{-1}+z^{-2}}$	$\lvert z\rvert > 1$
$r^n[\cos n\omega_0 T]u(n)$	$\dfrac{1-r(\cos\omega_0 T)z^{-1}}{1-2r(\cos\omega_0 T)z^{-1}+r^2z^{-2}}$	$\lvert z\rvert > \lvert r\rvert$
$r^n[\sin n\omega_0 T]u(n)$	$\dfrac{r(\sin\omega_0 T)z^{-1}}{1-2r(\cos\omega_0 T)z^{-1}+r^2z^{-2}}$	$\lvert z\rvert > \lvert r\rvert$

Linearity

The z transform of a weighted sum of sequences equals the corresponding weighted sum of z transforms. That is,

$$w(n) = ax(n) + by(n)$$

implies that

$$W(z) = aX(z) + bY(z), \qquad R_w \supset (R_x \cap R_y), \tag{3.4.1}$$

where set notation has been used to state that the region of convergence for $W(z)$ contains the intersection, at least, of those for $X(z)$ and $Y(z)$. R_w is larger than $R_x \cap R_y$ only if a pole on the boundary of R_x or R_y is cancelled by a zero produced in the weighted sum.

Delay or Advance

For $w(n) = x(n - n_d)$,

$$W(z) = z^{-n_d} X(z) \tag{3.4.2}$$

with R_w the same as R_x except, possibly, at $z = 0$ or $z = \infty$. Since a delay of $n_d = 1$ causes $X(z)$ to be multiplied by z^{-1}, or vice versa, z^{-1} is sometimes referred to as the *unit delay operator*. Likewise, an advance by n_a produces

$$W(z) = z^{n_a} X(z), \tag{3.4.3}$$

and z is sometimes called the *unit advance operator*.

Convolution of Sequences

If $w(n) = x(n) * y(n)$, i.e.,

$$w(n) = \sum_{k=-\infty}^{\infty} x(k)\, y(n - k),$$

then (see problem 3.2)

$$W(z) = X(z)\, Y(z), \qquad R_w \supset (R_x \cap R_y). \tag{3.4.4}$$

The region of convergence for $W(z)$ is larger than the intersection of those for $X(z)$ and $Y(z)$ only if a pole on the boundary of one is cancelled by a zero of the other.

Multiplication of Sequences

If two sequences are multiplied to produce

$$w(n) = x(n)\, y(n),$$

the corresponding z transform is given by

$$W(z) = \frac{1}{2\pi j} \oint_\Gamma Y\left(\frac{z}{v}\right) X(v)\, v^{-1}\, dv \tag{3.4.5}$$

with a region of convergence that includes at least

$$r_{x_-} r_{y_-} < |z| < r_{x_+} r_{y_+} \qquad \text{(see problem 3.2).}$$

From the theory of continuous-time systems, we expect that multiplication in one domain implies convolution in the other, and we saw above that this was indeed true for the convolution of sequences. We can, in fact, express (3.4.5) as a form of convolution by the change of variables

$$v = \varrho e^{j\phi} \quad \text{and} \quad z = re^{j\theta}$$

with the radii ϱ and r lying in R_w. In particular, if R_w contains the unit circle, we may choose $\varrho = r = 1$; and (3.4.5) becomes

$$W(e^{j\theta}) = \frac{1}{2\pi}\int_{-\pi}^{\pi} Y(e^{j(\theta-\phi)})\, X(e^{j\phi})\, d\phi, \tag{3.4.6}$$

which is a convolution of $X(e^{j\theta})$ and $Y(e^{j\theta})$ considered as functions of θ. Since $e^{j\theta}$ is periodic in θ with period 2π, $X(e^{j\theta})$ and $Y(e^{j\theta})$ are also, and (3.4.6) is often called a *periodic convolution*. Alternatively, since (3.4.6) is a form of (3.4.5) carried out on a circle in the z plane, it is also referred to as a *circular convolution*.

Complex Conjugation

For $y(n) = x^*(n)$, the definition of $Y(z)$ is easily manipulated to show that

$$Y(z) = X^*(z^*), \qquad R_y = R_x. \tag{3.4.7}$$

This property is useful in deriving other important properties, including the following.

Parseval's Relation

The total *energy* in a sequence $x(n)$ is defined to be

$$E = \sum_{n=-\infty}^{\infty} |x(n)|^2. \tag{3.4.8}$$

Letting $w(n) = x(n)x^*(n) = |x(n)|^2$, we have immediately that if E is finite, $W(z)$ must converge for $z = 1$ since $E = W(1)$. But from (3.4.6) and (3.4.7), we then have

$$\begin{aligned} E = W(1) &= \frac{1}{2\pi}\int_{-\pi}^{\pi} X^*(e^{j\phi})\, X(e^{j\phi})\, d\phi \\ &= \frac{1}{2\pi}\int_{-\pi}^{\pi} |X(e^{j\phi})|^2\, d\phi. \end{aligned} \tag{3.4.9}$$

Combining (3.4.9) and (3.4.8), we produce Parseval's Relation

$$\sum_{n=-\infty}^{\infty} |x(n)|^2 = \frac{1}{2\pi}\int_{-\pi}^{\pi} |X(e^{j\phi})|^2\, d\phi. \tag{3.4.10}$$

Additional properties of the z transformation are given in problem 3.6.

PROBLEMS

3.1 Let the z transform of $h(n)$ be

$$H(z) = \frac{z^{-1} - z^{-2}}{1 + z^{-1}}, \qquad |z| > 1.$$

a. Sketch the pole/zero diagram.
b. Find $h(n)$ for all n.
c. Is this filter stable?

3.2 Show the following z transform relationships:
a. For the convolution $w(n) = x(n) * y(n)$, $W(z) = X(z)Y(z)$.
b. For the product $w(n) = x(n)y(n)$, $W(z)$ is given by (3.4.5).

3.3 Find $X(z)$ for $x(n) = a^{|n|}$, including the region of convergence. What constraint is required on a for $X(z)$ to exist?

3.4 Find $X(z)$ for $x(n) = e^n u(-n)$, including the region of convergence. Repeat for $x(n) = e^n u(-n + 1)$.

3.5 Find $H(z)$ for $h(n) = Ar^n \cos(n\omega_0 T + \theta)u(n)$. Plot the pole/zero diagram for $0 < r < 1$, and show the region of convergence.

3.6 Show the following:
a. The z transform of $x^*(n)$ is $X^*(z^*)$.
b. The z transform of $x(-n)$ is $X(1/z)$.
c. The z transform of $a^n x(n)$ is $X(z/a)$. Determine how the poles and zeros of $X(z)$ are moved.
d. The z transform of $nx(n)$ is $-z \dfrac{dX(z)}{dz}$.
e. The initial value $x(0)$ of a causal sequence $x(n)$ is given by $\lim_{z \to \infty} X(z)$.

3.7 The *autocorrelation function* for a deterministic sequence $x(n)$ is defined as

$$r(n) = \sum_{k=-\infty}^{\infty} x(k)x^*(k + n).$$

a. Find $R(z)$ in terms of $X(z)$.
b. What must be the form of the region of convergence for $R(z)$ if it exists?
c. Show that the *energy spectrum* $R(e^{j\theta})$ is given by

$$R(e^{j\theta}) = |X(e^{j\theta})|^2.$$

d. Evaluate the total energy $E = r(0)$ in terms of $R(e^{j\theta})$ by Parseval's Relation.

3.8 Find $X(z)$ for the finite-duration sequence $x(n) = a^n[u(n) - u(n - N]$. Note a possible pole/zero cancellation when determining the region of convergence.

3.9 Evaluate $\dfrac{1}{2\pi}\displaystyle\int_{-\pi}^{\pi} |1 - ae^{-j\theta}|^{-2}\, d\theta$, $|a| < 1$, by Parseval's Relation. (*Hint*: Consider the form of the integrand when $z = e^{j\theta}$.) Repeat for $|a| > 1$.

3.10 Find $h(n)$ for the following cases of $H(z)$:

$$H(z) = \frac{2 - z^{-1}}{1 - z^{-1} - 0.75z^{-2}}.$$

a. $H(z)$ convergent for $z = 0$.
b. $H(z)$ convergent for $z = 1$.
c. $H(z)$ convergent for $z = \infty$.

3.11 The following second-order filter has a double pole at $z = a$ and is said to be *critically damped*:

$$H(z) = \frac{1}{(1 - az^{-1})^2}, \qquad |z| > |a|, \qquad |a| < 1.$$

Find and sketch the corresponding $h(n)$. (*Hint*: Use the relation in problem 3.6d).

4. INPUT/OUTPUT RELATIONSHIPS

4.0 INTRODUCTION

There are a number of different, but equivalent ways to describe the relationship between the input and output of a discrete-time filter, including the impulse response, the system function, the frequency response, difference equations, and state variables. Each has its own particular advantages in certain derivations and calculations. We have previously introduced and used the impulse response in chapter 2; in this chapter we will define and study the other four. We will also discuss the distinction between finite-impulse-response and infinite-impulse-response filters and recursive and nonrecursive filter implementations.

4.1 SYSTEM FUNCTION AND FREQUENCY RESPONSE

In section 2.2 we found that for a linear time-invariant filter with impulse response $h(n)$, the output $y(n)$ for an arbitrary input $x(n)$ is given by the convolution

$$y(n) = \sum_{k=-\infty}^{\infty} x(n-k)h(k). \tag{4.1.1}$$

But from (3.4.4) we can now express the same relationship in terms of the corresponding z transforms as

$$Y(z) = X(z)H(z), \tag{4.1.2}$$

where $H(z)$ is, therefore,

$$H(z) = \sum_{n=-\infty}^{\infty} h(n) z^{-n} \tag{4.1.3}$$

and $R_y \supset (R_x \cap R_h)$. $H(z)$ is called the *system function* of the discrete filter; from (4.1.2), it can also be written as

$$H(z) = \frac{Y(z)}{X(z)}. \tag{4.1.4}$$

EXAMPLE. We will rederive the output of a discrete-time filter with impulse response $h(n) = a^n u(n)$ to a unit-step input (from section 2.2). The corresponding z transforms are

$$H(z) = \frac{1}{1 - az^{-1}}, \qquad |z| > |a|$$

and

$$X(z) = \frac{1}{1 - z^{-1}}, \qquad |z| > 1.$$

Assuming that $|a| < 1$ for stability, we have from (4.1.2)

$$Y(z) = \frac{1}{(1 - z^{-1})(1 - az^{-1})}, \qquad |z| > 1$$

or, expanding in a partial-fraction expansion,

$$Y(z) = \frac{1/(1-a)}{1 - z^{-1}} + \frac{-a/(1-a)}{1 - az^{-1}}, \qquad |z| > 1.$$

Therefore,

$$y(n) = \frac{1}{1-a}[1 - a^{n+1}]u(n),$$

which is the same as our previous result.

The system function $H(z)$ will be the key function in most of our analysis and synthesis of discrete and digital filters. One reason is the simplicity of the relation in (4.1.2), as opposed to the convolution in (4.1.1). Another is that the frequency response of the filter is readily obtained from $H(z)$, as follows: Assume that the steady-state input to a linear time-invariant filter is the complex sinusoid

$$x(n) = e^{j\omega nT}, \qquad -\infty < n < \infty.$$

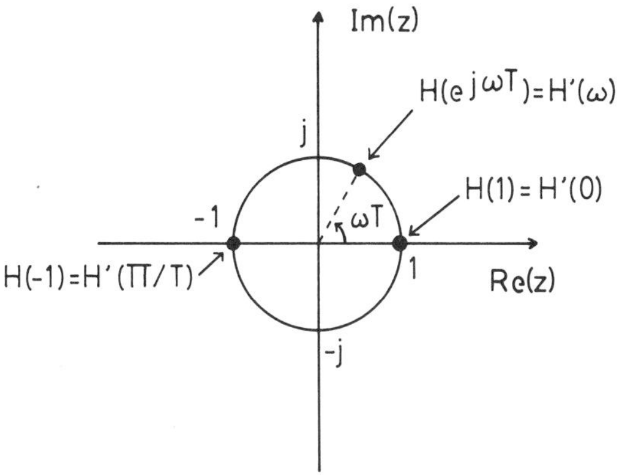

Figure 4.1. The unit circle in the z plane.

Then, from (4.1.1),

$$\begin{aligned} y(n) &= \sum_{k=-\infty}^{\infty} h(k) e^{j\omega T(n-k)} \\ &= e^{j\omega nT} \sum_{k=-\infty}^{\infty} h(k) e^{-j\omega kT} \\ &= x(n) H(z)\big|_{z=e^{j\omega T}} \end{aligned}$$

and the output thus equals the input multiplied by the complex quantity

$$H(e^{j\omega T}) = H'(\omega). \tag{4.1.5}$$

The function $H'(\omega)$ is the *frequency response* of the discrete filter.

Note that the frequency response is given by $H(z)$ evaluated on the unit circle in the z plane since $|z| = |e^{j\omega T}| = 1$. In particular, the dc or zero-frequency response is given by $H'(0) = H(1)$, and the response at the *Nyquist frequency* $\omega = \pi/T$ is given by $H'(\pi/T) = H(-1)$. This is depicted in figure 4.1. Since $e^{j\omega T}$ is periodic in ω with period $2\pi/T$, we have immediately from (4.1.5) that $H'(\omega)$ is also periodic with the same period. In addition, for $h(n)$ real, it follows from $h(n) = h^*(n)$ and (3.4.7) that

$$H'(\omega) = [H'(-\omega)]^*. \tag{4.1.6}$$

Hence, the *magnitude response* $|H'(\omega)|$ is an even function of ω for $h(n)$ real; while the *phase response* $\angle H'(\omega)$ is an odd function of ω. These properties are illustrated in figure 4.2.

The important concepts of stability and causality, defined in section 2.3, are readily and conveniently described in terms of the system function, as follows.

STABILITY. A linear time-invariant filter was previously shown to be stable if, and only if,

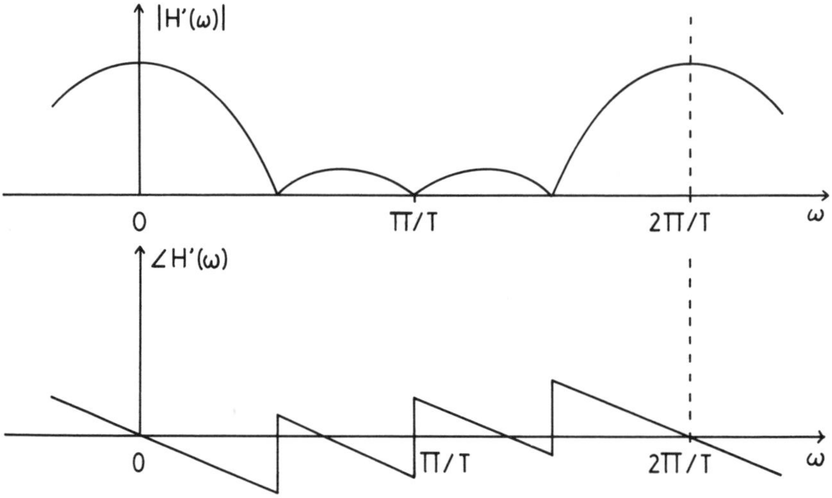

Figure 4.2. The magnitude response $|H'(\omega)|$ is a periodic and even function of ω; while the phase response $\angle\, H'(\omega)$ is periodic and odd.

$$\sum_{n=-\infty}^{\infty} |h(n)| < \infty.$$

But this implies that $H(z)$ must converge on the unit circle since

$$|H(e^{j\omega T})| \le \sum_{n=-\infty}^{\infty} |h(n)e^{-j\omega n T}| = \sum_{n=-\infty}^{\infty} |h(n)| < \infty.$$

Hence, for a stable system, R_h must contain the unit circle.

CAUSALITY. If $h(n)$ is causal, R_h must include $z = \infty$, and is thus of the form $|z| > r$, where r is the largest radius of the poles of $H(z)$. Hence, $H(z)$ converges everywhere outside a circle of radius r.

If the filter is both stable and causal, all of the poles of $H(z)$ must lie inside the unit circle since R_h then contains the unit circle as well as $z = \infty$.

4.2 DIFFERENCE EQUATIONS

For a large and important class of linear time-invariant discrete filters, the input and output sequences satisfy difference equations of the form

$$\sum_{k=0}^{N} a_k y(n-k) = \sum_{m=0}^{M} b_m x(n-m), \tag{4.2.1}$$

where the b_m and a_k are constant coefficients. However, (4.2.1) alone is not sufficient to completely specify the filter; additional information concerning causality and initial conditions is required.

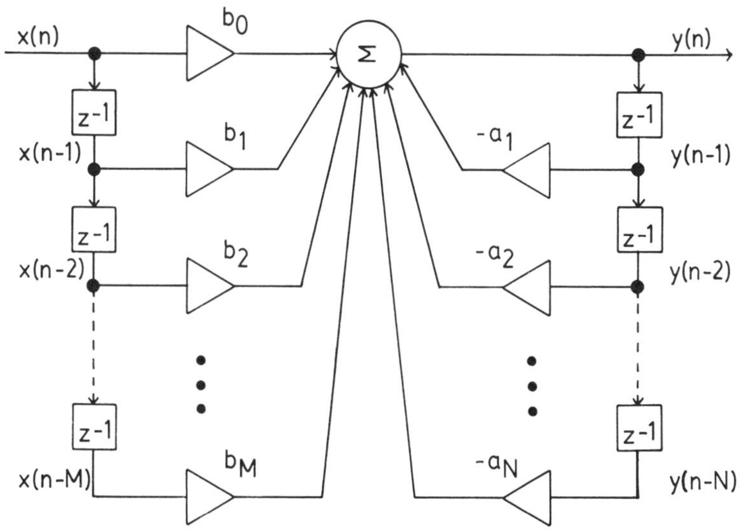

Figure 4.3. A direct implementation of equation (4.2.2).

Assuming that the filter is causal, and scaling the coefficients such that $a_0 = 1$, equation 4.2.1 can be rewritten in the form

$$y(n) = \sum_{m=0}^{M} b_m x(n - m) - \sum_{k=1}^{N} a_k y(n - k), \tag{4.2.2}$$

which shows that the present output value $y(n)$ can be computed from the present and M past input values and N past output values. This may be done directly as expressed by (4.2.2), or in other equivalent computational forms. If past output values (intermediate or final) are actually used in the computation of the present output, i.e., if the filter implementation contains feedback, then the implementation is said to be *recursive*. Otherwise, the filter implementation is *nonrecursive*.

A block diagram of one filter implementation may be produced directly from (4.2.2), as shown in figure 4.3. The unit delays are denoted by the corresponding z-transform operator z^{-1}, and the constant coefficient multipliers b_m and a_k are shown as *gain* factors. Each delay is realized by some form of storage element (register, memory location, switched capacitor, etc.) whose present output equals its preceding input. In general, a discrete-time or digital filter consists of these three basic components, namely: adders, multipliers, and delays.

The system function corresponding to (4.2.1) is readily derived by taking the z transform of both sides of that difference equation to produce

$$\sum_{k=0}^{N} a_k Z[y(n - k)] = \sum_{m=0}^{M} b_m Z[x(n - m)], \tag{4.2.3}$$

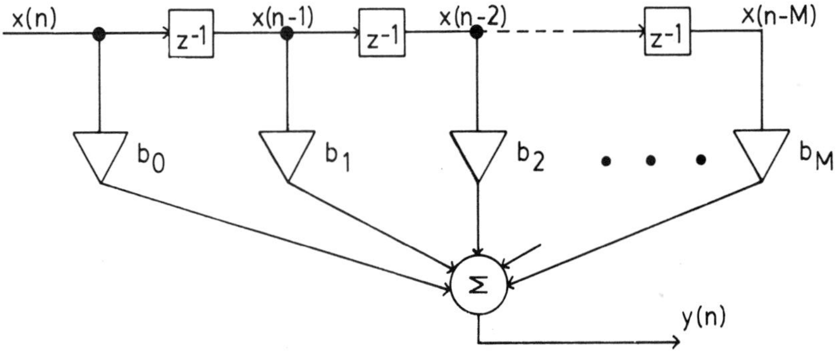

Figure 4.4. A nonrecursive implementation of an FIR filter.

where $Z[\ ]$ denotes the z transform and we have utilized (3.4.1). But from (3.4.2), we then have

$$\sum_{k=0}^{N} a_k z^{-k} Y(z) = \sum_{m=0}^{M} b_m z^{-m} X(z) \tag{4.2.4}$$

and thus

$$H(z) = \frac{Y(z)}{X(z)} = \frac{\Sigma_{m=0}^{M} b_m z^{-m}}{\Sigma_{k=0}^{N} a_k z^{-k}}. \tag{4.2.5}$$

Therefore, the system function $H(z)$ corresponding to (4.2.1) is a rational function of the form we have previously investigated.

EXAMPLE—Averaging. A common technique for smoothing a data sequence is to take a simple weighted average of $M + 1$ adjacent input values to produce each output value. A causal version of this filtering operation is thus described by the difference equation

$$y(n) = \sum_{m=0}^{M} b_m x(n - m)$$

and can be implemented nonrecursively as shown in figure 4.4. The corresponding system function is simply

$$H(z) = \sum_{m=0}^{M} b_m z^{-m};$$

and the impulse response $h(n)$ is obtained directly from $H(z)$ or from the block diagram as

$$h(n) = \begin{cases} b_n, & n = 0, 1, 2, \ldots, M \\ 0, & \text{otherwise.} \end{cases}$$

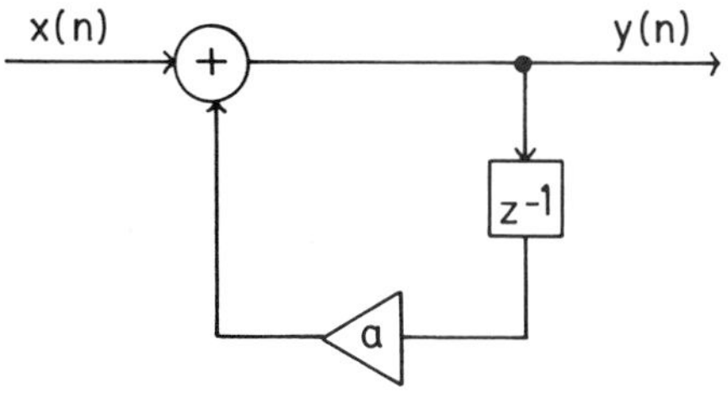

Figure 4.5. A "leaky" accumulator is a recursive IIR filter.

The impulse response of this filter has nonzero values only for a finite duration; such filters will be called *finite-impulse-response* (FIR) filters. Usually, FIR filters will be implemented nonrecursively, as in this case, but recursive implementations can also be generated. Hence, we will maintain this distinction, reserving *FIR* to describe the filter type and *nonrecursive* to describe the filter implementation.

EXAMPLE—Accumulation. A common recursive technique for smoothing a sequence is to accumulate (integrate) the sequence with a decay or *leak* in the accumulator as shown in the block diagram of figure 4.5. The corresponding difference equation is thus

$$y(n) \quad = \quad x(n) + a\,y(n-1)$$

and the system function is

$$H(z) \quad = \quad \frac{1}{1 - az^{-1}}.$$

The region of convergence for $H(z)$ is not obvious at first since we didn't sum anything to obtain it. But the filter is causal and there is a pole at $z = a$, which means that R_h must be of the form $|z| > |a|$. The impulse response is thus the familiar exponential sequence

$$h(n) \quad = \quad a^n u(n).$$

For stability, as we have seen previously, $|a| < 1$.

Note that $h(n)$ for positive n is nonzero for an infinite duration, and this is, therefore, an *infinite-impulse-response* (IIR) filter. As opposed to the FIR case, IIR filters will usually be implemented recursively; but since an IIR filter can have a predominantly nonrecursive implementation (involving recursion only at the beginning of each block of output data [5]), we will maintain the distinction between an *IIR* filter and a *recursive* implementation.

4.3 GEOMETRIC EVALUATION OF $H(z)$ AND $H'(\omega)$

For filters with a rational system function as in (4.2.5), the roots of the numerator and denominator polynomials are called the zeros and poles, respectively, of the filter. Factoring these polynomials into corresponding first-order factors, we produce

$$H(z) = \frac{b_0 \prod_{m=1}^{M} (1 - z_m z^{-1})}{a_0 \prod_{k=1}^{N} (1 - p_k z^{-1})} \tag{4.3.1}$$

or

$$H(z) = \frac{b_0 z^{N-M} \prod_{m=1}^{M} (z - z_m)}{a_0 \prod_{k=1}^{N} (z - p_k)}, \tag{4.3.2}$$

where the z_m are the zeros of $H(z)$ and the p_k are the poles.

The form of $H(z)$ in (4.3.2) is especially convenient for calculating or estimating the value of $H(z)$ from the geometry of the pole/zero diagram in the z plane. In particular, the complex quantity $(z - z_m) = B_m e^{j\theta_m}$ may be thought of as a vector from the point z_m in the z plane to the arbitrary point z. Likewise, $(z - p_k) = A_k e^{j\phi_k}$ may be considered to be a vector from p_k to z. The magnitudes B_m and A_k are the lengths of the corresponding vectors, and the angles θ_m and ϕ_k are the angles of the vectors measured counterclockwise from the positive real axis. Thus, since

$$H(z) = z^{N-M} \frac{b_0 (\Pi_m B_m) \exp(j \Sigma_m \theta_m)}{a_0 (\Pi_k A_k) \exp(j \Sigma_k \phi_k)}, \tag{4.3.3}$$

the product of the vector lengths and the sum of the vector angles are conveniently employed in the evaluation of the magnitude and angle, respectively, of $H(z)$.

This technique is particularly useful in the calculation or estimation of the frequency response $H'(\omega)$. In this case, we wish to evaluate $H(z)$ on the unit circle since $H'(\omega) = H(e^{j\omega T})$, and hence from (4.3.3),

$$|H'(\omega)| = \frac{b_0 \prod_{m=1}^{M} B_m}{a_0 \prod_{k=1}^{N} A_k} \tag{4.3.4}$$

and

$$\angle H'(\omega) = \sum_{m=1}^{M} \theta_m - \sum_{k=1}^{N} \phi_k + (N - M)\omega T. \tag{4.3.5}$$

The final term in (4.3.5) is a linear phase term and thus represents pure delay. An example of this case is shown in figure 4.8, which we will later analyze in detail.

We can often make significant qualitative observations concerning the frequency response from (4.3.4) and (4.3.5) without having to perform any actual computation. For example, if there is a zero directly on the unit circle at $e^{j\omega_0 T}$, then $|H'(\omega)|$ will go to zero at $\omega = \omega_0$, producing a notch in the frequency response. Also at that frequency, there will be an abrupt 180° phase shift in $\angle H'(\omega)$ as the vector for this zero reverses direction. If, however, the zero is only close to the unit circle (and there is no compensating pole), there will be a dip in $|H'(\omega)|$ that does not go to zero, and $\angle H'(\omega)$ will change rapidly, but not abruptly, by almost 180°. Likewise, if there is a pole near the unit circle (and no compensating

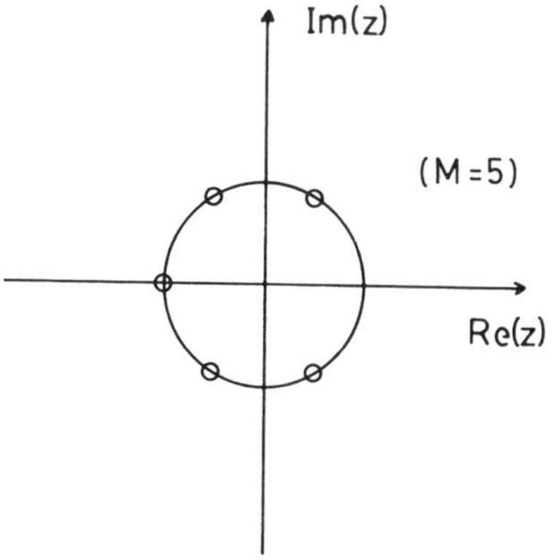

Figure 4.6. Zero locations for a simple-averaging filter.

zero), $|H'(\omega)|$ will tend to peak at the corresponding frequency as the vector to this pole becomes shorter, and $\angle H'(\omega)$ will change rapidly in a negative direction by almost $-180°$. As the pole is moved closer to the unit circle, the peak in $|H'(\omega)|$ becomes sharper until, in the limit, $|H'(\omega)|$ becomes infinite at that frequency and the filter becomes unstable.

EXAMPLE—Simple Averaging. Probably the simplest discrete-time filter of all is the simple average with uniform weights, defined by

$$y(n) = \frac{1}{M+1} \sum_{m=0}^{M} x(n-m),$$

with the corresponding system function

$$H(z) = \frac{1}{M+1} \sum_{m=0}^{M} z^{-m} = \frac{1 - z^{-(M+1)}}{(M+1)(1-z^{-1})}. \tag{4.3.6}$$

The zeros of $H(z)$ occur for values of z where

$$z^{-(M+1)} = 1,$$

and are thus given by the $(M+1)^{\text{st}}$ roots of unity

$$z_m = e^{j2\pi m/(M+1)}, \qquad m = 1, \ldots, M.$$

The root for $m = 0$ is not included because this zero at $z = 1$ is cancelled by the denominator factor $(1 - z^{-1})$; and there are, therefore, M zeros and no poles in $H(z)$ except at $z = 0$. This is illustrated in figure 4.6 for $M = 5$.

The magnitude of the frequency response is easily sketched on the basis of a pole/zero diagram such as figure 4.6. In this example, the maximum value of $|H'(\omega)|$ must occur

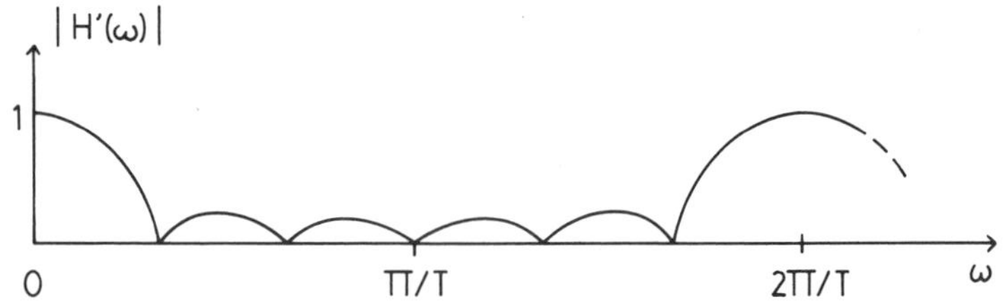

Figure 4.7. Magnitude response of a simple-averaging filter.

for $\omega = 0$ since the product of the vector lengths from the zeros to the point $z = 1$ will be greater than for any other point on the unit circle. $|H'(\omega)|$ will decrease as ω increases from $\omega = 0$ until it equals zero at $\omega = 2\pi/6T$, then increase to a relative maximum near $\omega = \pi/2T$, then decrease again to zero at $\omega = 2\pi/3T$, and so forth. We thus produce the sketch in figure 4.7.

It is readily shown (see problem 4.1) that $|H'(\omega)|$ is, in fact, given by

$$|H'(\omega)| = \left| \frac{\sin (M+1)\dfrac{\omega T}{2}}{(M+1)\sin \dfrac{\omega T}{2}} \right| . \tag{4.3.7}$$

Note that while $M + 1$ input values are averaged to produce each output, there are M zeros in $H(z)$, and accordingly this is an example of an Mth-order filter. This terminology is reasonable in view of figure 4.4, where we have seen that M delays are sufficient to realize this filter. In general, the *order* of a filter is the larger of the number of poles or the number of zeros; and this cannot exceed the number of delays in the filter implementation.

EXAMPLE—Unity-Gain Resonator. Another useful example of frequency response estimation is a second-order resonator with maximum gain at resonance of approximately unity. The appropriate system function is

$$H(z) = \frac{b_0(1 - z^{-2})}{1 + a_1 z^{-1} + a_2 z^{-2}}; \tag{4.3.8}$$

and hence there are zeros at $z_1 = 1$ and $z_2 = -1$ and a pair of poles somewhere inside the unit circle as determined by a_1 and a_2. We want the poles to be complex, and thus, since a_1 and a_2 are real, the poles are complex conjugates of the form

$$p_1, p_2 = -\frac{a_1}{2} \pm j\sqrt{a_2 - a_1^2/4}\,.$$

Rewriting the denominator of $H(z)$ as

$$(1 - p_1 z^{-1})(1 - p_1^* z^{-1}) = 1 - 2\,\mathrm{Re}(p_1) z^{-1} + |p_1|^2 z^{-2},$$

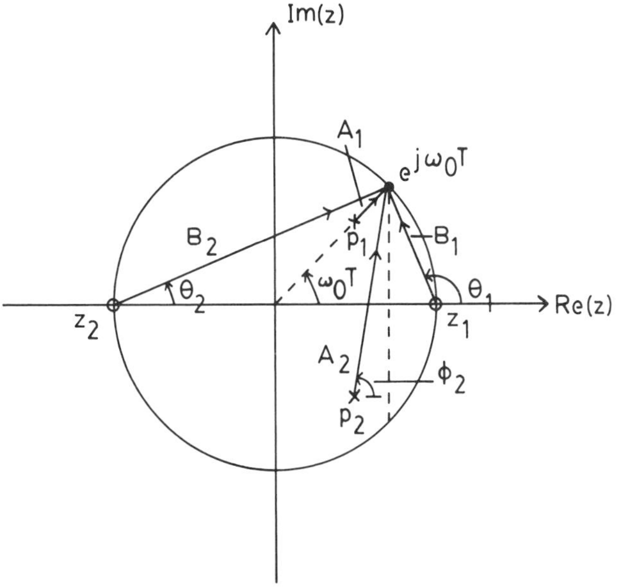

Figure 4.8. Pole/zero diagram for the unity-gain resonator.

we see that a_1 and a_2 are given by

$$a_1 \;=\; -2\,\mathrm{Re}\,(p_1) \quad \text{and} \quad a_2 \;=\; |p_1|^2 > a_1^2/4.$$

The pole/zero diagram showing the vectors for determining $H'(\omega)$ is shown in figure 4.8. Our objective is to determine b_0 such that the gain at the resonant frequency ω_0 is approximately unity, i.e.,

$$|H'(\omega_0)| \;\approx\; 1.$$

We note first that the maximum gain will occur approximately where $e^{j\omega T}$ is closest to p_1 (or p_2) and thus $\omega_0 T \;\approx\; \angle\, p_1$. We also note that the vectors from z_1 and z_2 form a right triangle with the real axis, and hence

$$B_1 \;=\; 2\sin\theta_2 \quad \text{and} \quad B_2 \;=\; 2\cos\theta_2.$$

At the resonant frequency ω_0, the vector from p_1 to $e^{j\omega_0 T}$ will have a length of

$$A_1 \;=\; 1 - |p_1| \;=\; 1 - \sqrt{a_2},$$

which, for p_1 near the unit circle, can be approximated by

$$A_1 \approx \frac{1 - a_2}{2}.$$

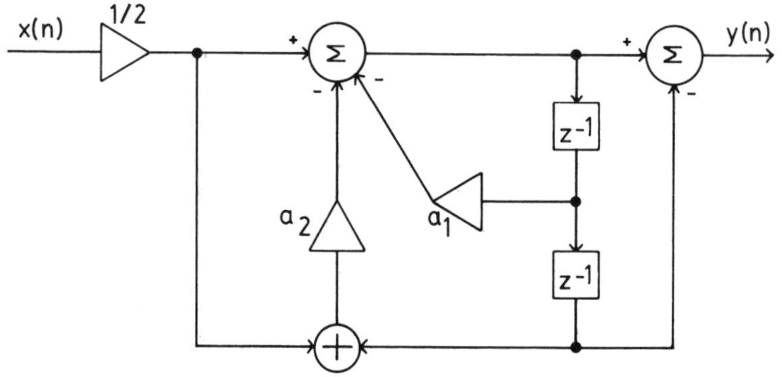

Figure 4.9. Unity-gain resonator implementation.

And then, since $p_2 = p_1^*$ is also near the unit circle, we can approximate the length of the vector from p_2 to $e^{j\omega_0 T}$ by that of the vector $e^{j\omega_0 T} - e^{-j\omega_0 T}$, shown as a dotted vertical line in figure 4.8, which gives

$$A_2 \approx 2B_2 \sin \theta_2 = B_1 B_2.$$

Therefore, from (4.3.4), we have

$$|H'(\omega_0)| = \frac{b_0 B_1 B_2}{A_1 A_2} = \frac{b_0}{A_1} \approx \frac{2b_0}{1 - a_2},$$

and for unity gain at ω_0

$$b_0 = \frac{1 - a_2}{2}.$$

A filter implementation that automatically provides this unity-gain scaling is shown in figure 4.9, and the resulting frequency responses for a bank of discrete-time resonators with equally spaced center frequencies are plotted in figure 4.10.

4.4 STATE VARIABLES

The state-variable approach is a generalization of difference equations for causal filters to include a description of the internal *state* of the network, as well as the input/output relationship. Many additional insights into the dynamics of the system can be obtained by state-variable analysis. As an example, consider the second-order discrete-time filter depicted in figure 4.11. In addition to the input $x(n)$ and output $y(n)$, the two states stored in the delays are designated as $s_1(n)$ and $s_2(n)$. These states are available at the outputs of the delays, as indicated. Difference equations relating the states at the next time index $n + 1$ to those at the present index n are readily written from the network diagram, as follows:

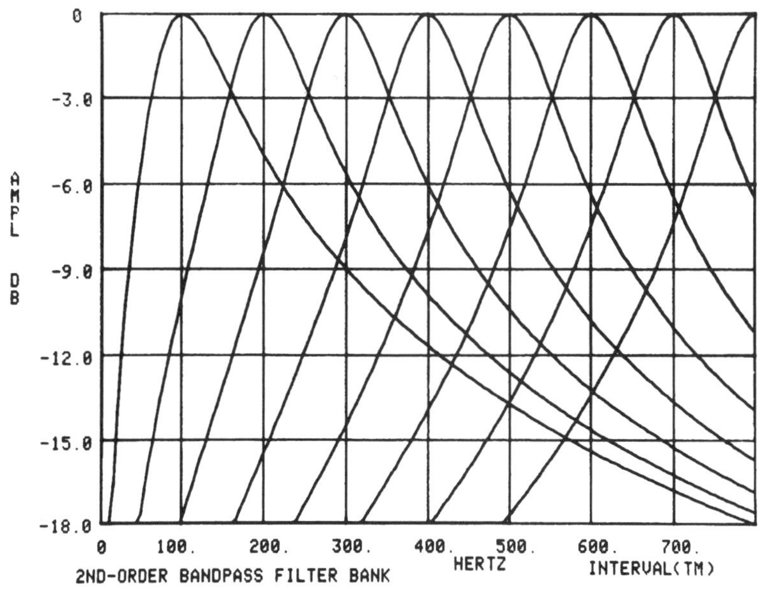

Figure 4.10. Magnitude responses for a bank of unity-gain resonators.

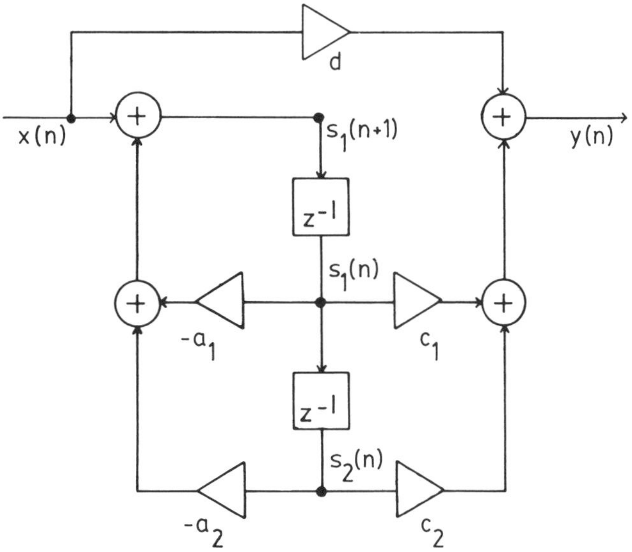

Figure 4.11. A second-order filter with states $s_1(n)$ and $s_2(n)$.

$$
\begin{aligned}
s_1(n+1) &= -a_1 s_1(n) - a_2 s_2(n) + x(n) \\
s_2(n+1) &= s_1(n).
\end{aligned}
\tag{4.4.1}
$$

In addition, the output is given by

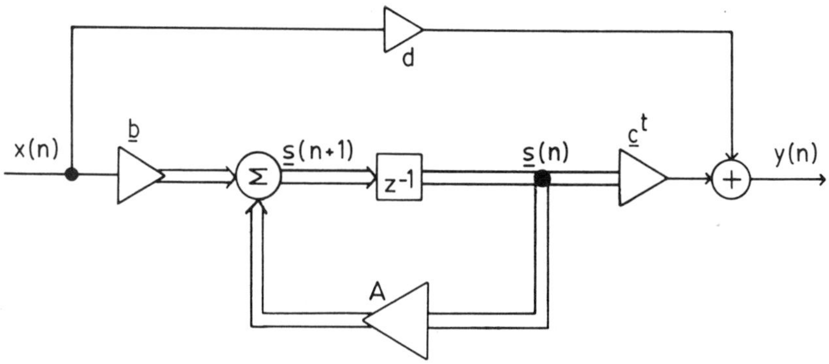

Figure 4.12. A filter block diagram in state-variable form.

$$y(n) = c_1 s_1(n) + c_2 s_2(n) + d\,x(n). \tag{4.4.2}$$

These are the *state equations* for the network.

Difference equations 4.4.1 and 4.4.2 are more conveniently written as vector equations of the form

$$\begin{aligned} \underline{s}(n+1) &= \mathbf{A}\underline{s}(n) + \underline{b}x(n), \\ y(n) &= \underline{c}^t\underline{s}(n) + dx(n), \end{aligned} \tag{4.4.3}$$

where, in this example,

$$\mathbf{A} = \begin{pmatrix} -a_1 & -a_2 \\ 1 & 0 \end{pmatrix} \qquad \underline{b} = \begin{pmatrix} 1 \\ 0 \end{pmatrix}$$

$$\underline{c}^t = (c_1 \quad c_2) \qquad \underline{s}(n) = \begin{pmatrix} s_1(n) \\ s_2(n) \end{pmatrix}.$$

One advantage of state-variable descriptions is that multiple-input, multiple-output systems are easily described and analyzed, in which case rectangular matrices **B**, **C**, and **D** are required in (4.4.3). However, because we will restrict our attention to single-input, single-output systems, we have chosen to write **B** as the column vector $\underline{b}$, **C** as the row vector $\underline{c}^t$, and **D** as the scalar d. A block diagram depicting the filter in state-variable form is given in figure 4.12, where double lines are used to indicate vector signals.

The state variables for a system are not unique. This flexibility to select among different networks for the same filter is explored further in chapters 5 and 11. One class of alternative networks is produced by linearly transforming the state vector $\underline{s}(n)$ to a new state vector

$$\hat{\underline{s}}(n) = \mathbf{T}\underline{s}(n) \tag{4.4.4}$$

where **T** is any nonsingular $N \times N$ matrix. From (4.4.3) and the inverse of (4.4.4), we then have

$$\begin{aligned}\hat{\underline{s}}(n+1) &= \mathbf{T}\underline{s}(n+1) = \mathbf{TA}\underline{s}(n) + \mathbf{T}\underline{b}x(n) \\ &= \mathbf{TAT}^{-1}\hat{\underline{s}}(n) + \mathbf{T}\underline{b}(x(n) \\ &= \hat{\mathbf{A}}\hat{\underline{s}}(n) + \hat{\underline{b}}x(n)\end{aligned} \tag{4.4.5}$$

and

$$\begin{aligned}y(n) &= \underline{c}^t\mathbf{T}^{-1}\hat{\underline{s}}(n) + dx(n) \\ &= \hat{\underline{c}}^t\hat{\underline{s}}(n) + dx(n),\end{aligned}$$

from which the state matrices for the transformed filter are seen to be

$$\begin{aligned}\hat{\mathbf{A}} &= \mathbf{TAT}^{-1} & \hat{\underline{b}} &= \mathbf{T}\underline{b} \\ \hat{\underline{c}}^t &= \underline{c}^t\mathbf{T}^{-1} & \hat{d} &= d.\end{aligned} \tag{4.4.6}$$

For example, let $\mathbf{T}$ be the scaling transformation

$$\mathbf{T} = \begin{pmatrix} g_1 & 0 \\ 0 & g_2 \end{pmatrix}.$$

Then $\hat{\underline{s}}(n)$ is the scaled state vector

$$\hat{\underline{s}}(n) = \begin{pmatrix} g_1 s_1(n) \\ g_2 s_2(n) \end{pmatrix}$$

and the state matrices for the scaled network are

$$\hat{\mathbf{A}} = \begin{pmatrix} a_{11} & \frac{g_1}{g_2} a_{12} \\ \frac{g_2}{g_1} a_{21} & a_{22} \end{pmatrix} \qquad \hat{\underline{b}} = \begin{pmatrix} g_1 b_1 \\ g_2 b_2 \end{pmatrix} \qquad \hat{\underline{c}}^t = \begin{pmatrix} \frac{c_1}{g_1} & \frac{c_2}{g_2} \end{pmatrix}.$$

By construction, the output $y(n)$ and thus the system function $H(z)$ are unchanged.

To determine the impulse response in terms of $\mathbf{A}$, $\underline{b}$, $\underline{c}^t$, d, let $x(n) = \delta(n)$ in (4.4.3) with zero initial conditions, i.e., $\underline{s}(0) = \underline{0}$. Clearly then, $\underline{s}(1) = \underline{b}$, $\underline{s}(2) = \mathbf{A}\underline{b}$, $\underline{s}(3) = \mathbf{A}^2\underline{b}$, etc., and in general,

$$\underline{s}(n) = \mathbf{A}^{n-1}\underline{b}, \qquad n \geq 1. \tag{4.4.7}$$

Similarly, the output $y(n) = h(n)$ is $h(0) = d$ and $h(n) = \underline{c}^t\underline{s}(n)$, $n \geq 1$, and thus

$$h(n) = d\delta(n) + \underline{c}^t\mathbf{A}^{n-1}\underline{b}u(n-1). \tag{4.4.8}$$

Taking the z transform, we have

$$H(z) = d + \sum_{n=1}^{\infty} \underline{c}^t \mathbf{A}^{n-1} \underline{b} z^{-n}$$

$$= d + z^{-1} \underline{c}^t \left[\sum_{k=0}^{\infty} \mathbf{A}^k z^{-k} \right] \underline{b}. \tag{4.4.9}$$

This summation looks very much like the familiar sum $\Sigma_{k=0}^{\infty} a^k z^{-k} = (1 - az^{-1})^{-1}$, and it can be shown that if all the eigenvalues of **A** have magnitudes less than $|z|$, the sum does indeed converge to the matrix inverse $(\mathbf{I} - z^{-1}\mathbf{A})^{-1}$. Hence,

$$H(z) = d + z^{-1} \underline{c}^t (\mathbf{I} - z^{-1}\mathbf{A})^{-1} \underline{b}. \tag{4.4.10}$$

Combining z^{-1} into the matrix inverse, we have the alternate expression

$$H(z) = d + \underline{c}^t (z\mathbf{I} - \mathbf{A})^{-1} \underline{b}. \tag{4.4.11}$$

As an exercise, the reader may want to check that $H(z)$ in (4.4.11) is not altered by the transformation in (4.4.6).

The poles of the filter are the eigenvalues of **A**. To see this, note from (4.4.11) that the denominator polynomial $D(z)$ will arise from the matrix inverse and will equal $|z\mathbf{I} - \mathbf{A}|$, where $|\cdots|$ denotes the determinant. Hence, the poles are the roots of the *characteristic equation*

$$D(z) = |z\mathbf{I} - \mathbf{A}| = 0. \tag{4.4.12}$$

On the other hand, the eigenvalues of **A** are the values satisfying

$$\mathbf{A}\underline{x} = \lambda \underline{x}$$

or (4.4.13)

$$(\mathbf{A} - \lambda \mathbf{I})\underline{x} = \underline{0},$$

where the solution vectors $\underline{x}$ are the corresponding eigenvectors. But this equation will have nontrivial solutions if, and only if, $(\mathbf{A} - \lambda\mathbf{I})$ is singular, or

$$|\mathbf{A} - \lambda \mathbf{I}| = 0. \tag{4.4.14}$$

Since this is the same characteristic equation (except for sign) as (4.4.12), the eigenvalues must equal the poles.

As an example of the ease with which the poles are found by state-variable analysis, we evaluate the characteristic polynomial for the earlier example:

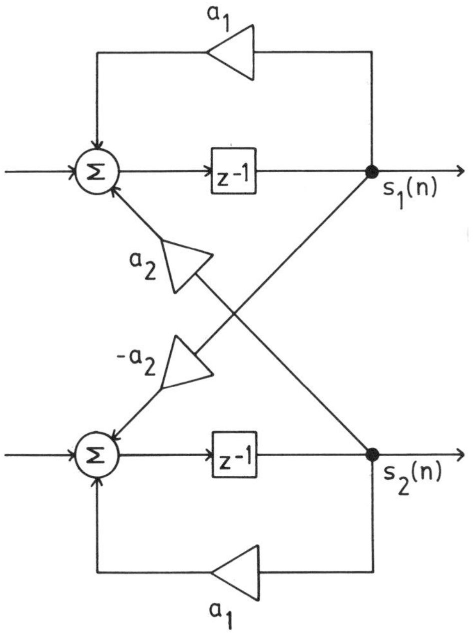

Figure 4.13. The recursive normal structure.

$$D(z) = |z\mathbf{I} - \mathbf{A}|$$

$$= \begin{vmatrix} z + a_1 & a_2 \\ -1 & z \end{vmatrix}$$

$$= z^2 + a_1 z + a_2.$$

The roots of this quadratic are the poles, as expected. However, to find the zeros, we must completely evaluate (4.4.11). (See problem 4.8.)

Another important example is provided by the *normal* matrix

$$\mathbf{A} = \begin{pmatrix} a_1 & a_2 \\ -a_2 & a_1 \end{pmatrix} \tag{4.4.15}$$

(also called the *coupled form*) which yields the characteristic polynomial

$$D(z) = (z - a_1)^2 + a_2^2. \tag{4.4.16}$$

Therefore, the poles are simply $a_1 \pm ja_2$. That is, a_1 specifies the real part of the poles, and a_2 the imaginary part. A network realizing this $\mathbf{A}$ matrix is shown in figure 4.13. Networks incorporating this recursive structure have reduced sensitivity to coefficient and signal quantization, as discussed in chapter 11.

PROBLEMS

4.1 Show that the system function $H(z)$ for the simple-averaging filter in (4.3.6) implies the magnitude response $|H'(\omega)|$ in (4.3.7).

4.2 Find the region of convergence and the impulse response $h(n)$ for

$$H(z) = \frac{1 - z^{-1}}{1 - 2z^{-1} + 0.75z^{-2}}$$

such that:

a. The filter is causal.

b. The filter is stable.

4.3 One simple, but quite useful technique for smoothing a noisy data sequence is called *averaging by 5's and 7's*. It consists effectively of a cascade of two simple-averaging filters, with one filter averaging five adjacent samples and the other filter averaging seven adjacent samples.

a. Give the impulse response $h(n)$ of the overall (causal) filter.

b. What is the order (M) of the overall filter?

c. What is the delay to the midpoint of the impulse response?

d. Sketch the pole/zero diagram for the overall filter.

e. Sketch the magnitude response of the overall filter. (*Note*: If you plot the magnitude response with a computer, you will find that it is almost *equiripple* in the stopband with about 30-dB attenuation.)

4.4 A first-order lowpass filter has the system function

$$H(z) = K\frac{1 + z^{-1}}{1 - cz^{-1}}, \qquad |z| > |c|.$$

a. Sketch the magnitude response of the filter.

b. At what frequency ω_0 does the peak gain occur? What scaling constant K is required to produce a peak gain of unity?

c. Find the coefficient c to produce a 3-db cutoff frequency of $\omega_c \approx 0.05\omega_s$, $\omega_s = 2\pi/T$. (*Hint*: Approximate the unit circle by a straight line tangent to it at $z = 1$.)

d. Give the impulse response of this filter.

4.5 Repeat problem 4.4 for the first-order highpass filter

$$H(z) = K\frac{1 - z^{-1}}{1 - cz^{-1}}, \qquad |z| > |c|.$$

4.6

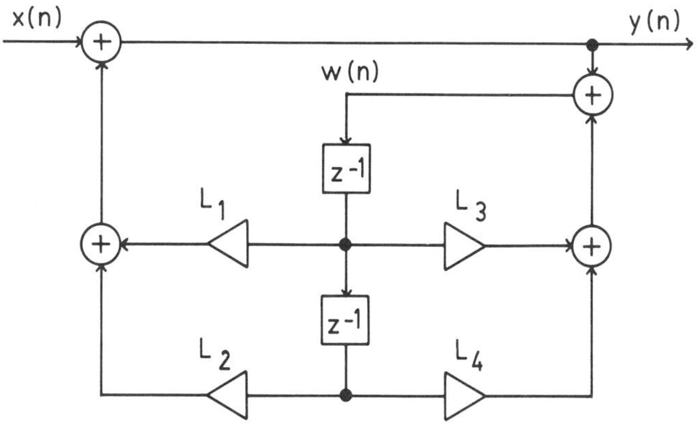

a. Write difference equations for the network shown in the figure relating $x(n)$, $w(n)$, and $y(n)$.
b. Find $H(z)$ by transforming the equations in (a) and eliminating $W(z)$.
c. Find the state matrices $\mathbf{A}$, $\underline{b}$, $\underline{c}^t$, d.
d. Find $H(z)$ from part (c) and check it with that from (b).

4.7 Show from Parseval's Relation in (3.4.10) that $h(n)$ and $H'(\omega)$ are related by

$$E_h = \sum_{n=-\infty}^{\infty} h^2(n) = \frac{1}{\omega_s} \int_0^{\omega_s} |H'(\omega)|^2 \, d\omega,$$

where $\omega_s = 2\pi/T$ is the radian sampling frequency.

4.8 Find $H(z)$ using (4.4.11) for the system in figure 4.11, which has the state matrices ($\mathbf{A}$, $\underline{b}$, $\underline{c}^t$, d) immediately following equation (4.4.3).

4.9 Find the approximate 3-dB bandwidth of the unity gain resonator in (4.3.8). (*Hint*: Approximate the unit circle by a straight line that is tangent to it at the point $e^{j\omega_0 T}$.)

4.10 A second-order *notch* filter has the system function

$$H(z) = \frac{1 - 2\cos\theta z^{-1} + z^{-2}}{1 - 2r\cos\theta z^{-1} + r^2 z^{-2}}$$

a. Sketch the pole/zero diagram for $\theta = 60°$ and $0 \ll r < 1$.
b. Sketch the magnitude response $|H'(\omega)|$.
c. At what frequency ω_0 does the maximum gain $|H'(\omega_0)|$ occur? Does this gain differ greatly from unity?
d. Find the approximate 3-dB bandwidth of the notch. (*Hint*: Approximate the unit circle by a straight line tangent to it at a zero.)

4.11 Consider the network in figure 4.13 to be a 2-input, 2-output filter with input vector $\underline{x}(n)$ and output vector $\underline{s}(n)$.

a. Write state equations for the network.
b. Find the state matrices **A**, **B**, **C**, **D** (all 2×2).
c. Find the matrix system function $\mathscr{H}(z)$ relating the vector z transforms $\underline{X}(z)$ and $\underline{S}(z)$, which is given by the generalization of (4.4.11), i.e.,

$$\mathscr{H}(z) = \mathbf{D} + \mathbf{C}(z\mathbf{I} - \mathbf{A})^{-1}\mathbf{B}.$$

5. DISCRETE-TIME NETWORKS

5.0 INTRODUCTION

To realize a discrete-time filter with either computer software or special-purpose hardware, a network must be specified describing the computations to be performed. For software realizations, the network corresponds to a flowchart of the filter algorithm; while for hardware realization, the network describes the actual circuit elements and their interconnection. Many important properties of the discrete-time filter are placed in evidence by the coefficients of certain network structures. Significant computational savings can also be achieved in many cases by the proper choice of the network. And finally, the performance of a digital implementation is affected very substantially by the choice of the network structure because of the quantization effects we will study in chapter 11.

5.1 FLOW GRAPH PROPERTIES

At least one implementation of a discrete-time filter is usually obvious from the form of its system function $H(z)$, and many others can be generated as we will see. Conversely, the system function $H(z)$ is readily deduced, in most cases, from a block diagram of some implementation of the filter. This is in marked contrast to the general case for passive continuous-time filters where quite complicated synthesis or analysis procedures are required to go from the system function to a circuit diagram, or vice versa.

The simple synthesis and analysis procedures described above for discrete-time networks are based on the following properties of linear flow graphs:

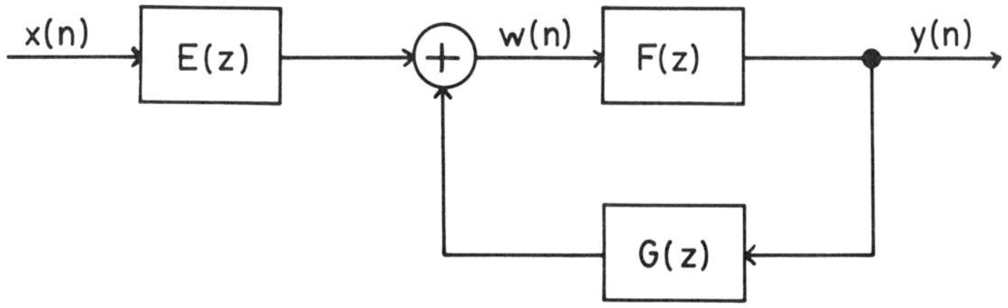

Figure 5.1. A general feedback configuration.

Parallel Networks

We saw in (2.2.6) that if two filters or filter elements with impulse response $h_1(n)$ and $h_2(n)$ are placed in parallel, the impulse response $h(n)$ of the combination equals $h_1(n) + h_2(n)$. Hence, the system function of the parallel network is given by

$$H(z) = H_1(z) + H_2(z). \tag{5.1.1}$$

Cascade Networks

If two filters or filter elements are placed in cascade, then from (2.2.5) we have $h(n) = h_1(n) * h_2(n)$, and thus the system function of the cascade network is simply

$$H(z) = H_1(z)H_2(z). \tag{5.1.2}$$

Feedback Networks

If three filter elements with individual system functions $E(z)$, $F(z)$, and $G(z)$ are arranged in the positive feedback configuration of figure 5.1, the overall system function of the feedback network is given by

$$H(z) = \frac{E(z)F(z)}{1 - F(z)G(z)}. \tag{5.1.3}$$

This is readily derived by relating $X(z)$, $Y(z)$ and $W(z)$, and is directly analogous to the corresponding Laplace transform property for continuous-time networks. In words, "the system function equals the feedforward transfer function divided by one minus the loop transfer function." (See problem 5.1.)

> **EXAMPLE.** As an example of the ease with which the above properties can usually be employed to obtain the system function from a block diagram of the filter, consider the discrete-time network of figure 4.3. This network is readily identified as being of the general form shown in figure 5.1 by using the preceding parallel and cascade network properties to show that

$$E(z) = \sum_{m=0}^{M} b_m z^{-m}, \qquad F(z) = 1, \qquad G(z) = -\sum_{k=1}^{N} a_k z^{-k}. \tag{5.1.4}$$

But then, from (5.1.3), we have immediately that

$$H(z) = \frac{\sum_{m=0}^{M} b_m z^{-m}}{1 + \sum_{k=1}^{N} a_k z^{-k}}, \tag{5.1.5}$$

which is the same as (4.2.5) since $a_0 = 1$ in the difference equation that originally led to this network.

Note also that since $z^{-1}\mathbf{A}$ is the loop gain of the network in state-variable form in figure 4.12, we can interpret (4.4.10) as being a generalization of (5.1.3) to include vector signals. In this case, however, we must take care to order the vectors and matrices correctly as in (4.4.10).

Another interesting and useful network property is that pertaining to *transpose* networks.

Transpose Networks

If the directions of all branches in the flow graph for a discrete-time filter are reversed, the system function of the resulting *transpose* network is the same as that of the original network. The input and output of the transpose network correspond, respectively, to the output and input of the original network. All *branch nodes* in the original network become *summation nodes* in the transpose network, and likewise summation nodes become branch nodes. This is illustrated in figure 5.2 for a second-order filter. The fact that the system function is unchanged by transposition can be proved using several approaches including Tellegen's theorem for discrete-time networks, Mason's rule for transfer function evaluation, or a state-variable formulation. We will use the last method to prove this result, as follows: Reversing the branch directions of a network in state-variable form as depicted in figure 4.12, we produce the state-variable description of the transpose network shown in figure 5.3. Note that $\underline{c}$ replaces $\underline{b}$ and $\underline{b}^t$ replaces $\underline{c}^t$. To see that $\mathbf{A}^t$ replaces $\mathbf{A}$, as indicated, note that the gain a_{ij} from $s_j(n)$ to $s_i(n+1)$ becomes the gain from $s_i(n)$ to $s_j(n+1)$ in the transpose network. Hence, a_{ji} must replace a_{ij}. The system function for the transpose network is, therefore,

$$H^t(z) = d + z^{-1}\underline{b}^t(\mathbf{I} - z^{-1}\mathbf{A}^t)^{-1}\underline{c}. \tag{5.1.6}$$

But this is just the matrix transpose of the 1×1 "matrix" $H(z)$, and thus $H(z) = H^t(z)$, which proves the assertion.

5.2 NETWORK STRUCTURES

The variety of possible structures for discrete-time networks is extremely wide and diverse, and the question of optimal digital-filter structures has attracted great research interest. The term *digital-filter* is used above in conjunction with optimal structures because it is only when the effects of quantization are considered that significant differences arise in the performance of different network structures. Therefore, we will defer most of our comparative analysis of network structures until quantization effects are studied in chapter 11, but the most common structures and their basic properties will be introduced in this section.

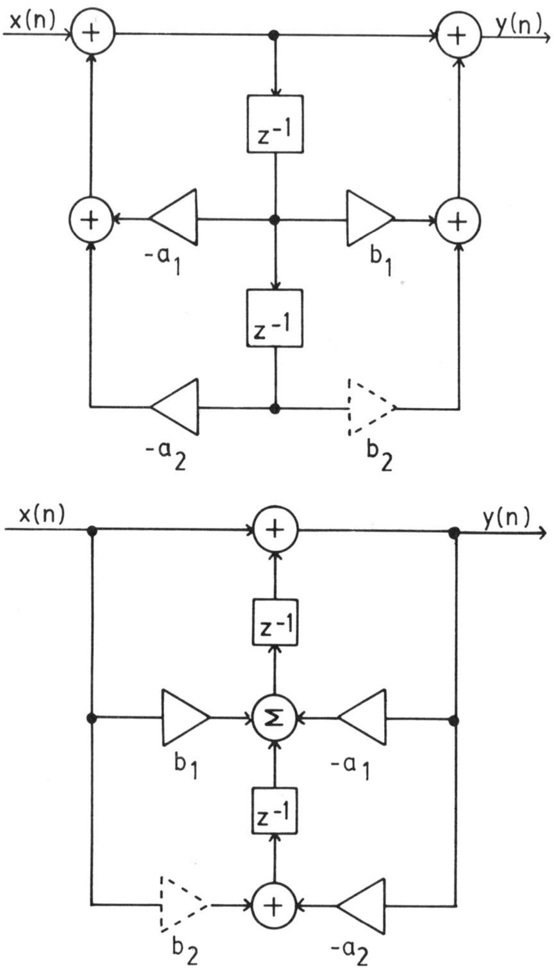

Figure 5.2. Transpose second-order networks.

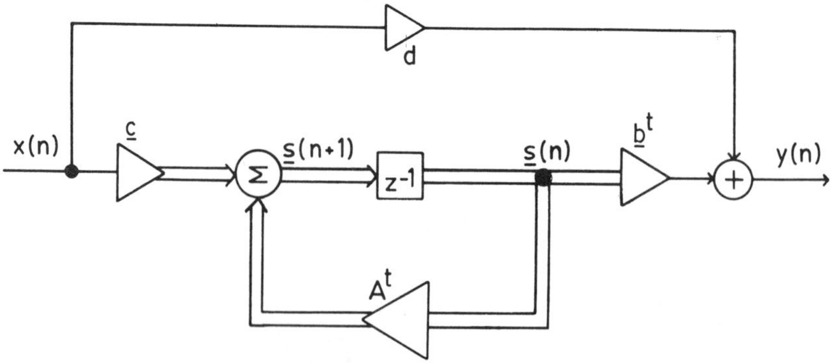

Figure 5.3. Transpose of the state-variable description in figure 4.12.

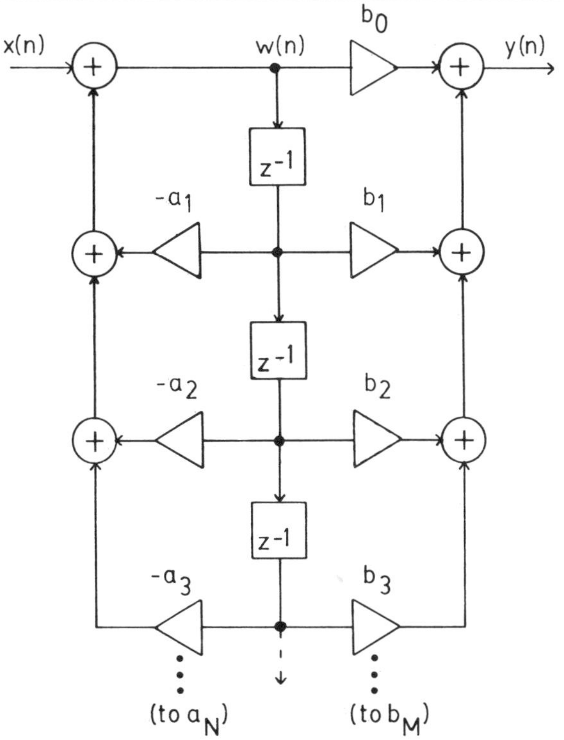

Figure 5.4. Direct form II structure.

The structure previously shown in 4.3 is often called the *direct form* of a discrete-time or digital filter because it is a direct implementation of the basic difference equation in (4.2.2). However, in the case of IIR filters, a more efficient structure in the sense of fewer delays can be generated by reversing the order of the two filter sections, which are effectively in cascade. That is, letting

$$H_1(z) = N(z) = \sum_{m=0}^{M} b_m z^{-m}$$

and

$$H_2(z) = \frac{1}{D(z)} = \frac{1}{\Sigma_{k=0}^{N} a_k z^{-k}}, \tag{5.2.1}$$

where $a_0 = 1$, figure 4.3 realizes the system function $H(z) = H_1(z)H_2(z)$ with $H_1(z)$ first, followed by $H_2(z)$. Reversing this order and eliminating the min (N, M) delays, which are thereby made redundant, we obtain the structure shown in figure 5.4 which we will call *direct form II*. Although direct form II still satisfies the difference equation in (4.2.2), it does not implement this difference equation directly. Rather, it is a direct implementation of a pair of equations relating $x(n)$, $y(n)$, and $w(n)$, as follows:

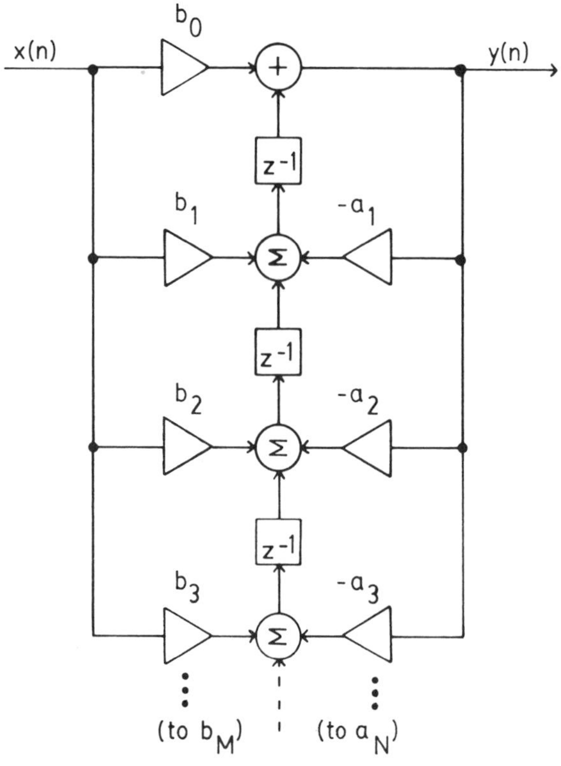

Figure 5.5. Direct form I structure.

$$w(n) = x(n) - \sum_{k=1}^{N} a_k w(n - k)$$

$$y(n) = \sum_{m=0}^{M} b_m w(n - m). \tag{5.2.2}$$

Note that this structure is still recursive, but in the intermediate output $w(n)$ and not the final output $y(n)$.

The transpose network corresponding to direct form II, which we will call *direct form I* is shown in figure 5.5. Since the order of $H_1(z)$ and $H_2(z)$ is reversed again by transposition in direct form I, this form is quite similar to the original direct form in figure 4.3. However, up to half of the delays have been eliminated by moving their location in the network structure.

Direct forms I and II are examples of *canonical* structures in that, in general, they realize the given system function with the smallest possible numbers of delays, adders, and multipliers. Assuming $N = M$, as is often the case for IIR filters, the number of each of these components required in a canonical implementation is as follows:

$$\begin{aligned} \text{number of delays} &= N, \\ \text{number of adders} &= 2N, \\ \text{number of multipliers} &= 2N + 1. \end{aligned} \tag{5.2.3}$$

By *adders* we mean two-input adders, and thus a summation node with J inputs implies $J - 1$ adders in the implementation. The reader can check that (5.2.3) is indeed satisfied by direct forms I and II when $N = M$.

The expansion of $H(z)$ in a partial-fraction expansion leads to another pair of canonical structures called the *parallel forms*. In order to produce real multiplier coefficients in the filter structure, the terms in the partial-fraction expansion corresponding to complex-conjugate pole pairs must be combined into second-order terms; and it is convenient for both notation and implementation reasons to combine the real poles in pairs, as well. We thereby produce the following form of the system function, assuming again that $N = M$:

$$H(z) = \gamma_0 + \sum_{i=1}^{L} \frac{\gamma_{1i} z^{-1} + \gamma_{0i}}{\alpha_{2i} z^{-2} + \alpha_{1i} z^{-1} + 1} \tag{5.2.4}$$

with

$$L = \left[\frac{N + 1}{2} \right]_{\text{int}}, \tag{5.2.5}$$

where $[\ ]_{\text{int}}$ denotes the *integer part of*. If N is odd, there is one first-order term in (5.2.4), say for $i = L$, and we then have $\gamma_{1L} = \alpha_{2L} = 0$. We will investigate additional properties of these filter coefficients in the next section.

Since the system function in (5.2.4) is expressed as a sum of second-order terms, the corresponding implementations must consist of parallel combinations of second-order networks or *sections*. Realizing each section as a direct-form-II network, we obtain *parallel form II*, which is shown in figure 5.6. Alternatively, we may employ direct-form-I networks for the second-order sections to produce *parallel form I*, shown in figure 5.7. Note that parallel forms I and II both satisfy (5.2.3) (keeping in mind that one section is first-order if N is odd), and hence these are also canonical forms. Moreover, note that parallel form II is, in fact, the transpose of parallel form I.

If, in addition to factoring the denominator polynomial $D(z)$ in (5.2.1) into second-order factors as for the parallel forms, the numerator polynomial $N(z)$ is similarly factored, the system function can be written as the product

$$H(z) = b_0 \prod_{i=1}^{L} H_i(z), \tag{5.2.6}$$

where

$$H_i(z) = \frac{\beta_{2i} z^{-2} + \beta_{1i} z^{-1} + 1}{\alpha_{2i} z^{-2} + \alpha_{1i} z^{-1} + 1}. \tag{5.2.7}$$

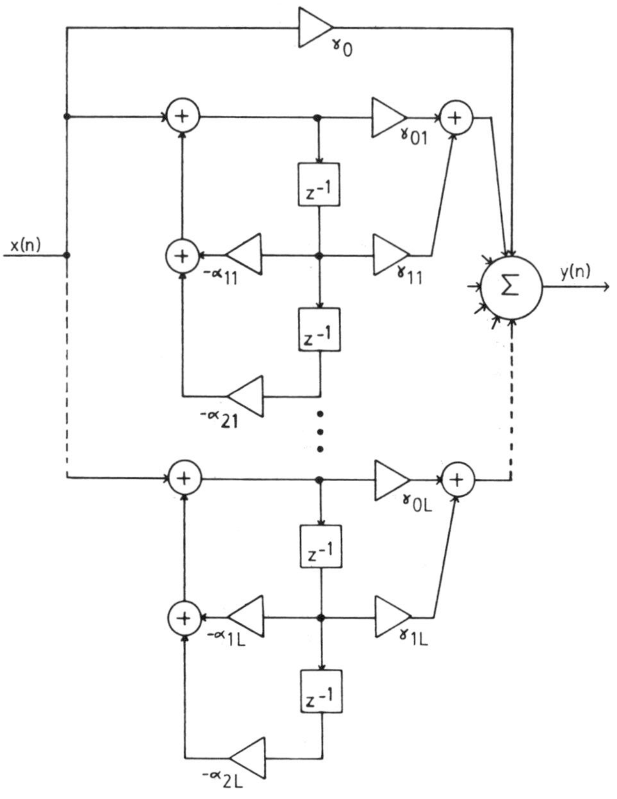

Figure 5.6. Parallel form II structure.

L is again given by (5.2.5), and $\alpha_{2L} = \beta_{2L} = 0$ if N is odd. Since $H(z)$ is formed as the product of the second-order functions $H_i(z)$, the corresponding filter structures must comprise a cascade of second-order sections. Implementing these sections in direct forms II and I, we produce *cascade forms II and I*, respectively, shown in figures 5.8 and 5.9. The reader can verify that these structures are also canonical and that they are related in topology by transposition. However, cascade form II is actually the transpose of form I only if the order of the filter coefficients is reversed from one form to the other.

Many other filter structures can obviously be produced using, for example, combinations of the parallel and cascade forms, various feedback arrangements, continued-fraction expansions of $H(z)$, and so forth. Many additional structures can be generated as analogs of classical continuous-time filters and have names such as *wave digital filters*, *ladder structures*, and *lattice structures*. A complete description of all of these structures is beyond the scope of this book, but several of them will be considered in subsequent chapters. By far the most common digital-filter structures are the direct, parallel, and cascade forms; and of these, the cascade forms are most often employed for IIR filters for reasons concerning quantization and implementation. In the case of FIR filters, only the direct and cascade forms are applicable, with the direct form being most commonly employed.

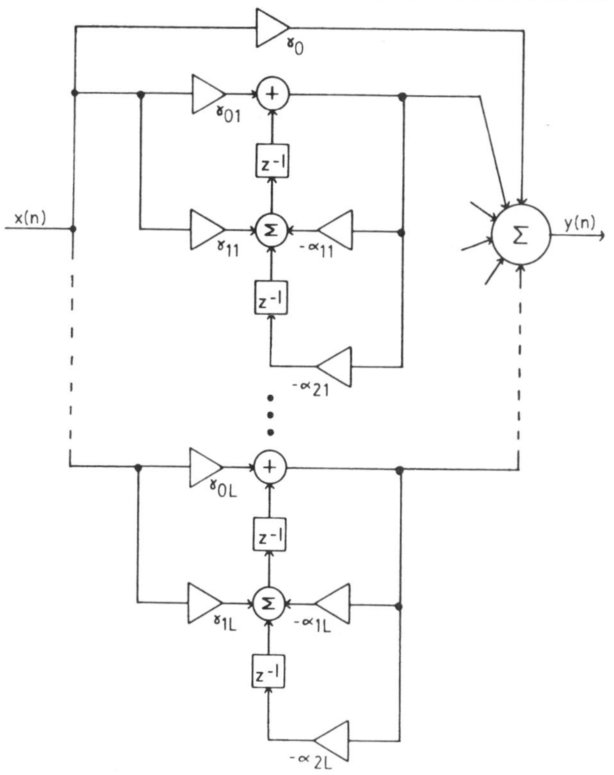

Figure 5.7. Parallel form I structure.

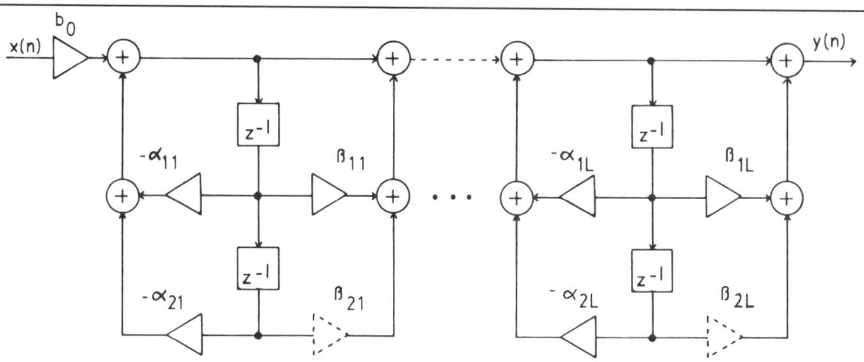

Figure 5.8. Cascade form II structure.

We will often refer in the remainder of this book to simply the "direct form", the "parallel form," or the "cascade form"; whereas we have actually defined two network structures in each case. This is done for convenience in view of the fact that each pair of structures implements the same form of the system function $H(z)$. Thus, in effect, we refer by this terminology to the form of the system function itself as expressed in (5.1.5), (5.2.4), or (5.2.6).

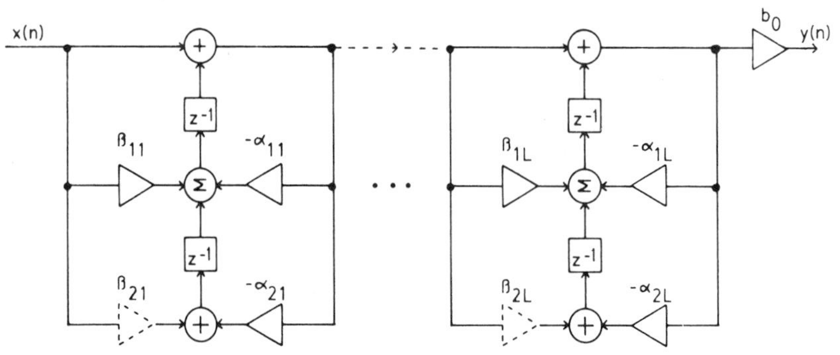

Figure 5.9. Cascade form I structure.

5.3 PROPERTIES OF NETWORK COEFFICIENTS

One can quickly determine a number of important properties of a discrete-time filter from the coefficients of certain network structures. In the case of IIR filters, information concerning such things as stability, real or complex singularities, transmission zeros, and quantization effects is readily obtained from the cascade-form coefficients or, to a lesser extent, the parallel-form coefficients. For FIR filters, such properties as linear phase, 90° phase shift, or symmetrical frequency response are apparent from the coefficients of the direct form or certain cascade structures.

The denominator factors of the parallel and cascade forms and the numerator factors of the cascade form are all quadratic functions of the form

$$\begin{aligned} F_i(z) &= 1 + c_{1i}z^{-1} + c_{2i}z^{-2} \\ &= (1 - q_{1i}z^{-1})(1 - q_{2i}z^{-1}), \end{aligned} \tag{5.3.1}$$

where the roots q_{1i} and q_{2i} are the corresponding singularity (pole or zero) locations in the z plane. The roots may be real or complex, but in either case

$$c_{1i} = -(q_{1i} + q_{2i}), \tag{5.3.2}$$

$$c_{2i} = q_{1i}q_{2i}. \tag{5.3.3}$$

If $c_{1i}^2 \geq 4c_{2i}$, the roots are real; while if $c_{1i}^2 < 4c_{2i}$, the roots are complex conjugates. Letting $q_{1i} = q_{2i}^* = q_i$ in the latter case, we have

$$c_{1i} = -2\,\mathrm{Re}\,(q_i) = -2r_i\cos\theta_i \tag{5.3.4}$$

$$c_{2i} = |q_i|^2 = r_i^2, \tag{5.3.5}$$

where r_i is the radius of the singularities in the z plane, and $\pm\theta_i$ are the corresponding angles. Therefore, the pole locations are quickly calculated from the coefficients of the parallel or cascade forms, and for the cascade form, the zeros are similarly determined.

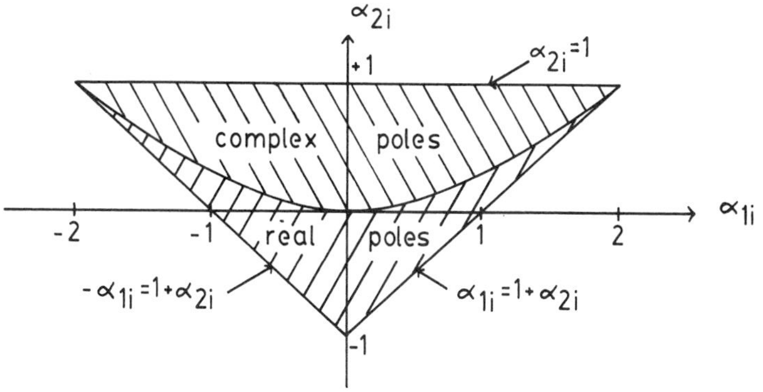

Figure 5.10. Region of coefficient values for which a second-order filter is stable.

As a result of the above, we can readily obtain conditions on the coefficients of the parallel or cascade forms to ensure the stability of the filter. Specifically, the second-order denominator factors of these forms are given by

$$D_i(z) = 1 + \alpha_{1i} z^{-1} + \alpha_{2i} z^{-2}$$
$$= (1 - p_{1i} z^{-1})(1 - p_{2i} z^{-1}),$$

and the poles must lie inside the unit circle for stability, i.e.,

$$|p_{1i}|, |p_{2i}| < 1.$$

Hence, from (5.3.3),

$$|\alpha_{2i}| = |p_{1i} p_{2i}| < 1. \tag{5.3.6}$$

The corresponding condition on α_{1i} is obtained from the expression

$$p_{1i}, p_{2i} = \frac{-\alpha_{1i} \pm \sqrt{\alpha_{1i}^2 - 4\alpha_{2i}}}{2}$$

(see problem 5.4) and is given by

$$|\alpha_{1i}| < 1 + \alpha_{2i}. \tag{5.3.7}$$

Conditions (5.3.6) and (5.3.7) are illustrated in figure 5.10, which shows the resulting *stability triangle* in the α_1, α_2 plane. That is, the second-order section is stable if, and only if, α_{1i} and α_{2i} define a point that lies inside this triangle. As previously noted, the poles are complex if $\alpha_1^2 < 4\alpha_2$, and real otherwise.

Very often, the zeros of $H(z)$ lie on the unit circle in the z plane. In fact, as we will see in chapter 8, this is always true for discrete-time filters derived by bilinear transformation from the classical continuous-time designs. If these zeros are complex conjugates (including two equal real zeros), then from (5.3.5) we see that the numerator coefficients β_{2i} of the cascade form must be unity. So, in this case,

$$\beta_{1i} = -2\cos\theta_i \quad \text{and} \quad \beta_{2i} = 1.$$

In particular, if the zeros are real and equal, we have

$$\beta_{1i} = \pm 2.$$

If a second-order section has unequal real zeros of unit magnitude, then they must be $z_{1i} = 1$ and $z_{2i} = -1$, and thus from (5.3.2) and (5.3.3)

$$\beta_{1i} = 0 \quad \text{and} \quad \beta_{2i} = -1.$$

If the numerator is actually first-order with only one zero, then of course

$$\beta_{1i} = \pm 1 \quad \text{and} \quad \beta_{2i} = 0.$$

Therefore, in all of the above cases of zeros on the unit circle, the β_{2i} coefficients of the cascade form are always the integers ± 1 or 0, and the β_{1i} are also integers (± 2, ± 1, or 0) if the zeros are real. For this reason, the β_{2i} multipliers were shown with dotted lines in figures 5.8 and 5.9 because actual multipliers are not required to implement these integer coefficients. This, in part, explains the popularity of the cascade form with designers since 25 percent to as much as 50 percent of the multiplication time or hardware can be saved in filters having zeros on the unit circle.

Turning to FIR filters, we will investigate what is perhaps their most important property, namely: that *the coefficients of an FIR filter are easily constrained to produce a linear phase response. The corresponding constraint is simply that the finite-duration impulse response have conjugate-even symmetry about its midpoint.* To see that this constraint ensures linear phase, consider the FIR system function

$$H(z) = \sum_{m=0}^{M} b_m z^{-m} \tag{5.3.8}$$

with

$$b_m = b^*_{M-m} = |b_m| e^{j\phi_m}.$$

If M is even, the coefficient $b_{M/2}$ is real and corresponds to the center of symmetry of $h(n)$; while if m is odd, there is no central coefficient. These two cases are illustrated in figure 5.11 for b_m real, where we then have

$$b_m = b_{M-m}.$$

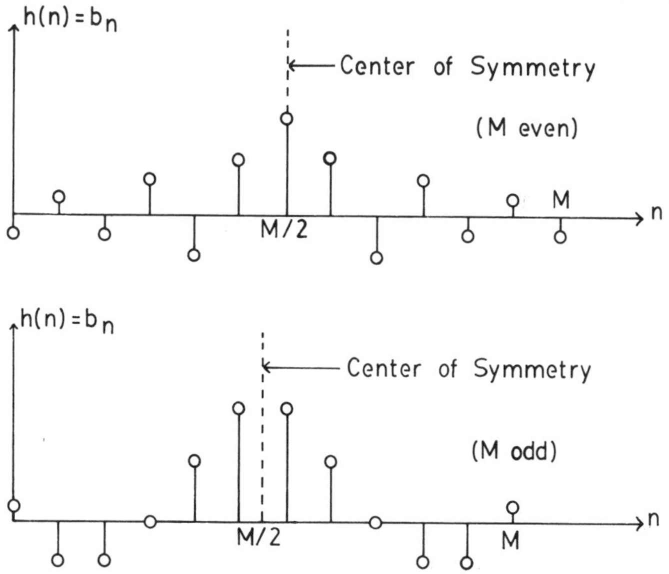

Figure 5.11. Two cases of FIR symmetry for linear phase.

Considering first the case of M even, we may rewrite (5.3.8) as

$$\begin{aligned} H(z) &= z^{-M/2}\left[\sum_{m=0}^{M} b_m z^{-m+M/2}\right] \\ &= z^{-M/2}\left[b_{M/2} + \sum_{m=0}^{M/2-1}(b_m z^{-m+M/2} + b_{M-m} z^{m-M/2})\right]. \end{aligned} \tag{5.3.9}$$

Substituting $z = e^{j\omega T}$ and taking into account that $b_m^* = b_{M-m}$, we find that the frequency response is given by

$$\begin{aligned} H'(\omega) &= e^{-j\omega TM/2}\left\{b_{M/2} + \sum_{m=0}^{M/2-1} 2|b_m|\cos\left[\left(\frac{M}{2} - m\right)\omega T + \phi_m\right]\right\} \\ &= e^{-j\omega TM/2} R(\omega), \end{aligned} \tag{5.3.10}$$

where $R(\omega)$ is purely real. If $R(\omega)$ is of constant sign for all ω, then $R(\omega) = \pm|H'(\omega)|$ and we indeed have the linear phase response

$$\angle\, H'(\omega) = -\omega\frac{TM}{2}.$$

If, however, there are sign changes in $R(\omega)$, there are corresponding 180° phase shifts in $\angle\, H'(\omega)$, and $\angle\, H'(\omega)$ is only piecewise linear. It is common practice, nonetheless, still to refer to the filter as having simply *linear phase*. This terminology is reasonable since we are actually trying to constrain the filter's *group delay* $D(\omega)$ to be constant, and since

$$D(\omega) = -\frac{d}{d\omega} \angle H'(\omega),$$

we have

$$D(\omega) = \frac{M}{2} T \tag{5.3.11}$$

except at those frequencies where $R(\omega)$ changes sign. But at those frequencies, $R(\omega) = |H'(\omega)| = 0$ and hence there is no output contribution anyway.

The implication of constant group delay is that all frequency components of an input sequence are similarly delayed in the output sequence. Hence, the symmetry (odd or even) of an input pulse sequence, for example, is preserved in the output sequence if the impulse response $h(n)$ has even symmetry. The delay between the corresponding centers of symmetry is just the group delay, and from (5.3.11) we see that this corresponds to the delay in $h(n)$ at its center of symmetry. For the above case of M even, the group delay equals an integer number of sampling periods T and thus is easily accounted for in subsequent data processing. The case of M odd leads to expressions similar to (5.3.9) and (5.3.10), and equation (5.3.11) still holds for the group delay in this case. However, since this delay is no longer an integer number of sampling periods for odd M, it can be more difficult to account for unless we wish to interpolate between samples.

The effect of the linear-phase constraint on the zeros of $H(z)$ is seen by noting from (5.3.8) that

$$z^M H(z) = H^*(1/z^*) \tag{5.3.12}$$

because of the conjugate-even symmetry in the b_m. Equation (5.3.12) implies that the zeros of $H(z)$ must also be zeros of $H^*(1/z^*)$, which means that if z_m is a zero of $H(z)$, then $1/z_m^*$ is also. Therefore, *the zeros of a linear-phase filter either must lie on the unit circle or must occur in pairs with reciprocal radii.* For b_m real, the zeros must also occur as complex conjugates; and thus in that case, those not lying in the unit circle will actually occur in quadruples, as illustrated in figure 5.12. It is readily shown (see problem 5.5) that if M is odd and the b_m are real, the odd zero must fall at $z = -1$.

Conjugate-odd symmetry in an FIR impulse response is also associated with an important class of discrete-time filters. These filters provide a 90° phase shift at all frequencies, in addition to constant group delay. This is seen by evaluating (5.3.9) for $z = e^{j\omega T}$ and $b_{M-m} = -b_m^*$ to obtain

$$\begin{aligned} H'(\omega) &= e^{-j\omega TM/2} \left\{ j \sum_{m=0}^{M/2-1} 2|b_m| \sin\left[\left(\frac{M}{2} - m\right)\omega T + \phi_m\right]\right\} \\ &= je^{-j\omega TM/2} R(\omega) \end{aligned} \tag{5.3.13}$$

because $j = e^{j\pi/2}$ implies a phase shift of $\pi/2$. $R(\omega)$ is again real, and we have noted by the upper limit on the sum that $b_{M/2}$ must equal zero (if M is even). Such FIR filters are useful, for example, in approximating the ideal differentiator

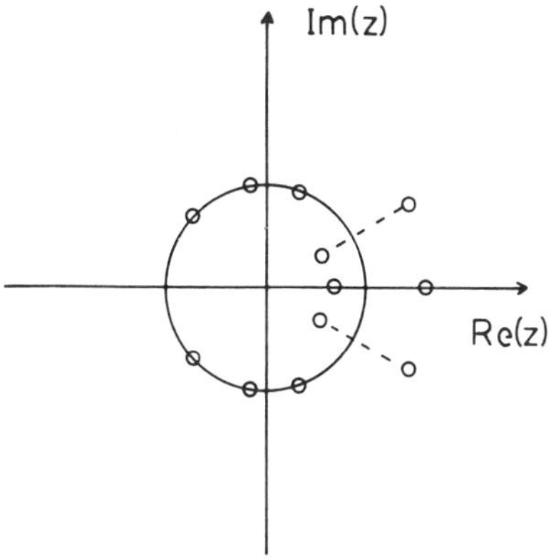

Figure 5.12. Zeros of a linear-phase FIR filter.

$$H'_D(\omega) \;=\; j\omega, \qquad -\frac{\pi}{T} < |\omega| < \frac{\pi}{T}. \tag{5.3.14}$$

As indicated above, the factor j in $H'_D(\omega)$ is realized exactly by the symmetry condition, and we design the coefficients b_m so that $R(\omega)$ approximates

$$R(\omega) \;\approx\; \omega, \qquad -\frac{\pi}{T} < |\omega| < \frac{\pi}{T}. \tag{5.3.15}$$

There is, of course, an additional linear-phase factor in (5.3.13), which can be eliminated for M even if the filter is made noncausal and is centered at $n = 0$.

Another example is the ideal Hilbert transformer

$$H'_H(\omega) \;=\; \begin{cases} -j, & 0 < \omega < \pi/T \\ +j, & -\pi/T < \omega < 0, \end{cases} \tag{5.3.16}$$

in which case the coefficients b_m are designed to give

$$R(\omega) \approx \begin{cases} -1, & 0 < \omega < \pi/T \\ +1, & -\pi/T < \omega < 0, \end{cases} \tag{5.3.17}$$

and there is, of course, a delay of $MT/2$. For real b_m, the odd symmetry condition implies, with M even or odd, that $H(1) = 0$ and thus there is a zero at $z = 1$. Otherwise, the zeros are located as in the case of linear-phase filters.

As a final point concerning real FIR networks with even or odd symmetry in the filter coefficients, we note that these networks can be implemented with only $[(M/2) + 1]_{\text{int}}$ multipliers. This results from writing $H(z)$ as

$$H(z) \;=\; b_0(1 \pm z^{-M}) + b_1(z^{-1} \pm z^{-M+1}) + \cdots + b_{M/2}z^{-M/2}$$

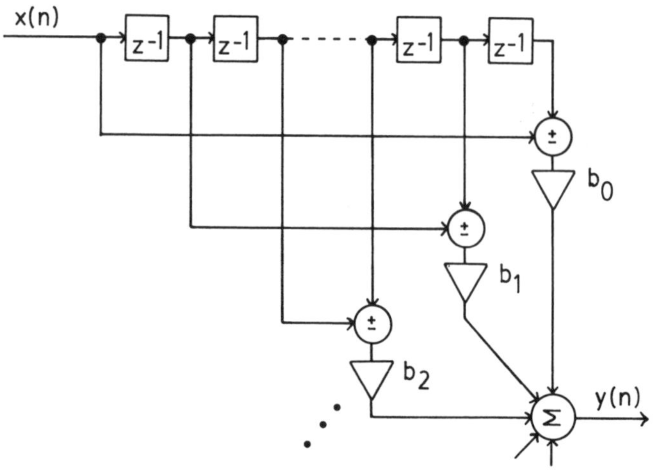

Figure 5.13. Structure that saves about half the multipliers for linear-phase FIR filters.

for M even, and likewise for M odd without the last term. The corresponding direct-form-I network is shown in figure 5.13.

5.4 SPECIAL DISCRETE-TIME NETWORKS

In addition to linear-phase and linear-plus-90°-phase filters, there are many other special discrete-time networks with interesting and useful properties. Those to be considered here are the allpass filter, the comb filter, and unity-gain filters.

Allpass Filters

An *allpass* filter has a magnitude response that is unity for all frequencies, i.e.,

$$|H'(\omega)| = 1, \qquad \text{for all } \omega. \tag{5.4.1}$$

Such filters are useful for phase equalization of IIR designs, and they also play a central role in discrete-time spectral transformations (see section 8.4). The system function of an allpass filter is of the form

$$H(z) = \frac{\Sigma_{k=0}^{N} a_k z^{-N+k}}{\Sigma_{k=0}^{N} a_k z^{-k}} = \frac{z^{-N} + a_1 z^{-N+1} + \cdots + a_N}{1 + a_1 z^{-1} + \cdots a_N z^{-N}}$$

$$= \prod_{i=1}^{L} \frac{z^{-2} + \alpha_{1i} z^{-1} + \alpha_{2i}}{\alpha_{2i} z^{-2} + \alpha_{1i} z^{-1} + 1}, \tag{5.4.2}$$

where all coefficients are real. Hence, the numerator and denominator coefficients are the same except that their order is reversed in both the direct and cascade forms. To see that (5.4.2) implies (5.4.1), we rewrite (5.4.2) as

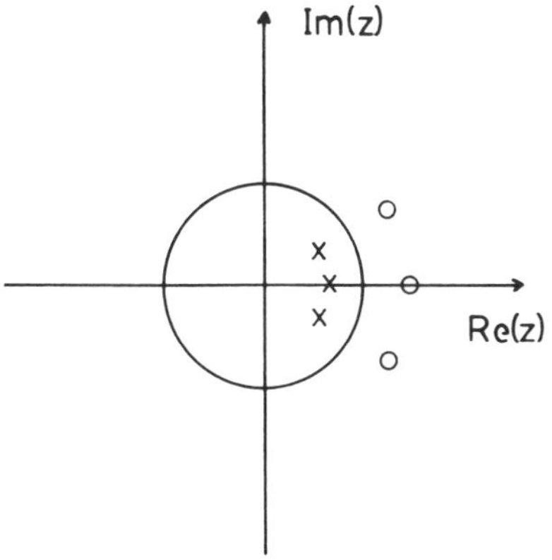

Figure 5.14. Sample allpass pole/zero diagram.

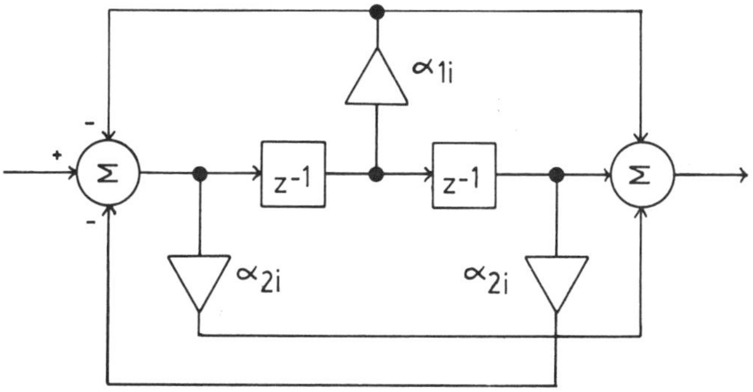

Figure 5.15. Cascade-form allpass section.

$$H(z) = \frac{z^{-N}\sum_{k=0}^{N} a_k z^k}{\sum_{k=0}^{N} a_k z^{-k}} = z^{-N}\frac{D(z^{-1})}{D(z)}. \tag{5.4.3}$$

It follows immediately that

$$|H'(\omega)| = \left|\frac{D'(-\omega)}{D'(\omega)}\right| = 1$$

because $|D'(\omega)|$ is an even function if $D(z)$ has real coefficients. From (5.4.3) we also note that the zeros of $H(z)$ are the reciprocals of its poles. A typical pole/zero diagram of an allpass filter is illustrated in figure 5.14.

A cascade-form allpass section having three multipliers is shown in figure 5.15. Cascade-form networks with only two multipliers per section can also be generated [3, 4].

Comb Filters

Given an arbitrary system function $H(z)$, consider the response of a filter with the system function $H(z^k)$. Since $H'(\omega)$ is periodic with period $\omega_s = 2\pi/T$, $H'(k\omega)$ must be periodic

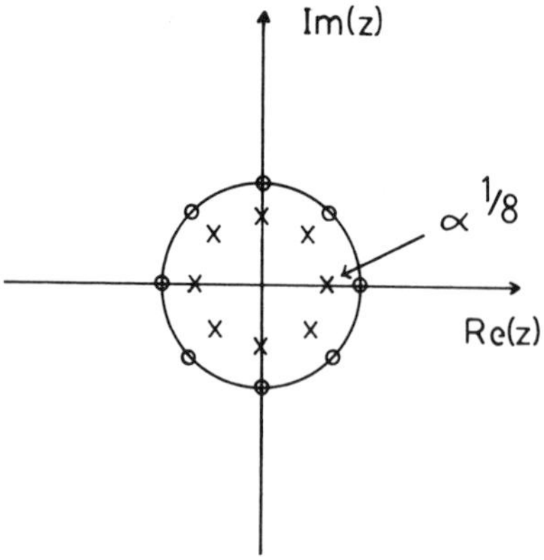

Figure 5.16. Sample comb-filter pole/zero diagram.

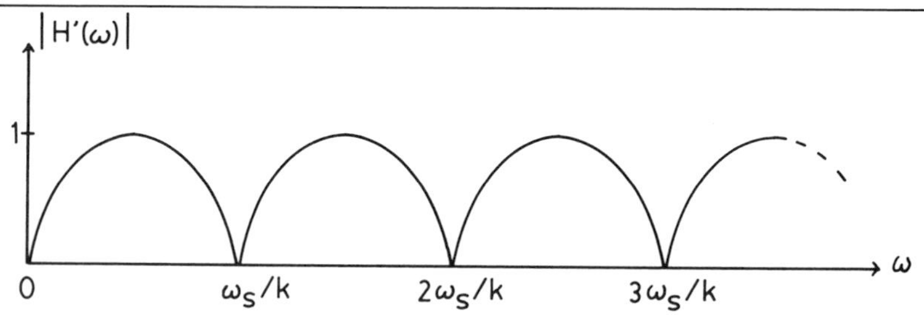

Figure 5.17. Sample comb-filter magnitude response.

with period ω_s/k. Hence, the frequency response corresponding to $H(z^k)$ is periodic within the Nyquist interval $0 \leq \omega \leq \omega_s/2$. Such filters are called *comb filters* and have various applications including the rejection of power-line harmonics and the suppression of *clutter* from fixed objects in moving-target-indicator (MTI) radars. They are also useful conceptually in the design of FIR filters via the frequency sampling technique (see section 9.2).

As an example of comb filters, we will transform the highpass function

$$H(z) = \frac{1 - z^{-1}}{1 - \alpha z^{-1}}$$

into a multi-notch comb filter by substituting z^k for z. The system function $H_k(z)$ for the comb filter is then

$$H_k(z) = H(z^k) = \frac{1 - z^{-k}}{1 - \alpha z^{-k}},$$

which has the pole/zero diagram shown in figure 5.16 (for $k = 8$). The corresponding frequency response is sketched in figure 5.17, and we see that the frequency ω_s/k and all its harmonics will be rejected by this filter (as well as $\omega = 0$, or dc). An implementation of this filter is diagrammed in figure 5.18.

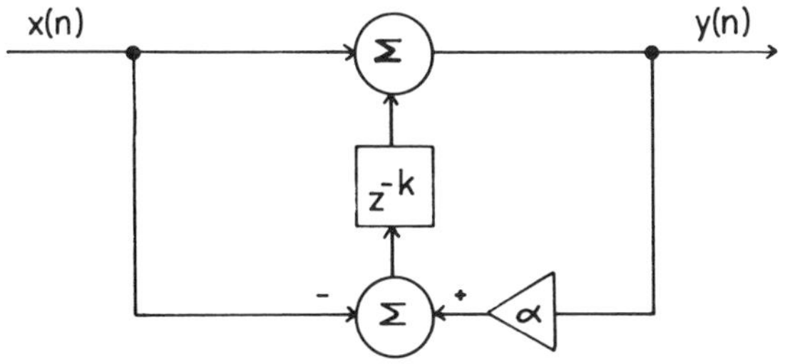

Figure 5.18. Comb filter implementation.

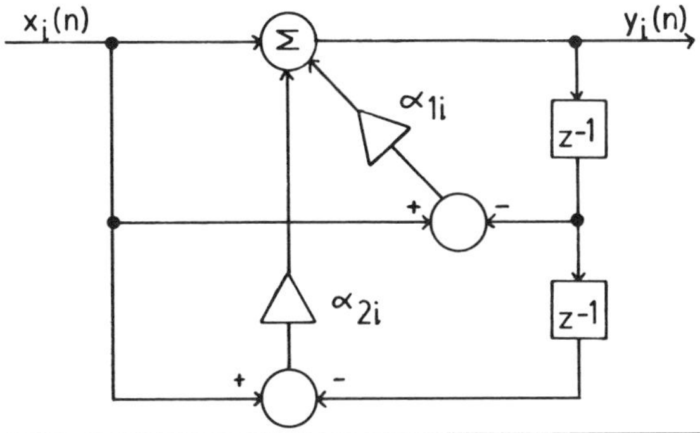

Figure 5.19. Second-order section with unity gain at dc.

Unity-Gain Filters

A unity-gain resonator was described at the end of section 4.3, where it was shown that a network could be generated to provide this feature automatically. A similar technique can be used to implement filters with unity gain at dc ($\omega = 0$), which is useful in lowpass filters and in speech (formant) synthesizers. Assume, for the moment, an all-pole filter in cascade form with the system function for each section given by

$$H_i(z) = \frac{A_i}{\alpha_{2i} z^{-2} + \alpha_i z^{-1} + 1},$$

where the gain A_i is chosen such that

$$H_i(1) = 1.$$

Obviously then, the required gain is

$$A_i = \alpha_{2i} + \alpha_{1i} + 1,$$

which can be incorporated into the section network as shown in figure 5.19. (To see that the desired A_i is achieved by this network, the reader can manipulate either the flow graph or the corresponding difference equations—see problem 5.6.)

If the filter also has zeros, this technique for unity-gain scaling is not generally applicable. However, in the frequently encountered case of all zeros at $z = -1$, the technique can still be used because each zero implies an additional gain factor of 2, which is easily compensated for (by a shift of one bit in a digital implementation).

PROBLEMS

5.1 Derive equation (5.1.3) for the system function $H(z)$ of a feedback network by relating $X(z)$, $Y(z)$, and $W(z)$ in figure 5.1.

5.2

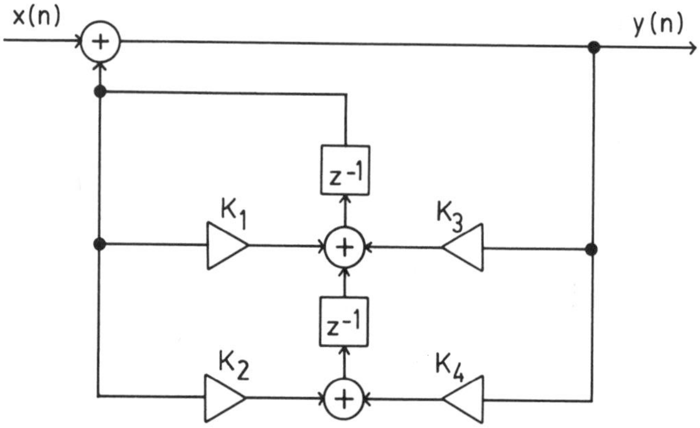

a. Find $H(z)$ for the network in the figure. (*Hint*: Identify the feedback network $G(z)$.)
b. What conditions on the parameters K_i ensure stability?
c. Draw the transpose network.
d. Give the state matrices $\mathbf{A}$, $\underline{b}$, $\underline{c}^t$, d.
e. Find $H(z)$ from (d) and check it with (a).
f. If the fractional parts of all multiplier coefficients are quantized to the same number of bits, is there any advantage to this configuration over the second-order direct-form structure with regard to the accuracy with which $H(z)$ can be realized? Why?

5.3 As a result of a particular computer-aided-design (CAD) algorithm, the following causal second-order filter was produced:

$$H(z) = \frac{1 + 2z^{-1} + z^{-2}}{1 - 2z^{-1} + 1.33z^{-2}}.$$

a. Is this filter stable?
b. If not, give a causal and stable $H(z)$ having the same magnitude response $|H'(\omega)|$.
c. Sketch $|H'(\omega)|$.

5.4 Show that equation (5.3.7) is required, along with (5.3.6), for the stability of a second-order section.

5.5 Show that for a linear-phase FIR filter with odd order M and real coefficients b_m, one zero must lie at $z = -1$. (*Hint*: Look at figure 5.11(b)).

5.6 Show that the network in figure 5.19 has unity gain at dc, i.e., $H_i(1) = 1$.

5.7 Sketch the pole/zero diagram and the magnitude response for the comb filter

$$H(z) = K\frac{1 + z^{-4}}{1 - az^{-4}}.$$

Find K such that the peak gain is unity, and design a structure that incorporates the scaling constant K without an additional multiplier.

5.8 Find $H(z)$ and $h(n)$ for the network in the figure.

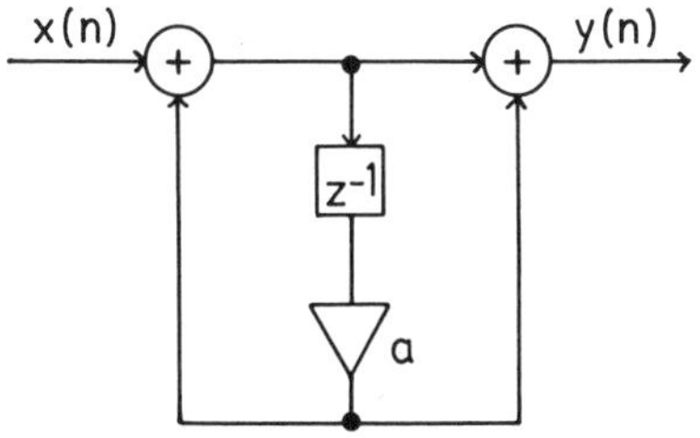

5.9 Sketch the pole/zero diagram and magnitude response for the network in the figure.

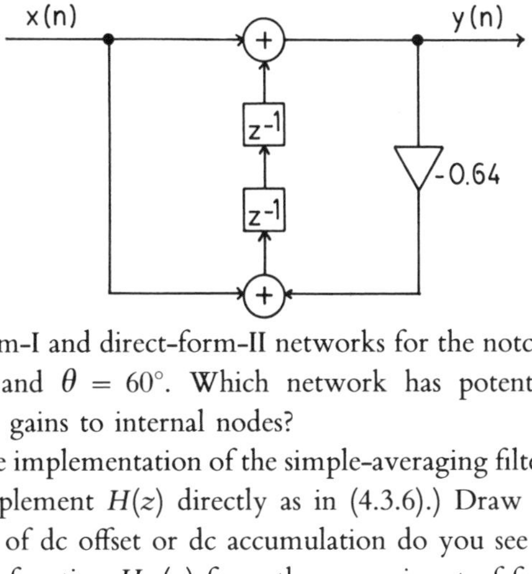

5.10 Draw direct-form-I and direct-form-II networks for the notch filter in problem 4.10 with $r = 0.95$ and $\theta = 60°$. Which network has potential overflow problems because of large gains to internal nodes?

5.11 Draw a recursive implementation of the simple-averaging filter incorporating a comb filter. (*Hint*: Implement $H(z)$ directly as in (4.3.6).) Draw the transpose network. What problems of dc offset or dc accumulation do you see in these networks?

5.12 Find the system function $H_{11}(z)$ from the upper input of figure 4.13 to the output $s_1(n)$ by feedback analysis. Repeat for $H_{21}(z)$ from the same input to output $s_2(n)$. You may wish to check your answers with those from problem 4.11(c).

5.13 A first-order allpass filter has the system function

$$H(z) = \frac{z^{-1} - a}{1 - az^{-1}}.$$

a. Draw form-I and form-II networks for this filter.
b. Sketch the phase response $\angle H'(\omega)$ of this filter.
c. When the coefficient a is quantized to a finite number of bits, its value will be perturbed. Will this perturb the magnitude and/or the phase responses of the filter?

5.14 Find the state matrices $\mathbf{A}$, $\underline{b}$, $\underline{c}^t$, d for a 4th-order filter in parallel form II. Repeat for cascade form II. (Be careful to include all paths to and from the states.)

5.15 A discrete-time oscillator has the impulse response $h_1(n) = \cos n\omega_0 T u(n)$. Draw a network with this impulse response. Repeat for $h_2(n) = \sin n\omega_0 T u(n)$. Incorporate both outputs into a single network to produce a *quadrature* oscillator.

5.16 Draw cascade- and parallel-form-I networks for the following filter using first-order sections:

$$H(z) = \frac{0.7(1 - z^{-2})}{1 - 0.3z^{-1} - 0.4z^{-2}}.$$

Compare the number of multipliers in the two networks.

6. SAMPLING CONTINUOUS-TIME SIGNALS

6.0 INTRODUCTION

Up to now, we have studied the fundamental definitions and properties of discrete-time signals and systems with only parenthetical reference to the analgous theory for continuous-time signals and systems. We have thereby demonstrated that the discrete-time theory stands on its own without the necessity of any direct reference to continuous-time theory, given that we assume the form of the discrete-time input sequences. It may well be that our input is digitally generated, for example, and that the output from our filter is to be processed or displayed without conversion to a continuous-time signal. In that case, we need not consider the relationship between discrete- and continuous-time signals. However, the real physical world is continuous in time and space, and our input data is usually derived from continuous-time (or equivalent) signals. The output data must often be returned to this form, as well. Therefore, in this chapter we will study the effects of conversion from continuous- to discrete-time signals, and vice versa, and will relate the z transform to the Fourier and Laplace transforms.

As discussed in chapter 1, the term *analog* is often used in place of *continuous-time* or other similarly precise terms. This is especially true for conversion equipment, where the terms *analog-to-digital converter* (ADC) and *digital-to-analog converter* (DAC) are almost universally employed. There is some variation in the exact usage of the term ADC, however, and we will describe this problem and the most commonly accepted terminology.

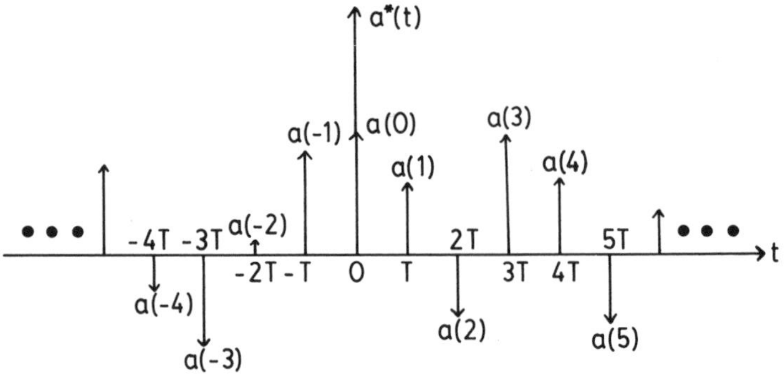

Figure 6.1. Ideal continuous-time signal $a^*(t)$ corresponding to the discrete-time sequence $a(n)$.

6.1 FOURIER TRANSFORM RELATIONSHIPS

We have called the function $H'(\omega) = H(e^{j\omega T})$ the *frequency response* of a discrete-time filter because it relates the amplitudes and phases of sinusoidal input and output sequences at the frequency ω. By linearity, it thus relates the individual frequency components of arbitrary input and output discrete-time sequences. However, for a *signal* $a(n)$ with z transform $A(z)$, what is the significance of the function $A'(\omega) = A(e^{j\omega T})$? We might be inclined to say that it's the *spectrum* of $a(n)$, or perhaps the Fourier transform, but what does it mean for a sequence of numbers to have a spectrum? And how can the Fourier transform, which is defined as a continuous-time integral, be applied to discrete-time signals?

In an attempt to answer the above questions, it is reasonable to investigate the continuous-time signal

$$a^*(t) = \sum_{n=-\infty}^{\infty} a(n)\, \delta(t - nT) \tag{6.1.1}$$

because, as illustrated in figure 6.1, it "looks" rather like our previous plots of discrete-time sequences. The Fourier transform of $a^*(t)$ is, by definition,

$$A^*(j\omega) = \int_{-\infty}^{\infty} a^*(t)\, e^{-j\omega t}\, dt, \tag{6.1.2}$$

which, after substituting (6.1.1) and reversing the order of summation and integration, becomes

$$\begin{aligned} A^*(j\omega) &= \sum_{n=-\infty}^{\infty} a(n) \int_{-\infty}^{\infty} e^{-j\omega t}\, \delta(t - nT)\, dt \\ &= \sum_{n=-\infty}^{\infty} a(n)\, e^{-j\omega n T} \\ &= A(z)\Big|_{z=e^{j\omega T}} = A'(\omega). \end{aligned} \tag{6.1.3}$$

Hence, we find that $A'(\omega)$ is, in fact, the Fourier transform of $a^*(t)$, and we will refer to it as the *spectrum* of $a(n)$ in this sense.

The signal $a^*(t)$ is an abstraction, of course, but a quite useful one, nonetheless. It is produced theoretically by "sampling" a continuous-time input signal $a_c(t)$ for which

$$a_c(nT) = a(n). \tag{6.1.4}$$

and thus the sampling operation may be modelled as the multiplication of $a_c(t)$ by a periodic impulse train. That is, letting

$$i(t) = \sum_{n=-\infty}^{\infty} \delta(t - nT), \tag{6.1.5}$$

we have

$$a^*(t) = a_c(t) \cdot i(t). \tag{6.1.6}$$

But for continuous-time signals, as for discrete-time signals, multiplication in the time domain implies convolution in the frequency domain, and thus

$$A'(\omega) = A^*(j\omega) = A_c(j\omega) * I(j\omega)/2\pi, \tag{6.1.7}$$

where $A_c(j\omega)$ and $I(j\omega)$ are the Fourier transforms of $a_c(t)$ and $i(t)$, respectively. $I(j\omega)$ is itself a periodic impulse train in the frequency domain and is given by

$$I(j\omega) = \omega_s \sum_{k=-\infty}^{\infty} \delta(\omega - k\omega_s), \tag{6.1.8}$$

where ω_s is the radian sampling frequency

$$\omega_s = 2\pi f_s = 2\pi/T. \tag{6.1.9}$$

Therefore, from (6.1.7) and (6.1.8) we find that

$$A'(\omega) = A^*(j\omega) = \frac{1}{T} \sum_{k=-\infty}^{\infty} A_c[j(\omega - k\omega_s)], \tag{6.1.10}$$

which is illustrated in figure 6.2. We knew from before that $A'(\omega) = A(e^{j\omega T})$ must be periodic with period ω_s since $e^{j\omega T}$ is itself periodic; and (6.1.10) now confirms this and shows the exact relationship between $A_c(j\omega)$ and $A'(\omega)$. Note that any frequency components in $A_c(j\omega)$ above $\omega_s/2$ are, in effect, "folded" about this frequency due to the overlap of the shifted versions of $A_c(j\omega)$ in $A'(\omega)$. This phenomenon is known as *aliasing* and is avoided only if $A_c(j\omega)$ is strictly bandlimited to a bandwidth of less than $\omega_s/2$, as illustrated in figure 6.3.

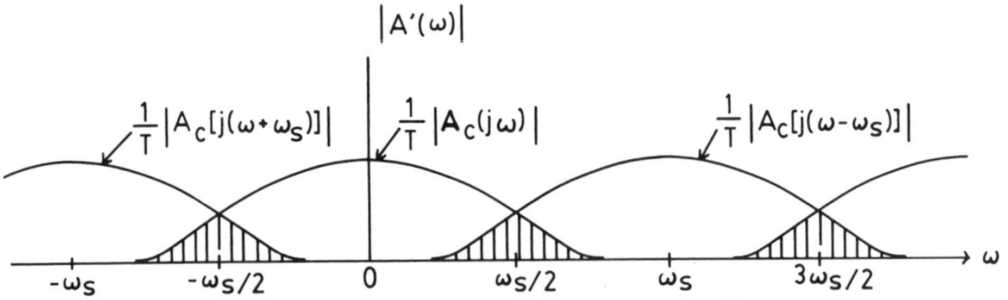

Figure 6.2. Relationship of $A'(\omega)$ to $A_c(j\omega)$, showing aliasing due to sampling.

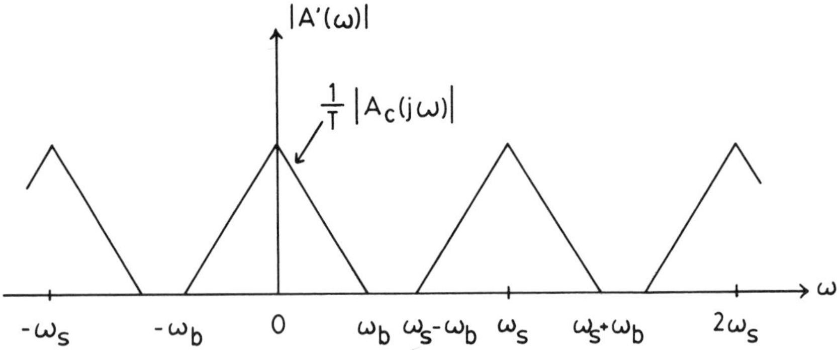

Figure 6.3. An ideal sampling case with no aliasing.

The latter observation forms the basis of the *Sampling Theorem* (also known as Nyquist's or Shannon's Theorem) which states that *a continuous-time signal can be completely recovered from its samples if, and only if, the sampling rate is greater than twice the signal bandwidth.* In the case of a lowpass signal with bandwidth ω_b, as depicted in figure 6.3, the continuous-time signal $a_c(t)$ can be recovered with an ideal lowpass filter having a cutoff frequency between ω_b and $\omega_s - \omega_b$ and a gain of T. (See problem 6.8.) For a bandpass signal with bandwidth ω_b, frequency translation may be required before sampling and after filtering to avoid aliasing, and the recovery filter can be either lowpass or bandpass depending upon the translation scheme used. The frequency $\omega_s/2$ (or $f_s = \omega_s/4\pi$) is called the *Nyquist frequency*, and the minimum allowable sampling rate ($2f_b$) is called the *Nyquist rate*.

6.2 LAPLACE TRANSFORM RELATIONSHIPS

The Fourier transform derivation for $a^*(t)$ in (6.1.3) is readily generalized to obtain the corresponding Laplace transform

$$A^*(s) = \int_{-\infty}^{\infty} a^*(t)e^{-st}\,dt, \tag{6.2.1}$$

which is thereby found to be

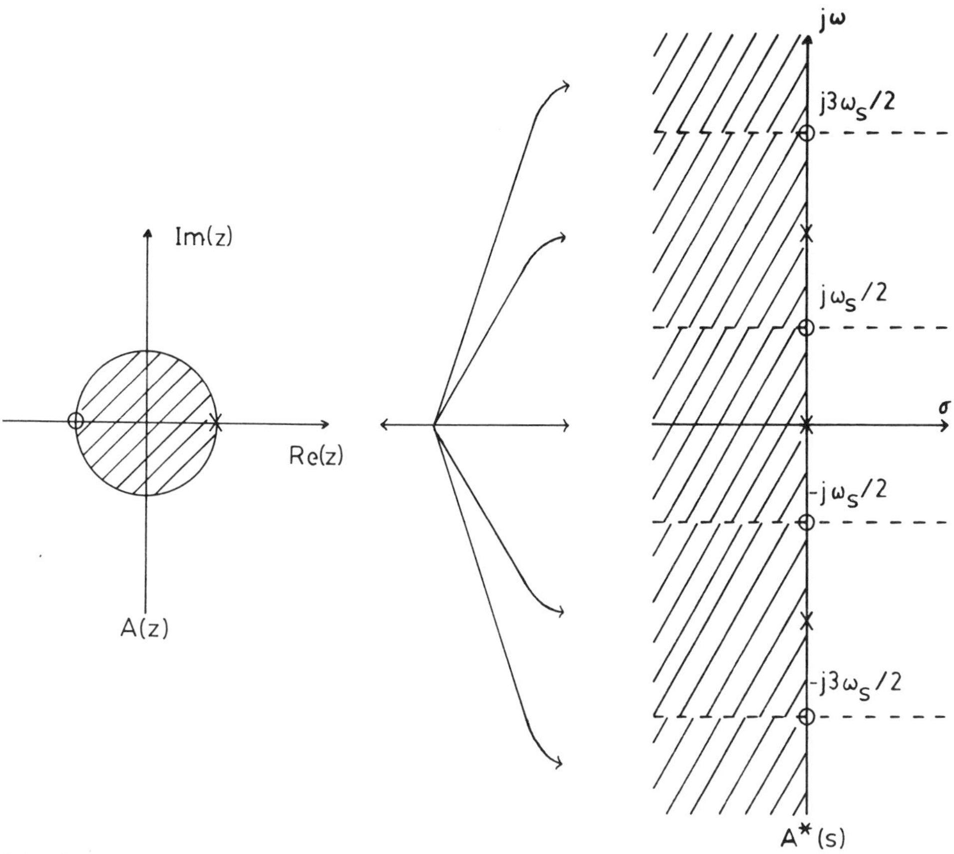

Figure 6.4. Mapping of $A^*(s)$ to and from $A(z)$.

$$A^*(s) \quad = \quad A(z)\big|_{z=e^{sT}} \tag{6.2.2}$$

The difference between (6.1.3) and (6.2.2) is, of course, that we are interested in the properties of $A^*(s)$ over all values of the complex variables $s = \sigma + j\omega$; while the Fourier transform $A^*(j\omega)$ restricts our attention to imaginary s axis alone. Equation (6.2.2) defines a mapping from the z plane to the s plane, which is one-to-many because $e^{(\sigma+j\omega)T}$ is periodic in ω with period $\omega_s = 2\pi/T$. Hence, the form of $A^*(s)$ in any horizontal *strip* of width ω_s in the s plane must be repeated in all other strips displaced from the first by integer multiples of $j\omega_s$. This is illustrated in figure 6.4 for the strip $-\omega_s/2 \le \omega \le \omega_s/2$.

The relationship of $A^*(s)$ to $A_c(s)$ is obtained by analytic continuation of (6.1.10), which yields

$$A^*(s) \quad = \quad \frac{1}{T}\sum_{k=-\infty}^{\infty} A_c\,(s - jk\omega_s). \tag{6.2.3}$$

This transformation from $A_c(s)$ to $A^*(s)$ is *many-to-many* in the sense that for any point $s_0 - jn\omega_s$, the value of $A^*(s_0 - jn\omega_s)$ equals $1/T$ times the sum of the values

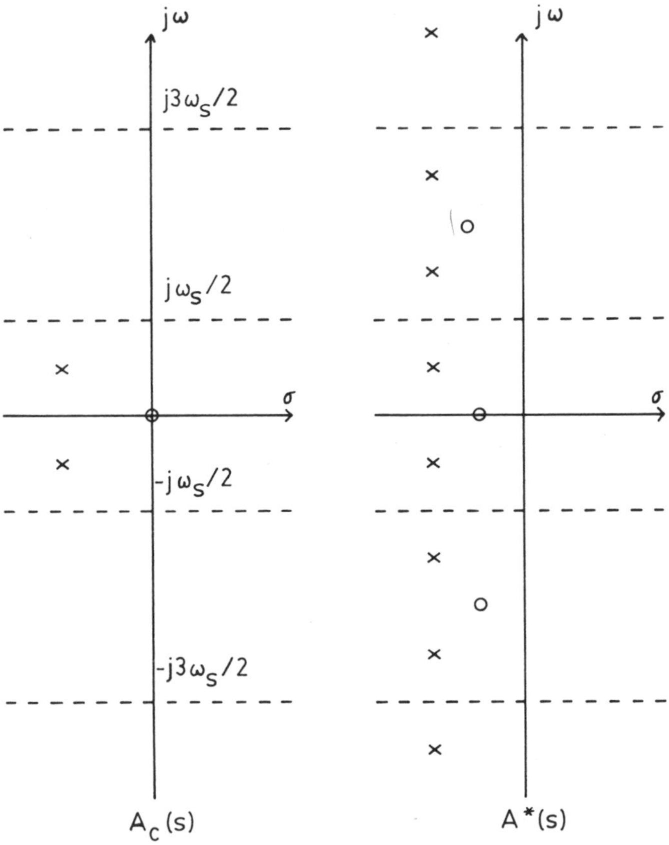

Figure 6.5. Example of corresponding $A_c(s)$ and $A^*(s)$.

$A_c(s_0 - jk\omega_s)$ for all k, where n and k are integers. This again explains the aliasing phenomenon, but in more general terms. Note that the poles of $A_c(s)$ are directly manifested by a corresponding (infinite) set of poles for $A^*(s)$, unless cancelled by zeros produced in the summation. However, the zeros of $A^*(s)$ can differ significantly from those of $A_c(s)$, due also to the effects of the summation. An example is shown in figure 6.5.

The transformation from $A_c(s)$ to $A(z)$, corresponding to the sampling of $a_c(t)$ to produce the sequence $a(n)$, can thus be viewed as a two-step process: first, the transformation from $A_c(s)$ to $A^*(s)$ described by (6.2.3), and second, the transformation from $A^*(s)$ to $A(z)$ given by the inverse of (6.2.2). Alternatively, we can view the sampling transformation as a single step directly from $A_c(s)$ to $A(z)$, as follows: Let $A_{ck}(s)$ be the function obtained by taking the kth strip of the s plane for $A_c(s)$ and repeating it for all other strips, i.e.,

$$A_{ck}(s) = A_c[s - (n - k)j\omega_s], \qquad \omega_s(n - \tfrac{1}{2}) < \omega < \omega_s(n + \tfrac{1}{2}) \tag{6.2.4}$$

for all integers k and n. Since $A_{ck}(s)$ is, therefore, periodic in ω with period ω_s, we can define a corresponding z transform by the mapping $z = e^{sT}$, i.e.,

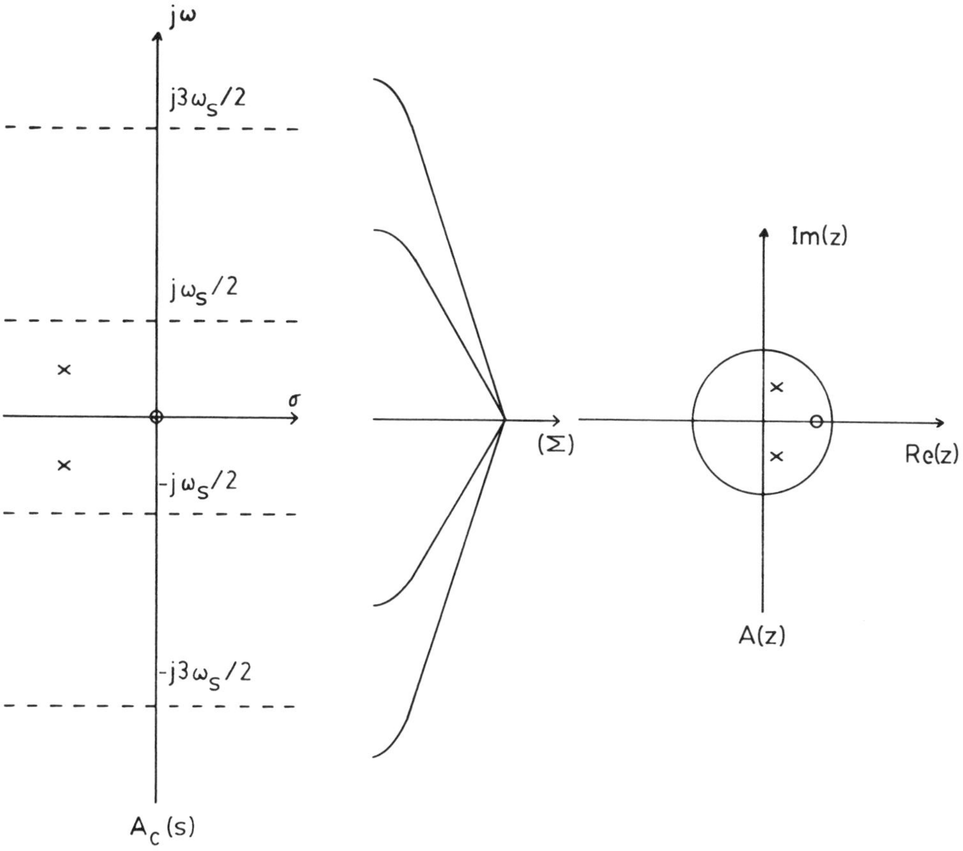

Figure 6.6. Mapping from $A_c(s)$ to $A(z)$.

$$A_k(z)\big|_{z=e^{sT}} = A_{ck}(s). \tag{6.2.5}$$

Manipulating (6.2.3) and (6.2.4), it is readily shown that

$$A^*(s) = \frac{1}{T}\sum_{k=-\infty}^{\infty} A_{ck}(s); \tag{6.2.6}$$

and thus from (6.2.2) and (6.2.5) we have that

$$A(z) = \frac{1}{T}\sum_{k=-\infty}^{\infty} A_k(z). \tag{6.2.7}$$

Hence, $A(z)$ can be thought of as the summation of the functions $A_k(z)$ derived by mapping the individual strips of the s plane for $A_c(s)$ to the z plane via $z = e^{sT}$. The inverse transformation from $A(z)$ to $A_c(s)$ does not exist unless the contributions from the individual strips are nonoverlapping, which is another explanation of the aliasing phenomenon. The transformation from $A_c(s)$ to $A(z)$ is depicted in figure 6.6, which also corresponds to figure 6.5.

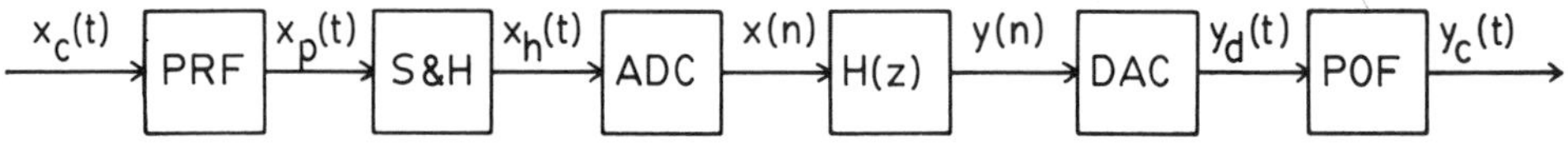

Figure 6.7. Stages in converting continuous-time signal to discrete-time and vice versa.

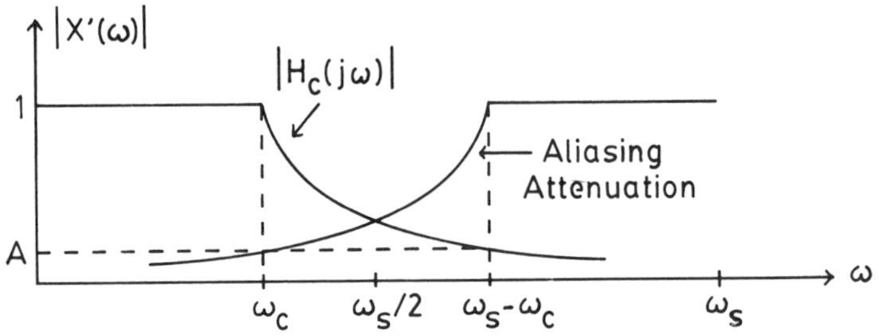

Figure 6.8. Effect of the prefilter $H_c(j\omega)$ in suppressing aliasing.

6.3 PREFILTERS, POSTFILTERS, AND D/A CONVERTERS

We have seen that to prevent aliasing, the continuous-time signal must be bandlimited before sampling. However, real-world signals are not strictly bandlimited, and often we wish to process only a portion of the bandwidth. Hence, we usually require an analog *prefilter* (PRF) before the ADC (analog-to-digital converter). Likewise, the output of the DAC (digital-to-analog converter) is not bandlimited because of the spectral images produced by sampling, and we often employ an analog *postfilter* (POF) to attenuate these high-frequency images. These operations are indicated in figure 6.7. Also shown is a *sample-and-hold* (S&H) unit to sample $x_p(t)$ at the times $t = nT$ and hold each value $x_p(nT)$ constant from nT to $(n + 1)T$ in order for the ADC to measure and encode it as $x(n)$. Sometimes, the term *analog-to-digital converter* is applied to the combined S&H and ADC operations as defined here; but because separate circuit modules are typically designed and purchased to perform these operations, we will distinguish them as above. The reader should be aware of this potential confusion in product terminology and literature.

Complete coverage of the properties and limitations of these analog circuits is beyond the scope of this book, and we will restrict our attention to those aspects relating to signal spectra. Specification of the prefilter characteristics is determined by what level of aliasing is acceptable over the bandwidth of interest. For example, in sampling telephone speech at a rate of 8 kHz, we may require that the aliased components be at least 30 dB below the signal components for frequencies up to 3 kHz. Since 5 kHz folds over to 3 kHz in this case, this specification determines the attenuation needed at 5 kHz and above. Of course, the speech spectrum itself rolls off with frequency at about 6 dB/octave, so the prefilter need not provide all of this attentuation.

The diagram in figure 6.8 is useful in determining the filter specifications for wideband input signals having negligible rolloff. Let the band of interest be 0 to ω_c. Clearly, by symmetry, the attenuation A of the aliased components at ω_c is given by

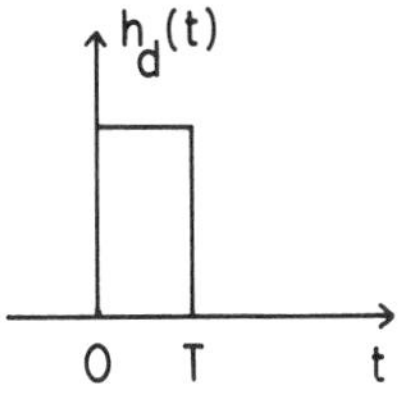

Figure 6.9. Effective impulse response of the DAC.

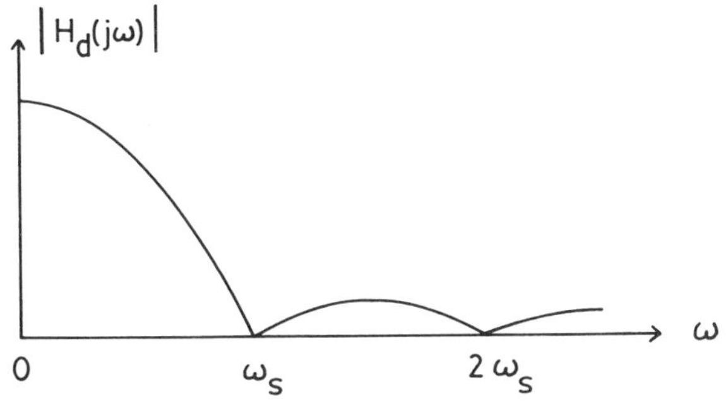

Figure 6.10. Effective magnitude response of the DAC.

$$A = |H_c(j(\omega_s - \omega_c))|; \tag{6.3.1}$$

thus it is the attenuation at $(\omega_s - \omega_c)$, not $\omega_s/2$, which we must specify. For example, assume that the specified value of A is 48 dB and we have an 8th-order analog prefilter with 48 dB/octave rolloff. Hence, A is achieved in one octave, i.e., at $2\omega_c$, and we require that

$$\omega_s - \omega_c \geq 2\omega_c \qquad \text{or} \qquad \omega_s \geq 3\omega_c.$$

That is, using this prefilter, we must sample at 3 times the bandwidth to satisfy the aliasing specifications. In practical systems, values of 2.5 to 4 are common for the ratio of sampling rate to bandwidth.

The specification of the postfilter is similar, but we must first take into account the filtering action of the DAC. To think of the DAC as an analog filter, we will consider its input $y(n)$ to be an impulse train $y^*(t)$ such as that illustrated in figure 6.1. The output $y_d(t)$ of the DAC is ideally the piecewise-constant signal

$$y_d(t) = y(n), \qquad nT < t < (n + 1)T. \tag{6.3.2}$$

Thus, the effective impulse response of the DAC is simply the square pulse depicted in figure 6.9, and its magnitude response is the sinc function shown in figure 6.10.

Therefore, the spectral images in $Y'(\omega)$ centered at $\pm\omega_s$, $\pm 2\omega_s$, etc., will be attenuated by the response of the DAC, but residual images will remain in $Y_d(j\omega)$ as illustrated in

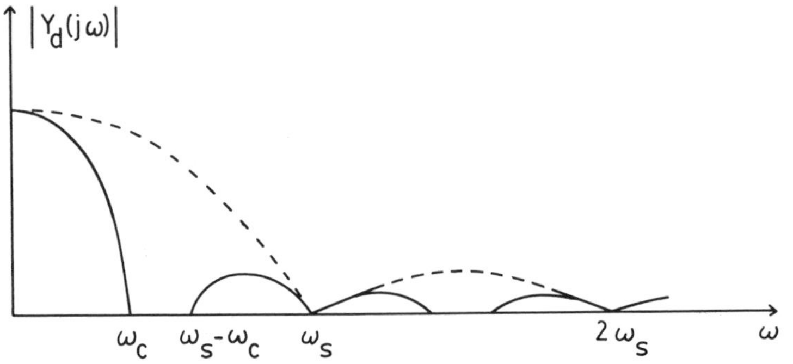

Figure 6.11. Effect of the DAC magnitude response on the spectrum of the continuous-time output $y_d(t)$.

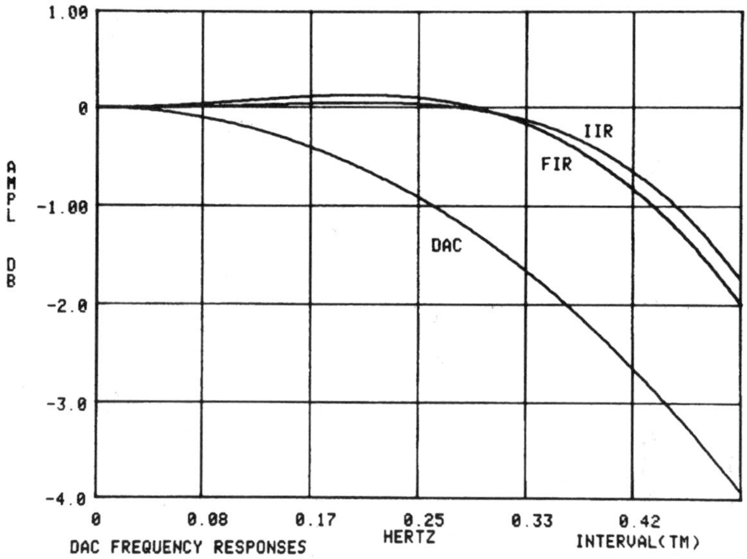

Figure 6.12. Effective DAC magnitude response with IIR compensation, FIR compensation, and no compensation.

figure 6.11. It is these residual images which the postfilter must further attenuate. Hence, as for the prefilter, the bandwidth ω_c of the signal determines the frequency $(\omega_s - \omega_c)$ above which attenuation specifications are needed for the postfilter.

In systems requiring precise spectral fidelity, it may be necessary to compensate for the *droop* in the magnitude response of the DAC up to the frequency $\omega_s/2$. For example, at $\omega_s/4$ the attenuation from the sinc function is 0.91 dB; while at $\omega_s/3$, it is 1.66 dB. If such compensation is required, it may be included in the analog postfilter or in the digital filter preceding the DAC. A simple linear-phase FIR compensation filter is

$$H_1(z) = -\tfrac{1}{16} + \tfrac{9}{8}z^{-1} - \tfrac{1}{16}z^{-2} \qquad (6.3.3)$$

and a simple IIR one is

$$H_2(z) = \frac{\frac{9}{8}}{1 + \frac{1}{8}z^{-1}}. \tag{6.3.4}$$

The coefficients of $H_1(z)$ and $H_2(z)$ have been selected for easy digital implementation. The compensated and uncompensated DAC responses are shown in figure 6.12. Note that this FIR compensation reduces the deviation of the DAC from unity gain to only ± 0.15 dB up to $\omega_s/3$; while with the IIR compensation, the deviation is even less.

PROBLEMS

6.1 Sketch the magnitude spectra $|A'(\omega)|$ for sampled sinusoids of the following frequencies:

a. $\omega_0 = \omega_s/8$
b. $\omega_0 = \omega_s/4$
c. $\omega_0 = \omega_s/2$
d. $\omega_0 = 3\omega_s/4$
e. $\omega_0 = \omega_s$
f. $\omega_0 = 0$ (dc)

6.2 A continuous-time square wave $a_c(t)$ is not a bandlimited signal and, therefore, will produce aliasing when sampled. Sketch the resulting sequence $a(n)$ and its magnitude spectrum $|A'(\omega)|$ when the fundamental frequency ω_0 of the square wave has the following values:

a. $\omega_0 = \omega_s/4$
b. $\omega_0 = \omega_s/10$
c. $\omega_0 = \omega_s/7$
d. $\omega_0/\omega_s \neq$ rational fraction.

6.3 A prefilter is required for a broadband analog input $x_c(t)$ to attenuate aliased components in the band of interest by at least 60 dB after analog-to-digital conversion.

a. If the sampling rate is 3 times the filter bandwidth, what rolloff in dB/octave is required in the magnitude response of the prefilter?
b. Repeat (a) if the sampling rate is 5 times the filter bandwidth.

6.4 A first-order analog lowpass filter has the impulse response

$$h_c(t) = \begin{cases} e^{-t/\tau}, & t \geq 0 \\ 0, & t < 0. \end{cases}$$

Let a discrete-time filter $h(n)$ be designed by sampling $h_c(t)$, i.e., $h(n) = h_c(nT)$.

a. Find the magnitude response $|H'(\omega)|$ and sketch it for $T = \tau$ and $T = \tau/4$.
b. Find the dc gain $H'(0)$ in each of the cases in (a).

6.5 An accurate sinusoid generator or *frequency synthesizer* can be implemented by first generating the discrete-time samples $y(n)$ of the desired sinusoid and then converting the signal to continuous-time form via a DAC. Assume that frequencies (ω_0) from 0 to $\omega_s/4$ are to be produced.

a. Sketch the magnitude spectrum $|Y_d(j\omega)|$ of the DAC output.
b. Sketch the DAC output signal $y_d(t)$ for $\omega_0 = \omega_s/4$.
c. What is the minimum attenuation (in dB) provided by the DAC of the first spectral image at $\omega_s - \omega_0$?
d. Sketch an appropriate postfilter magnitude response for the frequency synthesizer.

6.6 The DAC characteristic described by the equivalent impulse response $h_d(t)$ in figure 6.9 is sometimes called a *zero-order hold.* A *first-order hold,* on the other hand, corresponds to linear interpolation between the sample values $y(n)$, i.e.,

$$y_d(t) = y(n-1) + \left(\frac{t-(n-1)T}{T}\right)[y(n) - y(n-1)],$$

for $(n-1)T \leq t \leq nT$.

a. Sketch a typical output signal $y_d(t)$ from a first-order hold.
b. Find and sketch the equivalent impulse response $h_f(t)$ of the first-order hold.
c. Find and sketch the equivalent magnitude response $|H_f(j\omega)|$ of the first-order hold. (*Hint*: Express $h_f(t)$ in terms of $h_d(t)$.)
d. Comment on the attenuation of the spectral image around ω_s due to the first-order hold versus the droop from 0 to $\omega_s/2$.

6.7 Draw implementation of the DAC compensation filters $H_1(z)$ and $H_2(z)$ in (6.3.3) (6.3.4) showing the coefficient multipliers as simple binary shifts and/or additions.

6.8 Assuming an ideal lowpass filter with cutoff frequency $\omega_s/2$, show that a bandlimited signal $a_c(t)$ may be recovered from its samples $a(n) = a_c(nT)$ by

$$a_c(t) = \sum_{n=-\infty}^{\infty} a(n) \operatorname{sinc}[(\pi/T)(t - nT)],$$

where $\operatorname{sinc} x = (\sin x)/x$. Make a sketch to show how the sinc functions overlap to form $a_c(t)$.

7. DISCRETE FOURIER TRANSFORM

7.0 INTRODUCTION

In chapter 6, we showed that $A'(\omega) = A(e^{j\omega T})$ can be interpreted as the Fourier transform of the sequence $a(n)$ via the impulse train $a^*(t)$. That is, the z transform $A(z)$ evaluated on the unit circle $z = e^{j\omega T}$ is taken to be the Fourier transform of $a(n)$ (if the region of convergence contains the unit circle). However, much of the practice of digital signal processing is done in computers where we cannot evaluate a continuum of frequencies ω, nor can we input and store an infinite-duration sequence $a(n)$. Hence, for actual data sequences, as opposed to theoretically defined signals, we cannot compute the Fourier transform, in general.

As a consequence of the above, we define a new transform, known as the *discrete Fourier transform* (DFT), for N samples of $a(n)$ and at N uniformly spaced frequencies ω_k around the unit circle. This widely used transform is invertible and has other desirable properties. Multiplication of two DFT's corresponds to convolution of the corresponding sequences, as expected, but it is circular convolution and must be suitably employed. Most practical spectral-analysis methods are based upon the DFT, either directly or indirectly. Fast algorithms exist for its computation—most notably the *fast Fourier transform* (FFT) algorithm—and fast convolution algorithms based upon the FFT can be used to implement FIR filters as described in chapter 12.

7.1 DERIVATION AND PROPERTIES OF THE DFT

Given the sequence $a(n)$ for all n, its Fourier transform is defined to be

$$A'(\omega) = A(e^{j\omega T}) = \sum_{n=-\infty}^{\infty} a(n)\, e^{-jn\omega T}. \tag{7.1.1}$$

If, however, we take only the N samples $a(n)$ for $n = 0, 1, \ldots, N - 1$, and compute only N samples of $A'(\omega)$ for $\omega = k\omega_0$, $k = 0, 1, \ldots, N - 1$, and

$$\omega_0 = \frac{\omega_s}{N} = \frac{2\pi}{NT}, \tag{7.1.2}$$

we define a discrete transform of the form

$$A'_N(k\omega_0) = \sum_{n=0}^{N-1} a(n)\, e^{-jnk\omega_0 T} \tag{7.1.3}$$

or, in more convenient notation,

$$\tilde{A}(k) = \sum_{n=0}^{N-1} a(n)\, W^{kn}, \qquad k = 0, 1, \ldots, N - 1, \tag{7.1.4}$$

where

$$W = e^{-j2\pi/N}. \tag{7.1.5}$$

This is the discrete Fourier transform (DFT). Note that neither the sampling frequency ω_s nor the sampling period T appear in (7.1.4) or (7.1.5), and thus the DFT $\tilde{A}(k)$ is simply an N-point sequence computed from an N-point sequence, with a dimensionless independent variable (index) k. A familiar frequency scale is provided only if we plot $\tilde{A}(k)$ as a function of $k\omega_0$ or $kf_0 = k/NT$.

If possible (and appropriate), it is desirable that any transform be invertible, and since N samples of $a(n)$ produce N samples $\tilde{A}(k)$, it is reasonable to expect that this might be true for the DFT. Such is indeed the case, and the inverse DFT is given by

$$\tilde{a}(n) = \frac{1}{N} \sum_{k=0}^{N-1} \tilde{A}(k)\, W^{-kn}, \qquad n = 0, 1, \ldots, N - 1, \tag{7.1.6}$$

where $\tilde{a}(n) = a(n)$ over the interval $n = 0, 1, \ldots, N - 1$. This is readily proved by substitution of (7.1.4) into (7.1.6). (See problem 7.1.) Hence, although we expect some problems to arise because of the finite duration of $\tilde{a}(n)$, invertibility of the DFT is not one of them. Rather, the problems have to do with relating $\tilde{a}(n)$ to $a(n)$ for all n, or equivalently, relating $\tilde{A}(k)$ to $A'(\omega)$.

The properties of the DFT parallel those for the z and Fourier transforms except that the shifts and convolutions are circular. Linearity holds, for example, assuming that both sequences are of the same length N. The shift property, however, is that

$$W^{km}\tilde{A}(k) \leftrightarrow \tilde{a}(n - m)_{\text{mod}N}, \tag{7.1.7}$$

where $\leftrightarrow$ denotes a transform pair and the modulo operation is $(r)_{\text{mod}N} = r \pm iN$ for integer i such that $0 \leq (r)_{\text{mod}N} < N$. (See problem 7.2.) From (7.1.5), we note that the left side of (7.1.7) is a linear phase shift (delay) of $\tilde{A}(k)$ by $(-2\pi m/N)k$; while the right side is a circular shift (delay) of $\tilde{a}(n)$ by m samples. For example, if $m = 1$, $\tilde{a}(N - 1)$ replaces $\tilde{a}(0)$, $\tilde{a}(0)$ replaces $\tilde{a}(1)$, etc. A dual property holds for a circular shift of the DFT coefficients: namely,

$$W^{-ln}\tilde{a}(n) \leftrightarrow \tilde{A}(k - l)_{\text{mod}N}. \tag{7.1.8}$$

Consider next the product of two DFT's of length N, e.g.,

$$\tilde{A}_0(k) = \tilde{A}_1(k)\tilde{A}_2(k). \tag{7.1.9}$$

We expect this to correspond to convolution of $\tilde{a}_1(n)$ and $\tilde{a}_2(n)$, and convolution involves time reversal and shifting of one of the sequences. Hence, since shifts are circular with DFT's, we should expect $\tilde{a}_0(n)$ to be given by

$$\tilde{a}_0(n) = \sum_{m=0}^{N-1} \tilde{a}_1(m)\tilde{a}_2(n - m)_{\text{mod}N}, \qquad n = 0, 1, \ldots, N - 1. \tag{7.1.10}$$

(See problem 7.3.) This is called *circular convolution*, as opposed to the familiar *linear convolution* resulting from the multiplication of z transforms or Fourier transforms. For example, in problem 7.4, one shows that circular convolution of the two square pulses in figure 7.1(a) and (b) results in the circularly "wrapped" and overlapped triangular pulse in figure 7.1(c). A dual property holds for the DFT of a product of N-point sequences.

7.2 ZERO PADDING

Zero padding refers to the operation of extending a sequence of length N_1 to a length $N_2 > N_1$ by appending $N_2 - N_1$ zero samples to the given sequence. There are two principal reasons for doing this:

1. Circular convolution can be used to implement linear convolution if both sequences contain sufficient zero samples to prevent circular wrap-around and overlap of the result.
2. The density of DFT samples of the spectrum over the interval $0 \leq k\omega_0 < \omega_s$ is increased from N_1 to N_2. Hence, the spectrum between the DFT samples can be interpolated to an arbitrary density by sufficient zero padding.

To study both of these applications, it is convenient to introduce the important concept of *windowing*. Defining the *rectangular window* of length N_1 by

$$w_R(n) = \begin{cases} 1, & n = 0, 1, \ldots, N_1 - 1 \\ 0, & \text{elsewhere,} \end{cases} \tag{7.2.1}$$

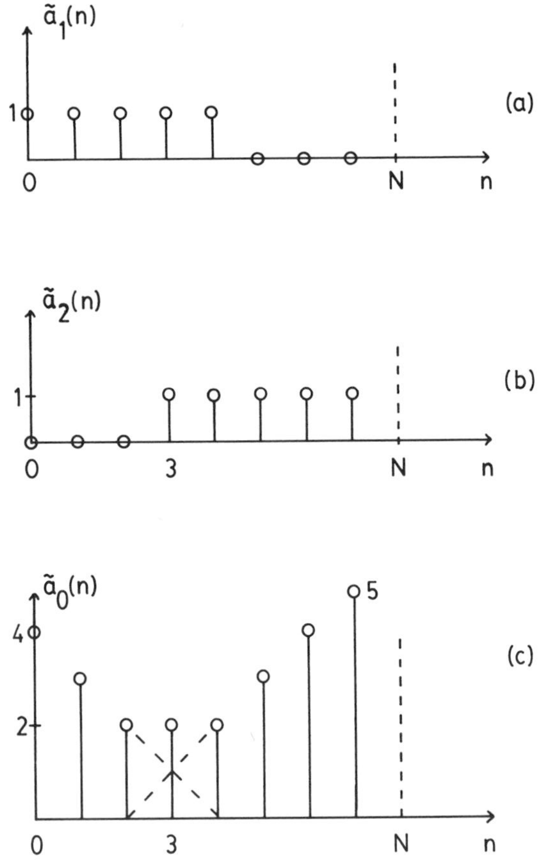

Figure 7.1. Circular convolution of N-point sequences in (a) and (b) results in (c).

it is clear that a finite-duration sequence $\tilde{a}(n)$ is related to the corresponding infinite-duration sequence $a(n)$ simply as

$$\tilde{a}(n) = w_R(n)a(n). \tag{7.2.2}$$

As seen in the last section, if we compute and multiply the N_1-point DFT's of two N_1-point sequences $\tilde{a}_1(n)$ and $\tilde{a}_2(n)$, and inverse transform the resulting DFT $\tilde{A}_0(k)$, we produce the circular convolution of $\tilde{a}_1(n)$ and $\tilde{a}_2(n)$. However, if we zero-pad $\tilde{a}_1(n)$ and $\tilde{a}_2(n)$ to a sufficient length $N_2 > N_1$ and use N_2-point DFT's, leaving the window length N_1 unchanged, there will be no circular wrap-around or overlap of the resulting sequence. It is important to recognize, however, that the linear convolution so effected is of $\tilde{a}_1(n)$ and $\tilde{a}_2(n)$, not $a_1(n)$ and $a_2(n)$. That is, we can produce the linear convolution only of windowed sequences via DFT's, and the effect of the windowing will appear explicitly in the result (as a general triangular envelope). Assuming that the N_1-point sequences are full length (i.e., have only occasional zero samples), any value of N_2 satisfying

$$N_2 \geq 2N_1 - 1 \tag{7.2.3}$$

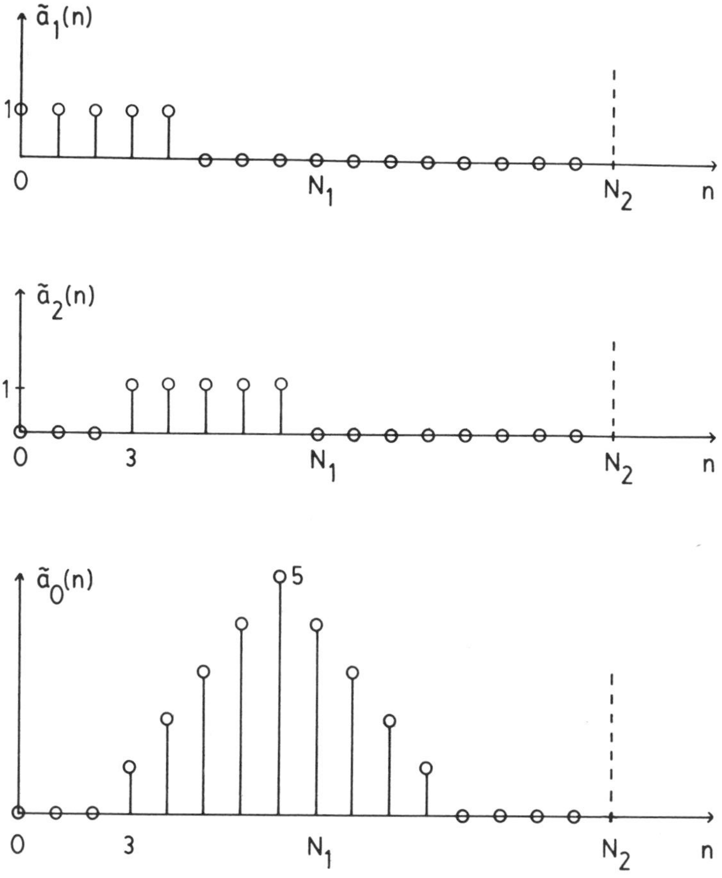

Figure 7.2. Zero padding the sequences in figure 7.1 makes circular convolution equivalent to linear convolution.

will produce linear convolution, but the commonly used value is simply $N_2 = 2N_1$. Overlap methods to produce linear convolution of an FIR response $h(n)$ with an infinite-duration sequence $a(n)$ via DFT's are presented in chapter 12.

As examples, consider first the previous example in figure 7.1. If these 8-point sequences are zero-padded to 16 points, the resulting circular convolution is shown in figure 7.2, where it is apparent that the linear convolution of the two square pulses has been realized. Hence, in the case of two FIR sequences, linear convolution can be realized directly by zero padding.

Consider next the convolution of a square pulse with a sinusoid. If we could use infinite-duration transforms, we know that the result would be simply a sinusoid of the same frequency, but different amplitude and phase, in general. However, with DFT's and zero padding, we get a result such as that in figure 7.3 (shown as a continuous function of time). The overlap methods in chapter 12 address and correct these end effects.

The second application of zero padding—interpolation of the DFT spectral samples—can also be understood readily in terms of the effects of windowing. Utilizing the

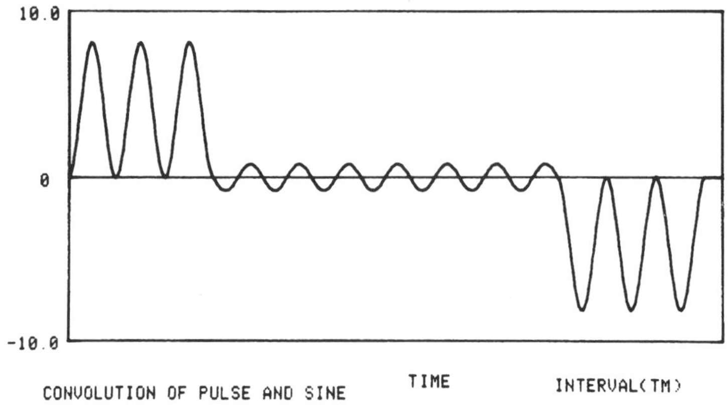

Figure 7.3. Result of convolution of square pulse and sinusoid after zero padding, showing end effects.

definition of the windowed sequence $\tilde{a}(n)$, we can write from (7.1.3) the N_2-point DFT of the zero-padded sequence as an infinite summation of the form

$$\tilde{A}(k) = A'_{N_2}(k\omega_0) = \sum_{n=-\infty}^{\infty} \tilde{a}(n)\, e^{-jnk\omega_0 T}, \qquad k = 0, 1, \ldots, N_2 - 1, \tag{7.2.4}$$

with $\omega_0 = \omega_s/N_2$, since the values of $\tilde{a}(n)$ outside the interval $n = 0, 1, \ldots, N_1 - 1$ are zero. That is, for $N_2 > N_1$, we have simply N_2 samples of the Fourier transform of the windowed sequence $\tilde{a}(n)$. But windowing is the multiplication of $a(n)$ by a square pulse $w_R(n)$, which produces a convolution of $A'(\omega)$ with a linear-phase transform having the magnitude

$$|W_R'(\omega)| = \left| \frac{\sin\left(\dfrac{\omega T N_1}{2}\right)}{\sin\left(\dfrac{\omega T}{2}\right)} \right|. \tag{7.2.5}$$

(See figure 4.7 or figure 7.6.) Hence, the spectral resolution that can be achieved by increasing N_2 is limited by the bandwidth of the window transform in (7.2.5) to something like $2\pi/N_1 T$. This can also be thought of as an interpolation of the basic N_1-point DFT for $\tilde{a}(n)$; and for integer multiples $N_2 = iN_1$, the interpolation produces $(i - 1)$ samples between each pair of N_1-point DFT samples (which are themselves unchanged).

The utility of this spectral interpolation is that we can locate the frequency of an isolated spectral peak, for example, as accurately as we wish by sufficient zero padding of the given N_1 samples before taking the DFT. We must distinguish this, however, from the resolution of two closely spaced spectral peaks which, as noted above, is limited to a spacing of about $2\pi/N_1 T$ by the implicit convolution of $A'(\omega)$ and $W_R'(\omega)$. For example, in figure 7.4(a) is shown the magnitude of the N_1-point DFT of a complex sinusoid whose frequency falls somewhere between two DFT samples (bins). From this DFT, we can not tell the exact

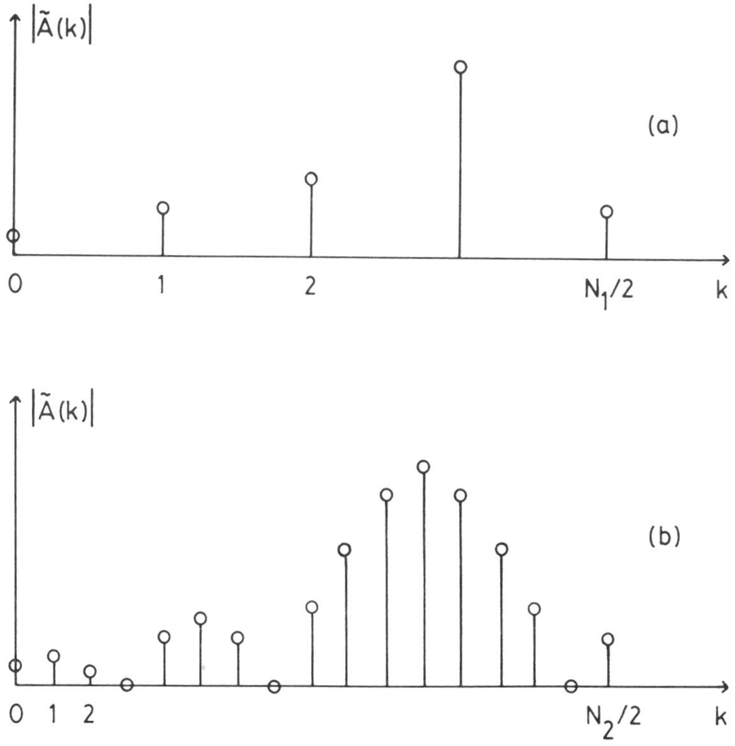

Figure 7.4. DFT's of a complex sinusoid before and after zero padding to $N_2 = 4N_1$.

frequency of the sinusoid. Figure 7.4(b) shows the N_2-point DFT magnitude after zero padding to $N_2 = 4N_1$, and we now see by the symmetry of the peak that the sinusoidal frequency is located three-quarters of the way between the two original bins. Note also that even the basic N_1-point DFT in figure 7.4(a) showed the effect of windowing because the sinusoidal frequency ω_1 was not centered in a bin and thus the DFT samples of $W_R'(\omega - \omega_1)$ were not zero.

To illustrate the spectral resolution of the DFT with zero padding, figure 7.5(a) shows the N_1-point DFT magnitude for two sinusoids located at the centers of adjacent bins, i.e., at $k\omega_0$ and $(k + 1)\omega_0$. Zero padding this data to N_2 points with $N_2 \gg N_1$ and connecting the DFT samples to form a continuous function in figure 7.5(b), we can just resolve the two spectral peaks. Any significant reduction of the spacing below this *Rayleigh limit*, however, would produce a single, broadened peak and no direct indication of the presence of two sinusoids. (The exact spacing for the resolution of two sinusoids depends on their relative phase, which in this example is 90°.)

7.3 WINDOWS IN SPECTRUM ANALYSIS

In this section, we consider alternatives to the rectangular window for application in spectrum analysis. The rectangular window inherent in the DFT has the major disadvantages that the peak *sidelobe* of $W_R'(\omega)$ in (7.2.5) is down only 13 dB from the *mainlobe* level,

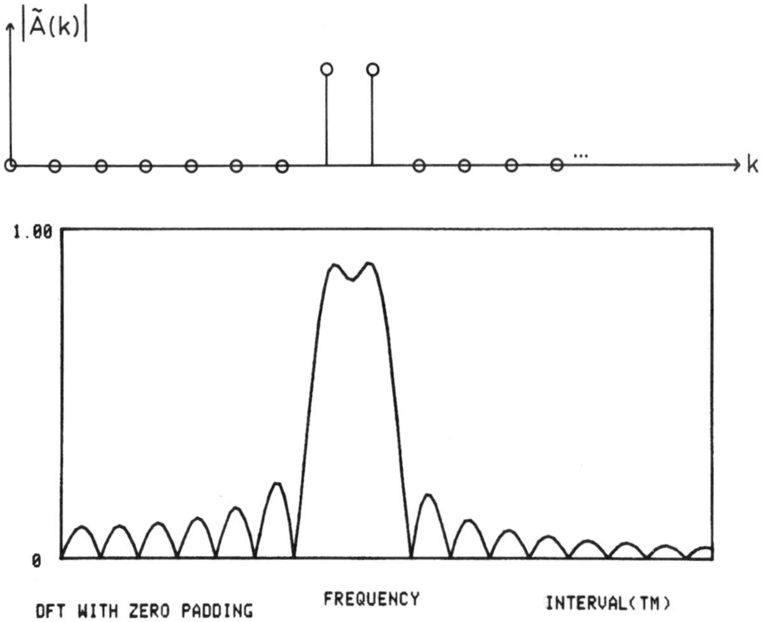

Figure 7.5. DFT's of two sinusoids before and after zero padding to $N_2 \gg N_1$.

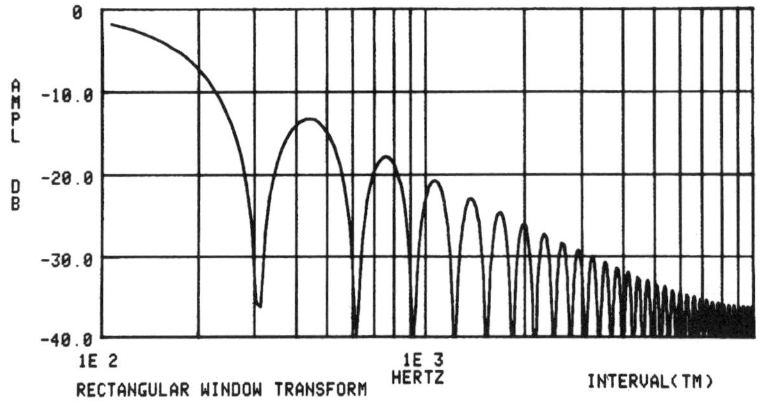

Figure 7.6. Magnitude of rectangular window transform $W_R'(\omega)$, showing -13-dB peak sidelobe and 6-dB/octave rolloff.

as shown in figure 7.6, while the sidelobes thereafter fall off by only 6 dB/octave. This severely limits the ability of the DFT to resolve a weak signal in the presence of a stronger one. The reason for the 6-dB/octave rolloff is readily seen in the z transform of $w_R(n)$, which is

$$W_R(z) = \frac{1 - z^{-N_1}}{1 - z^{-1}}. \tag{7.3.1}$$

This transform has N_1 zeros uniformly spaced around the unit circle and a pole at $z = 1$ (cancelling the zero at $z = 1$), which produces the 6-dB/octave rolloff. Figure 7.5(b) provides an example of the effect of this window on a narrowband spectrum.

The two most common window functions in spectrum analysis improve upon these disadvantages of the rectangular window in different ways, as follows:

1. The raised cosine, or Hanning window, $w_N(n)$ has an 18-dB/octave rolloff rate and a peak sidelobe level of -31 dB.
2. The Hamming window $w_M(n)$ reduces the peak sidelobe level to -41 dB, but the rolloff rate is only 6 dB/octave, as for the rectangular window.

(The similarity of the names is an unfortunate coincidence, but it is hoped that using the offending letters as the subscripts will help to distinguish them.) Window functions are employed, of course, by multiplying $a(n)$ by the window before the DFT computation to produce

$$\tilde{A}_w(k) = \sum_{n=0}^{N_1-1} w(n)a(n)W^{kn}, \qquad k = 0, 1, \ldots, N_2 - 1, \tag{7.3.2}$$

where

$$W = e^{-j2\pi/N_2}$$

to allow for the possibility of zero padding (if $N_2 > N_1$). Hence, the DFT provides N_2 samples of the convolved spectrum $A'(\omega) * W'(\omega)$.

The Hanning window is simply the raised cosine or sine-squared function

$$\begin{aligned} w_N(n) &= \tfrac{1}{2}\left[1 - \cos\frac{2\pi n}{N_1 - 1}\right] \\ &= \sin^2\frac{\pi n}{N_1 - 1}, \qquad n = 0, 1, \ldots, N_1 - 1, \end{aligned} \tag{7.3.3}$$

shown as a continuous function in figure 7.7, along with the other window functions to be considered. Note that it tapers smoothly to zero at each end and has a peak value of unity in the middle.

Since the end samples $w_N(0)$ and $w_N(N_1 - 1) = 0$, this window actually has a length of only $N_1 - 2$, and hence its z transform has $N_1 - 3$ zeros, not $N_1 - 1$ as for $W_R(z)$. These two missing zeros are cancelled by two additional poles, which accounts for the 18-dB/octave rolloff rate. Specifically, the numerator of $W_N(z)$ is of the form $(1 - z^{-(N_1-1)})(1 + z^{-1})$, and the three zeros at and nearest $z = 1$ are cancelled by poles. (See problem 7.5.) The resulting magnitude transform for the Hanning window is plotted in figure 7.8. Note the rolloff rate, the -31-dB peak sidelobe, and the effect of the extra zero at the Nyquist frequency ($z = -1$). Note also that the mainlobe width is

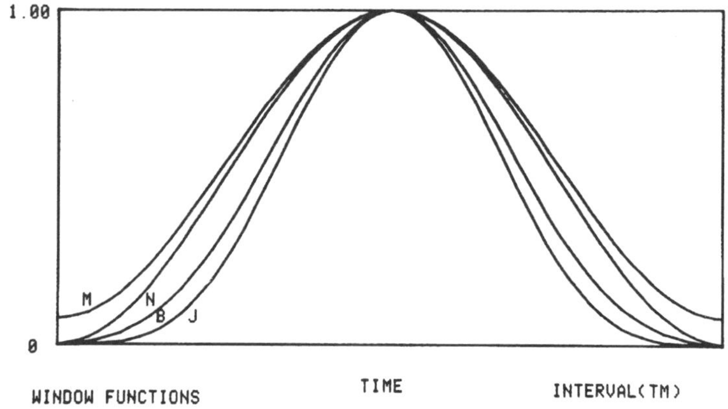

Figure 7.7. Hamming (M), Hanning (N), Blackman (B), and 30-dB/octave (J) window functions.

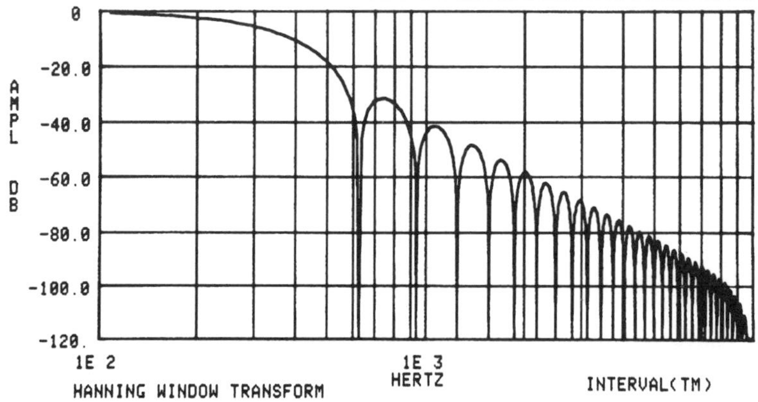

Figure 7.8. Magnitude of Hanning window transform $W_N'(\omega)$, showing -31-dB peak sidelobe and 18-dB/octave rolloff.

approximately twice that of the rectangular window transform in figure 7.6. Therefore, the price we have paid for these improvements is reduced spectral resolution. The Hanning window is probably the most widely used window function in spectrum analysis because of its excellent rolloff rate of 18 dB/octave.

Another advantage of the Hanning window is that if applied after taking the DFT, not before, it eliminates the N_1 implied multiplications by $w(n)$. That is, the windowed DFT can be computed from the unwindowed one by (see problem 7.6)

$$\tilde{A}_w(k) = -\tfrac{1}{4}\tilde{A}(k-1) + \tfrac{1}{2}\tilde{A}(k) - \tfrac{1}{4}\tilde{A}(k+1),$$

and these coefficients imply only binary shifts, not full multiplications.

The Hamming window is designed to minimize the peak sidelobe level while maintaining approximately the same mainlobe width as the Hanning window. This window is defined by

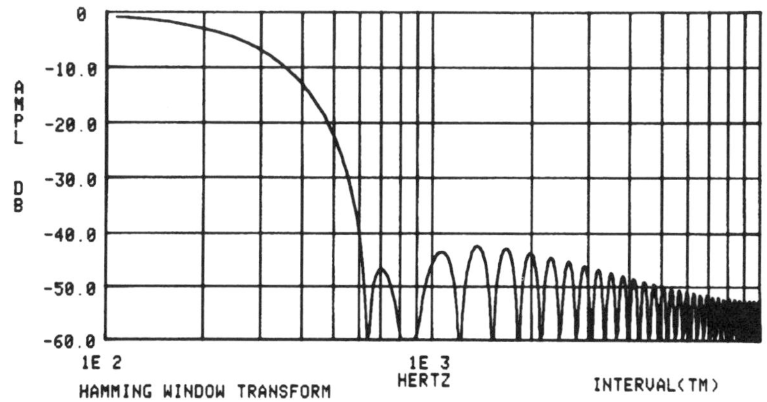

Figure 7.9. Magnitude of Hamming window transform $W'_M(\omega)$, showing -41-dB peak sidelobe and 6-dB/octave rolloff.

$$w_M(n) = 0.54 - 0.46\cos\frac{2\pi n}{N_1 - 1}, \qquad n = 0, 1, \ldots, N_1 - 1. \tag{7.3.4}$$

Note that the central value is again unity, but the ends taper to a value of 0.08, not zero. Hence, this window is full length, and its transform has $N_1 - 1$ zeros. Rather than cancelling the zeros to either side of $z = 1$ on the unit circle as with the Hanning window, the Hamming window is designed to move these zeros to the vicinity of the first sidelobe to minimize its level, as shown in figure 7.9. As a result, it is the third sidelobe, not the first, that is the peak sidelobe, and the level is reduced by 10 dB from the Hanning case to -41 dB. However, since we have simply moved these zeros, not cancelled them with poles, the rolloff rate is produced by only the one pole at $z = 1$ and is thus only 6 dB/octave. The bandwidth is about the same for both windows. As shown in chapter 9, the low peak sidelobe level of the Hamming window is of more importance in the design of FIR filters than we have found it to be here for spectrum analysis.

The ideas behind the Hanning and Hamming windows are combined in the Blackman window, given by

$$w_B(n) = 0.42 - 0.5\cos\frac{2\pi n}{N_1 - 1} + 0.08\cos\frac{4\pi n}{N_1 - 1}, \qquad n = 0, 1, \ldots, N_1 - 1. \tag{7.3.5}$$

Note that the central value is unity and that the ends taper to zero as for the Hanning window. Hence, two of the zeros to either side of $z = 1$ have been cancelled by poles to yield the steeper 18-dB/octave rolloff. Two more zeros have been moved to the vicinity of the first sidelobe, as in the Hamming case, to reduce its level. The result is plotted in figure 7.10. Note the rolloff rate, the -57 dB peak sidelobe, and the effect of the extra zero at $z = -1$. Note also that the mainlobe width is now three times that for the rectangular window, or 50% greater than for the Hanning and Hamming windows. Hence, use of this window further sacrifices spectral resolution to gain additional attenuation.

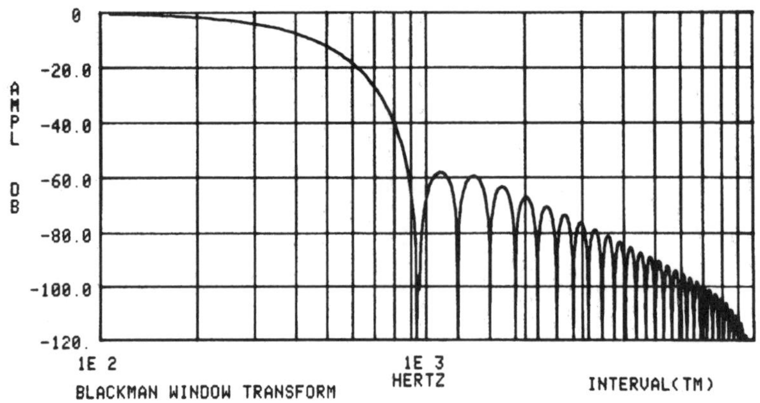

Figure 7.10. Magnitude of Blackman window transform $W_B'(\omega)$, showing -57-dB peak sidelobe and 18-dB/octave rolloff.

From the foregoing analyses, it is clear how to achieve an even steeper 30-dB/octave rolloff by altering the Blackman window coefficients. What we need to do is to cancel two pairs of zeros to either side of $z = 1$ by choosing the coefficients such that $w(0) = w(1) = w(N_1 - 1) = w(N_1 - 2) = 0$, while setting the central value to unity. That is, letting

$$w_J(n) = a + b\cos\frac{2\pi n}{N_1 - 1} + c\cos\frac{4\pi n}{N_1 - 1}, \qquad n = 0, 1, \ldots, N_1 - 1, \tag{7.3.6}$$

we choose a, b, and c to satisfy the linear equations

$$\begin{aligned} a - b + c &= 1 \\ a + b + c &= 0 \\ a + b\cos\frac{2\pi}{N_1 - 1} + c\cos\frac{4\pi}{N_1 - 1} &= 0. \end{aligned} \tag{7.3.7}$$

Note that the coefficient values are functions of the window length N_1, unlike the previous windows. However, for large values of N_1, which is the usual case of interest, these equations are ill-conditioned and their numerical solution is difficult. Rewriting the window function in the alternate form

$$w_J(n) = \sin^2\frac{\pi n}{N_1 - 1} + d\sin^2\frac{2\pi n}{N_1 - 1}, \tag{7.3.8}$$

we have no such numerical difficulties solving for d. Note that $w_J(0) = w_J(N_1 - 1) = 0$ and the central value $w_J[(N_1 - 1)/2]$ equals unity, as required. Hence, with the coefficient d, we need only satisfy the remaining constraint

$$w_J(1) = \sin^2(x) + d\sin^2(2x) = 0 \tag{7.3.9}$$

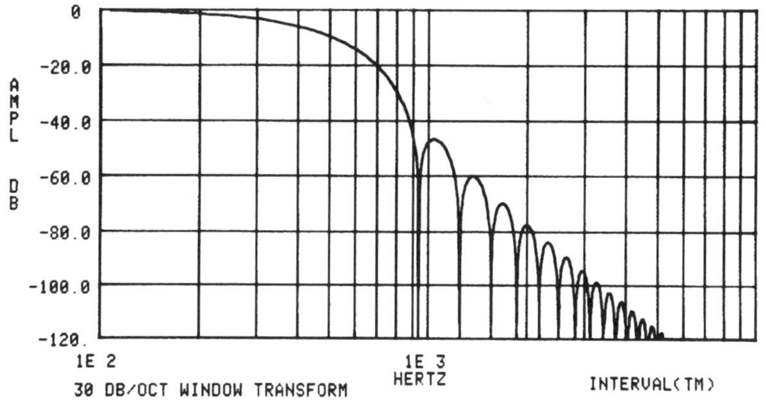

Figure 7.11. Magnitude of derived window transform $W_J'(\omega)$, showing -46-dB peak sidelobe and 30-dB/octave rolloff.

for $x = \pi/(N_1 - 1)$. The resulting window transform is shown in figure 7.11 and we observe that the peak sidelobe is -46 dB, the bandwidth is about the same as for the Blackman window, and the rolloff rate is indeed 30 dB/octave.

As the window length N_1 is increased, d and thus a, b, and c rapidly approach convenient asymptotic values. For N_1 large, x in (7.3.9) is small, and thus $\sin x \approx x$ and $\sin 2x \approx 2x$. Therefore, $d \approx -0.25$ and the asymptotic form of the window function becomes

$$\begin{aligned} w_J(n) &= \sin^2 \frac{\pi n}{N_1 - 1} - 0.25 \sin^2 \frac{2\pi n}{N_1 - 1} \\ &= 0.375 - 0.5 \cos \frac{2\pi n}{N_1 - 1} + 0.125 \cos \frac{4\pi n}{N_1 - 1} \qquad n = 0, 1, \ldots, N_1 - 1. \end{aligned} \tag{7.3.10}$$

Note the simple binary form of the asymptotic coefficients. This was previously noted by Marshall and Temes [6], who derived the window as the square of the Hanning window, i.e.,

$$w_J(n) = w_N^2(n) = \sin^4 \frac{\pi n}{N_1 - 1}. \tag{7.3.11}$$

They observed that this window, like the Hanning, is easily implemented in the frequency domain (DFT's) via

$$\tilde{A}_w(k) = \tfrac{1}{16}\tilde{A}(k-2) - \tfrac{1}{4}\tilde{A}(k-1) + \tfrac{3}{8}\tilde{A}(k) - \tfrac{1}{4}\tilde{A}(k+1) + \tfrac{1}{16}\tilde{A}(k+2).$$

The other windows that should be mentioned for application in spectrum analysis are the Dolph-Chebyshev window, which provides uniform (equiripple) sidelobes, and the Kaiser window, which provides selectable sidelobe levels and bandwidths (see section 9.1). Window functions are also employed in the design of antenna arrays, where their use is called *shading* and bandwidth implies angular resolution or *beamwidth*. A complete survey of window functions and their properties has been given by Harris [7].

7.4 FFT ALGORITHM

The widespread application of the DFT to convolution and spectrum analysis is due in large measure to the rediscovery in the mid-1960s by Cooley and Tukey of a fast algorithm for its computation, which is known as the *fast Fourier transform* (FFT) algorithm. Many variations exist in the formulation of this algorithm; we will consider only the basic approaches for $N = 2^r$ and integer r. The first approach—decimation in time—is based upon separating $\tilde{a}(n)$ into two sequences of length $N/2$ comprised of the even and odd-indexed samples, respectively, i.e.,

$$\begin{aligned}\tilde{A}(k) &= \sum_{n=0}^{N-1} \tilde{a}(n)\, W^{kn} \\ &= \sum_{n\,\text{even}} \tilde{a}(n)\, W^{kn} + \sum_{n\,\text{odd}} \tilde{a}(n)\, W^{kn} \qquad (7.4.1) \\ &= \sum_{m=0}^{(N/2)-1} \tilde{a}(2m)(W^2)^{km} + W^k \sum_{m=0}^{(N/2)-1} \tilde{a}(2m+1)(W^2)^{km}.\end{aligned}$$

Each of these summations is recognized as being simply an $N/2$-point DFT of the respective sequence because

$$W^2 = e^{-2j(2\pi/N)} = e^{-j2\pi/(N/2)}. \qquad (7.4.2)$$

Hence, if we compute the DFT $\tilde{A}_e(k)$ for the even-indexed sequence $\tilde{a}(0)$, $\tilde{a}(2), \ldots, \tilde{a}(N-2)$ and the DFT $\tilde{A}_0(k)$ for the odd-indexed sequence $\tilde{a}(1), \tilde{a}(3), \ldots,$ $\tilde{a}(N-1)$, we can find the overall DFT from (7.4.1) by combining them as

$$\tilde{A}(k) = \tilde{A}_e(k) + W^k \tilde{A}_0(k). \qquad (7.4.3)$$

The complex coefficients W^k are known as *twiddle factors*. We note that the $N/2$-point DFT's $\tilde{A}_e(k)$ and $\tilde{A}_0(k)$ are periodic in k with period $N/2$, and thus their values for $k \geq N/2$ need not be recomputed, given those for $0 \leq k < N/2$.

This basic idea is then applied again and again until we have only 2-point DFT's to compute. That is, each $N/2$-point DFT is computed by combining two $N/4$-point DFT's, each of which is computed by combining two $N/8$-point DFT's, etc., for r stages since $N = 2^r$. The initial 2-point DFT's require coefficients of only ± 1, as depicted in figure 7.12.

The overall decimation-in-time implementation of an 8-point FFT is shown in figure 7.13. Note the regularity of the twiddle factors W^k in successive DFT's. These complex coefficients are stored in memory (read-only memory in hardware implementations) and read out for multiplication as required. Note also that although the output DFT $\tilde{A}(k)$ is in natural order, the input array $\tilde{a}(n)$ is not. This ordering of the input is known as

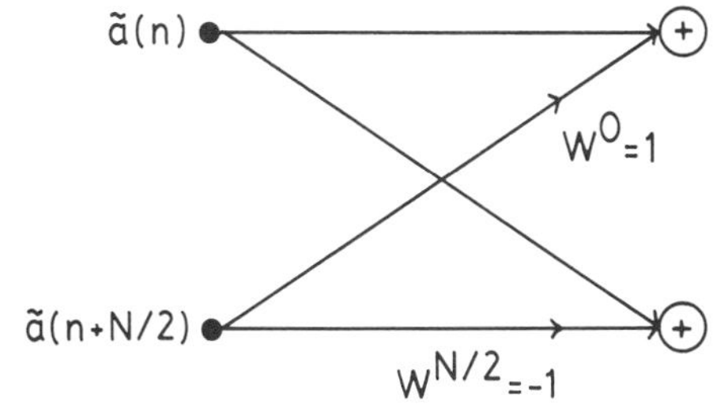

Figure 7.12. Diagram of initial two-point DFT's for decimation-in-time FFT.

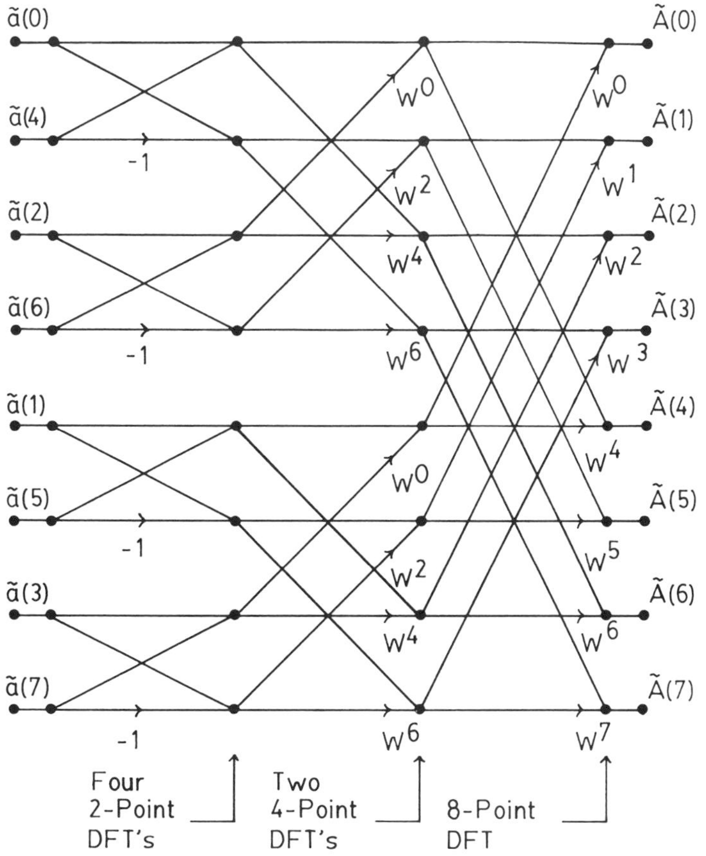

Figure 7.13. Diagram of eight-point decimation-in-time FFT, with inputs in bit-reversed order.

bit-reversed order because the index sequence n in binary form and natural order becomes this sequence when the bits representing each index value are reversed. For example, in this 8-point case, the natural order 000, 001, 010, 011, 100, 101, 110, 111 becomes 000, 100, 010, 110, 001, 101, 011, 111, or 0, 4, 2, 6, 1, 5, 3, 7, as shown in figure 7.13.

Let us estimate the computational saving provided by the FFT algorithm over a straightforward DFT implementation. By definition, each DFT sample $\tilde{A}(k)$ is a sum

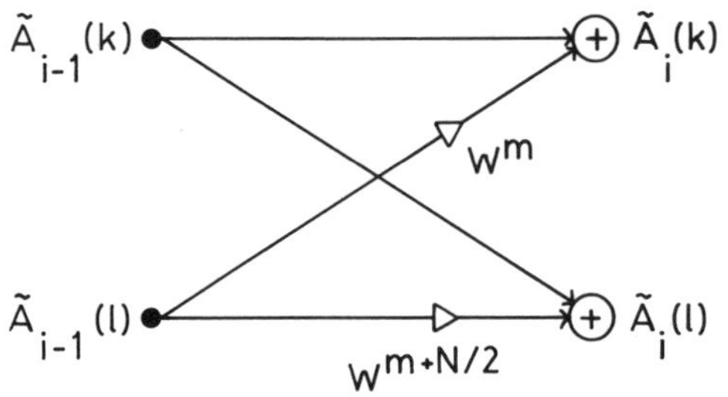

Figure 7.14. Decimation-in-time butterfly computation.

of N products, and there are N samples to compute, so direct DFT computation would require (at most) N^2 complex multiplications and additions. With the FFT algorithm, on the other hand, there are $r = \log_2 N$ stages, each of which requires up to N complex multiplications by twiddle factors and N complex additions. Hence, the number of complex multiplications and additions is of the order of $N \log_2 N$. For example, for $N = 2^{10} = 1024$, the FFT requires about 10^4 computations versus 10^6 for the direct DFT—a saving of a factor of 100! This sort of ratio can easily make the difference between a computationally viable program or system and one that is not. We note, of course, that the first stage in figure 7.13 contains coefficients of ± 1 and thus no actual multiplications, and that similar savings are readily found elsewhere. However, the computational order is still reasonably estimated as simply $N \log_2 N$.

An important aspect of the FFT algorithm in figure 7.13 is that it can be computed *in place* in memory. That is, if we do not need to retain the input array $\tilde{a}(n)$, we can overwrite it with the intermediate results of successive stages until it finally contains the DFT $\hat{A}(k)$. Hence, except for a few working registers, no additional memory is required. To see this, let the outputs of the ith stage be denoted by $\tilde{A}_i(k)$. Note then that the FFT in figure 7.13 consists of pairs of computations of the form

$$\begin{aligned} \tilde{A}_i(k) &= \tilde{A}_{i-1}(k) + W^m \tilde{A}_{i-1}(l) \\ \tilde{A}_i(l) &= \tilde{A}_{i-1}(k) + W^{m+N/2} \tilde{A}_{i-1}(l), \end{aligned} \tag{7.4.4}$$

where the initial inputs $\tilde{A}_0(k)$ are the $\tilde{a}(n)$ in bit-reversed order. This basic computational pair is shown in figure 7.14 and is known as the FFT *butterfly* computation. Clearly, after completing each butterfly, we can replace the input pair $\tilde{A}_{i-1}(k)$ and $\tilde{A}_{i-1}(l)$ in memory by the output pair $\tilde{A}_i(k)$ and $\tilde{A}_i(l)$ because we will not need those inputs in any further computation. Hence, the computation can proceed in place. The factor $W^{m+N/2}$ in (7.4.4) immediately suggests an additional saving of a factor of two in computation because $W^{N/2} = -1$. Therefore, each butterfly can be computed with only one complex multiplication as depicted in figure 7.15.

Many different variations of the FFT algorithm are possible depending upon whether we want the input or output in bit-reversed order, the need for in-place computation, and the associated bookkeeping complexity. For example, if we arrange the nodes of figure 7.13 so that the input array is in natural order and retain the use of in-place butterflies,

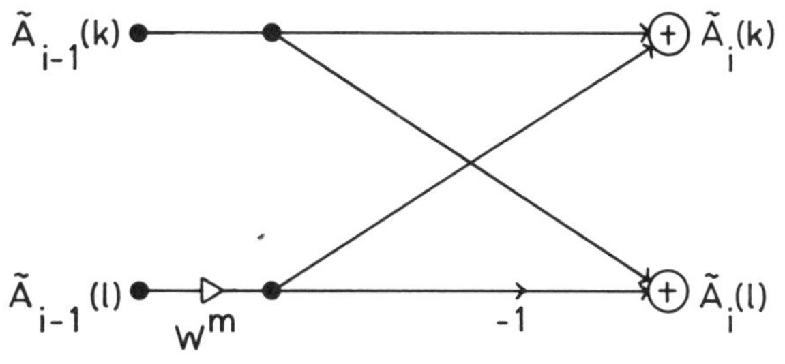

Figure 7.15. Decimation-in-time butterfly with only one complex multiplication.

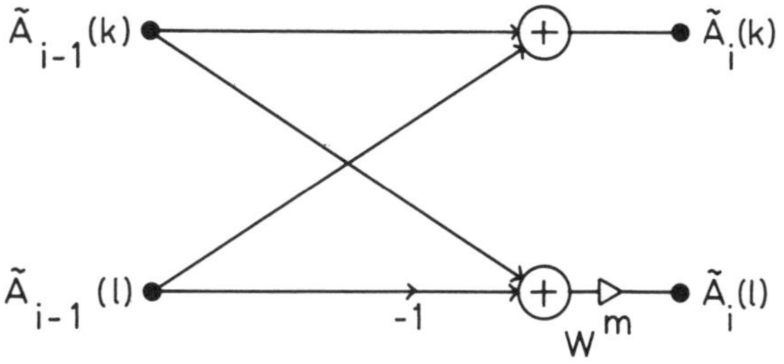

Figure 7.16. Decimation-in-frequency butterfly with only one complex multiplication.

the resulting output array is in bit-reversed order. (See Problem 7.11.) If we force both inputs and outputs to be in natural order, then in-place computation is destroyed, and the bookkeeping (control code or circuitry) is more complex. Another possibility is to form the transpose network for a particular decimation-in-time algorithm by reversing the direction of each branch in the network. This produces a new class of FFT algorithms known as the *decimation-in-frequency* algorithms. For example, the decimation-in-frequency butterfly is shown in figure 7.16, and the transpose of figure 7.13 is given in figure 7.17. (See Problem 7.14.)

Our purpose here is not to give a complete survey of the possible implementations of the FFT algorithm because most users will simply call an FFT subroutine or utilize an FFT circuit board or VLSI chip and will not need to know the details of the implementation. These users need simply to understand the basic approach, the various tradeoffs such as bit-reversed order and in-place computation, and the complexity estimate of $N \log_2 N$. A more complete survey is provided by Oppenheim and Schafer [2], including a description of algorithms for composite numbers N that are not powers of 2. Programs are also available in [16].

7.5 PERIODOGRAM

The examples of spectrum analysis given in section 7.2 were for deterministic signals, where the only effect to be considered was that of windowing. However, in practice, we usually process random signals, or combinations of deterministic and random signals, and this introduces additional statistical considerations into our choice of spectrum-analysis

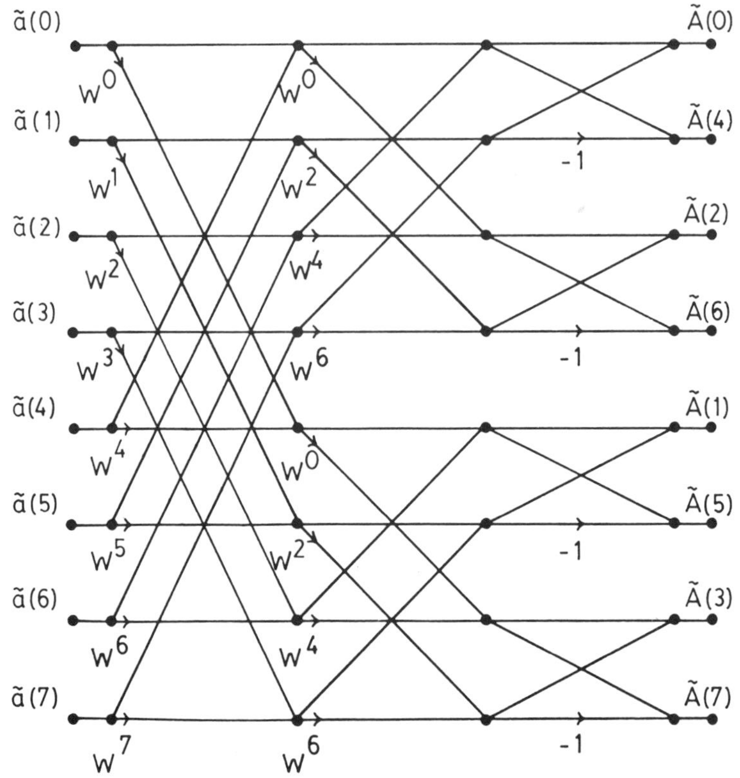

Figure 7.17. Diagram of eight-point decimation-in-frequency FFT.

methods. In this section, we give an overview of these considerations and the resulting *periodogram* methods based upon the DFT. This material assumes a basic knowledge of probability theory and random processes.

For random signals, the Fourier transform is not defined (because we don't know the signal for all time), and we seek instead to estimate the power spectrum. Defining the autocorrelation sequence for a stationary signal $a(n)$ to be

$$r(m) = E[a(n)a^*(n + m)] \tag{7.5.1}$$

with z transform $S(z)$, the power spectrum $S'(\omega)$ is the Fourier transform

$$S'(\omega) = \sum_{m=-\infty}^{\infty} r(m)e^{-j\omega mT}. \tag{7.5.2}$$

Given a finite data record $a(n)$, $n = 0, 1, \ldots, N - 1$, an unbiased estimate of $r(m)$ is simply

$$\hat{r}(m) = \frac{1}{N - |m|} \sum_{n=0}^{N-1-|m|} a(n)a^*(n + m) \tag{7.5.3}$$

for $|m| < N$ because then $E[\hat{r}(m)] = r(m)$. However, as $|m|$ approaches N, the variance of the estimate $\hat{r}(m)$ becomes large because the number $(N - |m|)$ of samples being averaged becomes small. Hence, the biased estimate

$$\tilde{r}(m) = \frac{1}{N} \sum_{n=0}^{N-1-|m|} a(n)a^*(n+m) \tag{7.5.4}$$

is commonly employed to avoid this excessive variance. The ˜ denotes the fact that $a(n)$ is effectively windowed in (7.5.4) by the rectangular window, i.e.,

$$\tilde{r}(m) = \frac{1}{N} \sum_{n=-\infty}^{\infty} \tilde{a}(n)\tilde{a}^*(n+m), \tag{7.5.5}$$

where $\tilde{a}(n) = w_R(n)a(n)$. Comparing (7.5.3) and (7.5.4), it is clear that the bias of the estimate $\tilde{r}(m)$ is given by

$$E[\tilde{r}(m)] = w_T(m)r(m), \tag{7.5.6}$$

where $w_T(m)$ is the triangular window $(N - |m|)/N$, $|m| < N$.

The Fourier transform of $\tilde{r}(m)$ is the *periodogram* and is readily derived from (7.5.5) to be

$$S'_R(\omega) = \frac{1}{N} |A'_R(\omega)|^2, \tag{7.5.7}$$

where $a_R(n) = \tilde{a}(n)$, or in terms of DFT's,

$$\tilde{S}_R(k) = \frac{1}{N} |\tilde{A}(k)|^2. \tag{7.5.8}$$

Hence, we can compute an estimate of the power spectrum directly from the DFT $\tilde{A}(k)$. The estimate is biased, however, since $\tilde{r}(m)$ is biased, and from (7.5.6) the bias is described by the convolution

$$E[S'_R(\omega)] = W'_T(\omega) * S'(\omega), \tag{7.5.9}$$

where

$$W'_T(\omega) = \frac{1}{N} [W'_R(\omega)]^2 = \frac{1}{N} \left(\frac{\sin(\omega TN/2)}{\sin(\omega T/2)} \right)^2. \tag{7.5.10}$$

Note also from (7.5.5) that $\tilde{r}(m)$ corresponds to a linear convolution of $\tilde{a}(n)$ and $\tilde{a}^*(-n)$ and thus is of length $2N - 1$. Hence, it cannot be recovered as the inverse DFT of (7.5.8) unless we have zero-padded $\tilde{a}(n)$ to at least this length before computing $\tilde{A}(k)$ and $\tilde{S}_R(k)$.

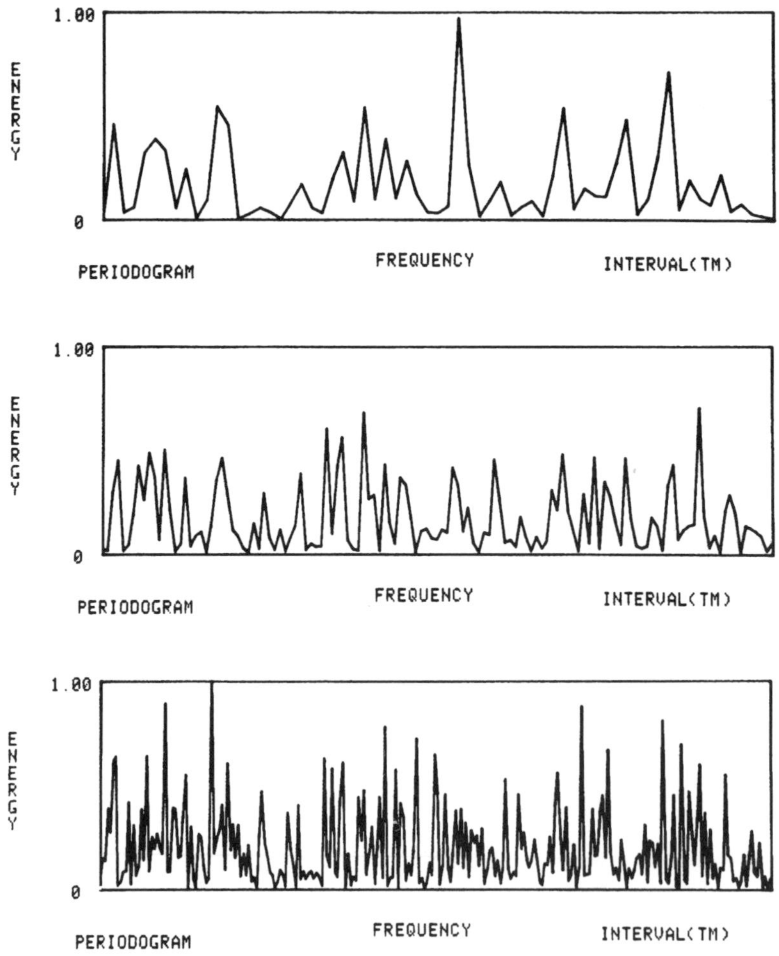

Figure 7.18. Periodograms of white noise for $N = 128$, 256, and 512.

As the length N of the sequence $\tilde{a}(n)$ is increased, the resolution of the spectral estimate $\tilde{S}_R(k)$ is increased, as expected. However, the variance of each spectral amplitude does not decrease with N, but rather remains approximately constant [2]. Therefore, the periodogram is not a consistent estimator of the power spectrum. This is illustrated in figure 7.18, which shows several spectral estimates for white noise with increasing values of N. Note that although the frequency resolution is increased, the range of spectral amplitudes does not decrease.

There are two basic approaches for dealing with this problem: averaging of multiple spectral estimates for shorter data segments, and windowing. Both approaches clearly sacrifice resolution to gain reduced variance. The best approach is often a combination of the two, as proposed by Welsh [8]. Dividing the data record into L nonoverlapping segments $\tilde{a}_i(n)$ of length $M = N/L$, we window each segment with an appropriate window (see section 7.3) and compute the L modified periodograms

$$ {}_i\tilde{S}_w(k) = \frac{1}{E_w}\left|\sum_{n=0}^{M-1} w(n)\tilde{a}_i(n)\,W^{kn}\right|^2 \qquad k = 0, 1, \ldots, M-1, \qquad (7.5.11)$$

$$ i = 1, 2, \ldots, L, $$

where E_w is the window energy

$$E_w = \sum_{n=0}^{M-1} w^2(n). \qquad (7.5.12)$$

Then, the overall spectral estimate is simply the average periodogram

$$\tilde{S}_w(k) = \frac{1}{L}\sum_{i=1}^{L} {}_i\tilde{S}_w(k). \qquad (7.5.13)$$

As before,

$$E[S'_w(\omega)] = \frac{1}{E_w}\,[W'(\omega)]^2 * S'(\omega), \qquad (7.5.14)$$

from which we can analyze the loss of spectral resolution, remembering that $w(n)$ is now of length M, not N. But, in compensation, the variance of the spectral estimate has been decreased by the factor $1/L$ [8].

It may bother the reader that a lot of data is largely discarded by being tapered toward zero by the window at the ends of each data segment. Certainly, if there is only a limited amount of data, one is reluctant to attenuate any of it. The spectral estimate can indeed be improved somewhat by overlapping the data segments [8]. In fact, using the Hanning window and 50-percent overlap (i.e., $2L - 1$ segments of the same length M as before, excluding two half segments at each end of the data record), every data sample except the $M/2$ at each end has exactly the same total weight in the spectral estimate in the sense that it is weighted by $\sin^2\theta$ in one segment and $\cos^2\theta$ in the overlapping one [see (7.3.3)].

PROBLEMS

7.1 Show the inverse DFT relationship in (7.1.6) from the definition of the DFT in (7.1.4).

7.2 Prove the DFT shift properties in (7.1.7) and (7.1.8).

7.3 Show that multiplication of DFT's as in (7.1.9) implies the circular convolution of the corresponding sequences in (7.1.10).

7.4 Circularly convolve the sequences in figure 7.1(a) and (b) to produce the result in figure 7.1(c). Compare this result with the results of linear convolution.

7.5 Derive the transform $W_N(z)$ for the Hanning window $w_N(n)$ and note the pole/zero cancellations.

7.6 Show that the windowed DFT $\tilde{A}_w(k)$ using the Hanning window can be computed from the unwindowed DFT $\tilde{A}(k)$ by

$$\tilde{A}_w(k) = -\tfrac{1}{4}\tilde{A}(k-1) + \tfrac{1}{2}\tilde{A}(k) - \tfrac{1}{4}\tilde{A}(k+1).$$

7.7 In this figure are shown several N-point sequences for $N = 8$.

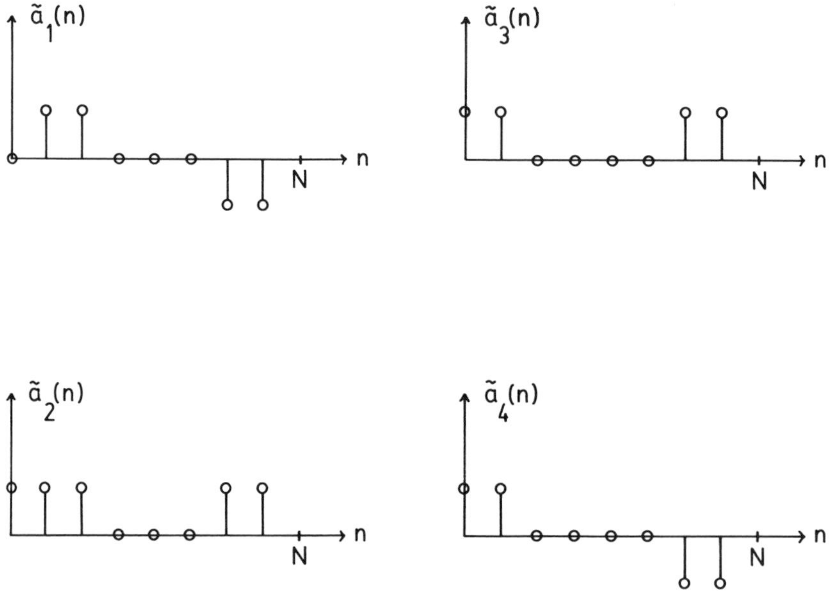

a. Which sequence has a real-valued DFT?
b. Which sequence has an imaginary-valued DFT?
c. For which sequence does $\tilde{A}(k) = 0$, $k = 2, 4, 6, \ldots$?
(*Hint*: Consider these finite-length segments to be single periods from corresponding periodic sequences $a_i(n)$.)

7.8 The DFT values $\tilde{A}(k)$ correspond to samples of the z transform of the (windowed) sequence $\tilde{a}(n)$ at uniformly spaced points around the unit circle, starting at $z = 1$. If, instead, we wish to sample the z transform on a circle of radius r, how can we modify the sequence $\tilde{a}(n)$ so that the resulting $\tilde{A}(k)$ are the desired samples?

7.9 Calculate the DFT's of the following sequences:
a. $\tilde{a}(n) = \delta(n)$.
b. $\tilde{a}(n) = \delta(n - n_0)$, $\quad 0 < n_0 < N$.
c. $\tilde{a}(n) = c^n$, $\quad n = 0, 1, \ldots, N-1$.

7.10 Derive Parseval's relation for the DFT, i.e.,

$$\sum_{n=0}^{N-1} |\tilde{a}(n)|^2 = \frac{1}{N} \sum_{k=0}^{N-1} |\tilde{A}(k)|^2.$$

7.11 Redraw the diagram of the decimation-in-time FFT algorithm in figure 7.13 for in-place computation with the inputs in natural order. Note that the outputs are now in bit-reversed order. Compare your diagram with that for the decimation-in-frequency algorithm in figure 7.17.

7.12 Redraw the diagram of the decimation-in-time FFT algorithm in figure 7.13, incorporating the butterfly computation in figure 7.15 to reduce the number of complex multiplications.

7.13 Derive a 9-point decimation-in-time FFT algorithm using basic 3-point DFT's, and draw a diagram of the algorithm.

7.14 Derive the decimation-in-frequency FFT algorithm as follows: Dividing the input sequence $\tilde{a}(n)$ into two halves, define sum and difference sequences of length $N/2$ by

$$f(n) = \tilde{a}(n) + \tilde{a}(n + N/2), \qquad 0 \leq n < N/2$$

$$g(n) = \tilde{a}(n) - \tilde{a}(n + N/2), \qquad 0 \leq n < N/2.$$

Show that the $N/2$-point DFT of $f(n)$ corresponds to $\tilde{A}(k)$ for $k = 0, 2, 4, \ldots, N-2$, and that the $N/2$-point DFT of $W^n g(n)$ corresponds to $\tilde{A}(k)$ for $k = 1, 3, 5, \ldots, N-1$. Likewise, each $N/2$-point DFT becomes two $N/4$-point DFT's, and so forth. Draw a diagram of a complete 8-point decimation-in-frequency FFT resulting from this derivation, and compare it with figure 7.17.

7.15 Derive a 16-point decimation-in-time FFT algorithm using four 4-point DFT's. Show the resulting *radix-4* butterflies with internal coefficients of ± 1 and $\pm j$ only. Noting that coefficients of ± 1 and $\pm j$ do not require actual multiplications, compare the number of complex multiplications in this radix-4 algorithm with the number for the *radix-2* algorithm based upon the butterfly of figure 7.15.

7.16 Show that the bias expression in (7.5.9) and (7.5.10) for the estimated power spectrum with rectangular windowing fits the general expression for arbitrary windowing in (7.5.14).

7.17 Show that (7.5.3) provides an unbiased estimate of the autocorrelation function $r(m)$; while (7.5.4) yields a biased estimate. That is, find $E[\hat{r}(m)]$ and $E[\tilde{r}(m)]$.

7.18 Show that the Fourier transform of $\tilde{r}(m)$ is the periodogram $S_R'(\omega)$ given by (7.5.7).

8. IIR FILTER DESIGN BY TRANSFORMATION

8.0 INTRODUCTION

The first techniques for discrete-time filter design we shall study involve transformation of the designs for continuous-time filters. The classical filter approximations are of particular interest because they are equally applicable, in most cases, to continuous or discrete-time design and because they are so widely known and extensively tabulated. Of necessity, discrete-time filters derived by transformation from continuous-time designs are IIR filters because the number of finite poles must equal or exceed the number of finite zeros in the continuous-time case in order to prevent infinite gain at infinite frequency.

The system function for a rational continuous-time filter is of the form

$$H_c(s) = \frac{\sum_{m=0}^{M_c} c_m s^m}{\sum_{k=0}^{N} d_k s^k}, \tag{8.0.1}$$

where $M_c \leq N$ to prevent poles at infinity, as mentioned above. As implied by the notation in (8.0.1), the order N of the denominator of the system function will, in general, be the same for corresponding continuous and discrete-time filters; whereas the numerator polynomials can be of different orders depending upon the transformation used.

Various transformations can be chosen to preserve various properties of the continuous-time filter, but we will generally want the $j\omega$-axis in the s plane to map into the unit circle in the z plane. We will also insist that the left-half plane map into the interior of the unit circle to preserve stability. We must emphasize at this point that although many different transformations can be used for *filter design* (as long as the above constraints are met), it is

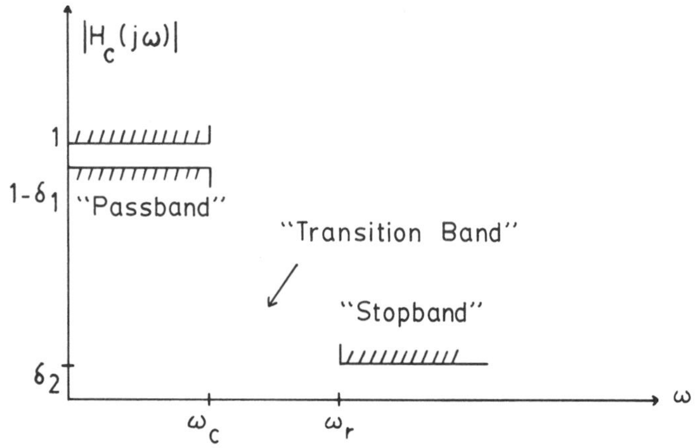

Figure 8.1. Classical lowpass filter specifications for the magnitude response.

usually meaningless to apply them to *signals*; only the standard z transform defined in chapter 3 is appropriate for signal analysis.

8.1 CLASSICAL FILTER DESIGNS

Most often, transformation techniques are applied to the design of discrete-time filters with classical continuous-time counterparts. The classical filter designs of primary interest to us —namely, Butterworth, Chebyshev, and elliptic—satisfy constraints on the magnitude of the frequency response $H_c(j\omega)$ of the form illustrated in figure 8.1 for the lowpass case. That is, in the *passband*, the frequency response is required to satisfy

$$1 \geq |H_c(j\omega)| \geq 1 - \delta_1, \qquad |\omega| \leq \omega_c \tag{8.1.1}$$

and in the *stopband*

$$|H_c(j\omega)| \leq \delta_2, \qquad |\omega| \geq \omega_r, \tag{8.1.2}$$

with the response unspecified in the *transition band* $\omega_c < |\omega| < \omega_r$. Analogous specifications are made in the case of highpass, bandpass, and bandstop filters, but we will first study the design of lowpass filters and then, in section 8.4, investigate spectral transformations to convert a lowpass design into one of these other types.

The specifications for and properties of the three principal classical designs are described below. Conspicuously absent from this group is the classical Bessel design, which approximates linear phase in the sense of maximally flat group delay in the passband. There are two reasons for this omission. First, as we have seen, FIR filters can easily be constrained to have exactly linear phase and are thus usually chosen if this is a requirement. Second, acceptable Bessel IIR designs are derived by transformation only for a relatively limited range of specifications—namely, sufficiently low cutoff frequencies ω_c. Thus, although satisfactory approximations to linear phase can indeed be obtained by transformation of

Bessel designs in some cases, this technique is not generally applicable, in contrast to transformation of the following filter types.

Butterworth Design

The Butterworth approximation results from the requirement that the magnitude response be *maximally flat* in both the passband and the stopband. That is, the first $(2N - 1)$ derivatives of $|H_c(j\omega)|^2$ are specified to equal zero at $\omega = 0$ and at $\omega = \infty$. The general specifications in (8.1.1) and (8.1.2) can be employed with arbitrary δ_1, δ_2, ω_c, and ω_r to determine the required filter order N, from which the filter design is readily obtained. Very often, however, the cutoff frequency ω_c of a Butterworth filter is chosen to correspond to the half-power or 3-dB point of $|H_c(j\omega)|$, in which case $1 - \delta_1 = 1/\sqrt{2}$. We will assume in the following that this is the case since the general properties of the Butterworth design are unaffected by this assumption.

The squared magnitude response resulting from the above specifications is of the form

$$|H_c(j\omega)|^2 = \frac{1}{1 + \left(\dfrac{\omega}{\omega_c}\right)^{2N}}. \tag{8.1.3}$$

Note that $|H_c(0)| = 1$ and $|H_c(j\omega_c)| = 1/\sqrt{2}$, as desired, and that $|H_c(j\omega)|$ falls off as $1/\omega^N$ for $\omega \gg \omega_c$. Note also that $|H_c(j\omega)|$ is monotonically decreasing for all ω and thus is monotonic in both the passband and the stopband.

The Butterworth magnitude response is shown in figure 8.2 for several values of the order N with both linear and logarithmic (dB) scales. The logarithmic scale is especially appropriate because for $\omega \gg \omega_c$, $|H_c(j\omega)|$ approaches an asymptotic rolloff of approximately $-6\,N$ dB per octave, as indicated in figure 8.2(b). A continuous-time Butterworth filter is often specified simply by the 3-dB cutoff frequency and the rolloff rate. A discrete-time Butterworth filter can also be specified in this way, but as we shall see, the rolloff can actually be greater (with the bilinear transformation) in the discrete-time case.

By analytic continuation of $|H_c(j\omega)|^2 = H_c(j\omega)H_c^*(j\omega) = H_c(j\omega)H_c(-j\omega)$, the system function of the continuous-time Butterworth filter must satisfy

$$H_c(s)H_c(-s) = \frac{1}{1 + \left(\dfrac{s}{j\omega_c}\right)^{2N}} \tag{8.1.4}$$

and the poles of $H_c(s)H_c(-s)$ are thus given by

$$s_k = (-1)^{1/2N}(j\omega_c). \tag{8.1.5}$$

These $2N$ poles are equally spaced around a circle of radius ω_c in the s plane and symmetrically located with respect to both the real and imaginary axes, as illustrated in figure 8.3 for $N = 4$ and $N = 5$. Note that there are poles on the real axis for N odd, and not for N even, but there are never poles on the $j\omega$ axis.

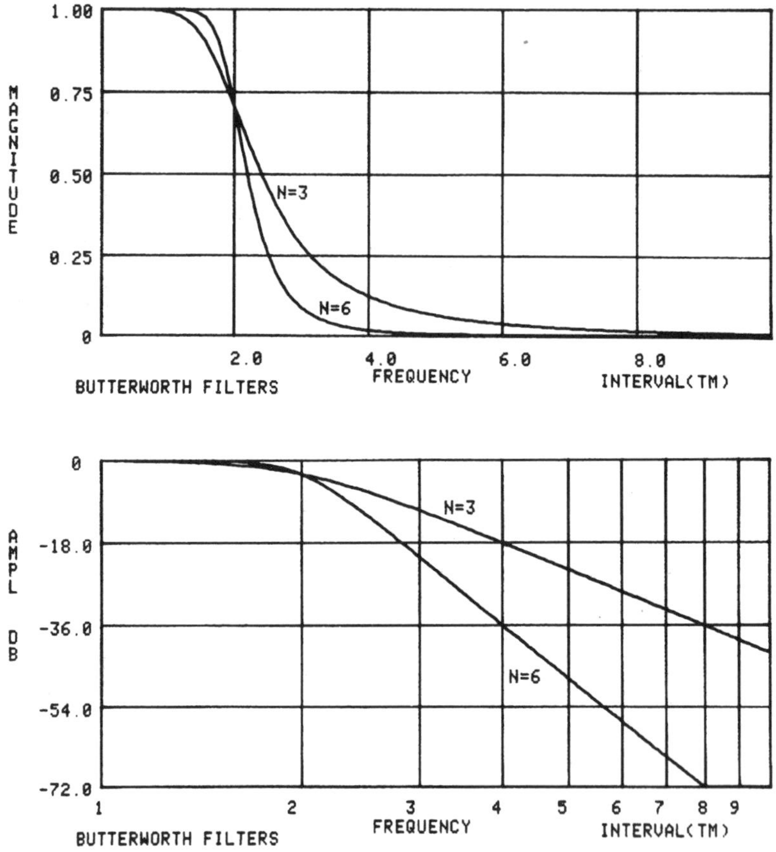

Figure 8.2. Magnitude responses of Butterworth LPF's for $N = 3$ and $N = 6$ on both linear and logarithmic scales.

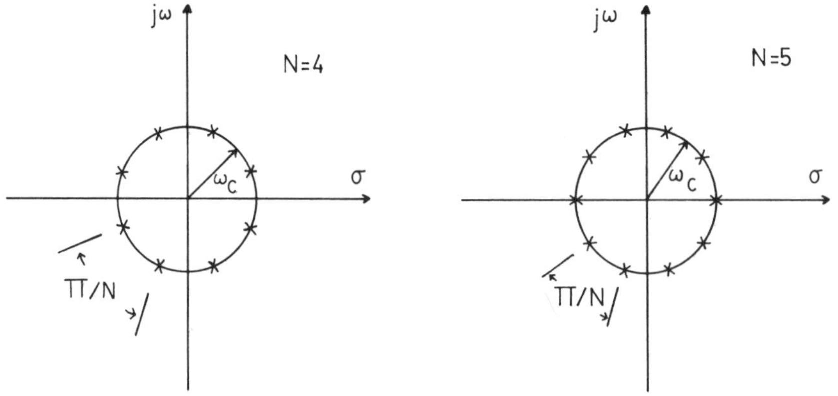

Figure 8.3. Pole locations of $H_c(s)H_c(-s)$ for fourth- and fifth-order Butterworth LPF's.

Since we are ultimately interested in the poles of $H_c(s)$ alone, we must select these N poles from the $2N$ poles of $H_c(s)H_c(-s)$. For a stable and causal filter, this is straightforward since we must choose the N poles lying in the left-half plane for $H_c(s)$, with $H_c(-s)$ then having the right-half-plane poles. Note that $H_c(s)$ has no zeros (except at infinity), and thus the Butterworth is one of the *all-pole* continuous-time designs.

EXAMPLE. As a simple example of a Butterworth design, we will derive the system function for a filter with cutoff frequency $\omega_c = 1$ and 24-dB attenuation at $\omega_r = 4$. Since ω_r is two octaves above ω_c, we need a rolloff of 12 dB/octave and thus $N = 2$. The two poles of $H_c(s)$ are located on a circle of radius 1 at angles of $\pm 45°$ from the negative real axis, or

$$s_1, s_2 = \frac{1}{\sqrt{2}}(-1 \pm j).$$

Hence,

$$H_c(s) = \frac{1}{\left(s + \dfrac{(1-j)}{\sqrt{2}}\right)\left(s + \dfrac{(1+j)}{\sqrt{2}}\right)}$$

$$= \frac{1}{s^2 + \sqrt{2}s + 1}.$$

Chebyshev Designs

A more rapid rolloff rate near the cutoff frequency than that of the Butterworth design can be achieved at the expense of a loss of monotonicity in the passband and/or the stopband. The Chebyshev designs (types I and II) maintain monotonicity in one band but are *equiripple* in the other band, as illustrated in figure 8.4.

The squared magnitude response for a Chebyshev-I design is of the form

$$|H_c(j\omega)|^2 = \frac{1}{1 + \varepsilon^2 T_N^2(\omega/\omega_c)}, \tag{8.1.6}$$

where $T_N(x)$ is the Nth-order Chebyshev polynomial defined by

$$T_N(x) = \cos(N \cos^{-1} x) = \cosh[N \cosh^{-1} x]. \tag{8.1.7}$$

The Chebyshev polynomials can be generated recursively by

$$T_{N+1}(x) = 2x T_N(x) - T_{N-1}(x)$$

with $T_0(x) = 1$, $T_1(x) = x$, and thus $T_2(x) = 2x^2 - 1$, etc. Since $T_N(1) = 1$ for all N, the squared magnitude response at cutoff equals $1/(1 + \varepsilon^2)$, and ε^2 is thus determined from the passband ripple δ_1 as

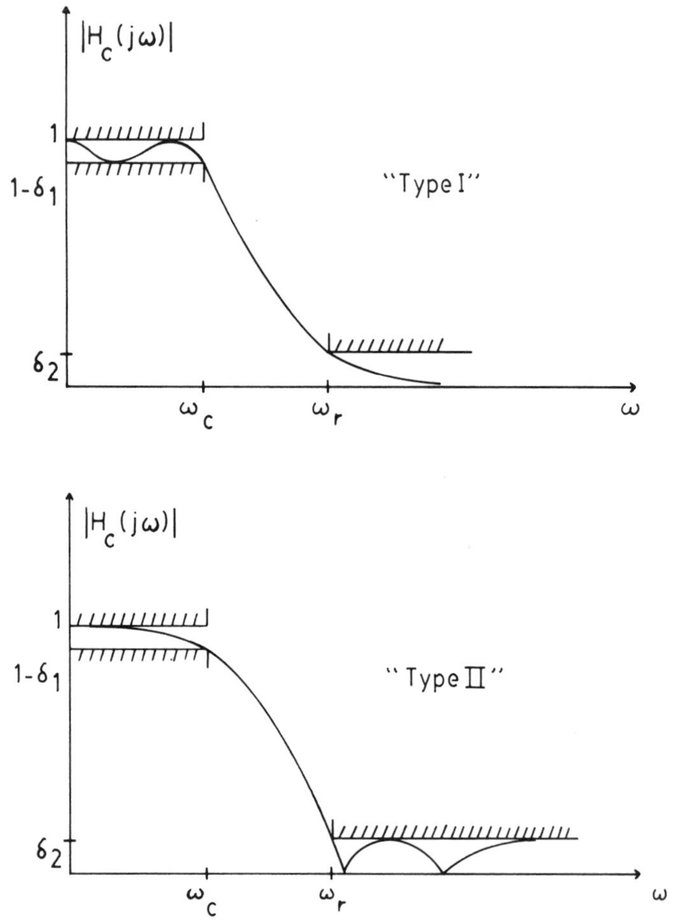

Figure 8.4. Sample Chebyshev magnitude responses for (a) Type-I LPF and (b) Type-II LPF.

$$\varepsilon^2 = \frac{1}{(1 - \delta_1)^2} - 1. \tag{8.1.8}$$

Therefore, a Chebyshev-I design is specified by the cutoff frequency, the passband ripple, and the order N, with the latter determining the stopband characteristics. The required value of N for given stopband specifications is obtained from (8.1.6) and (8.1.7) as

$$N \approx \frac{\cosh^{-1}(1/\delta_2\varepsilon)}{\cosh^{-1}(\omega_r/\omega_c)}. \tag{8.1.9}$$

The poles of $H_c(s)$ are found from those of $H_c(s)H_c(-s)$ as in the Butterworth case and lie on an ellipse in the s plane. The zeros of $H_c(s)$ are all at infinity for a type-I filter, and hence this is another all-pole continuous-time design.

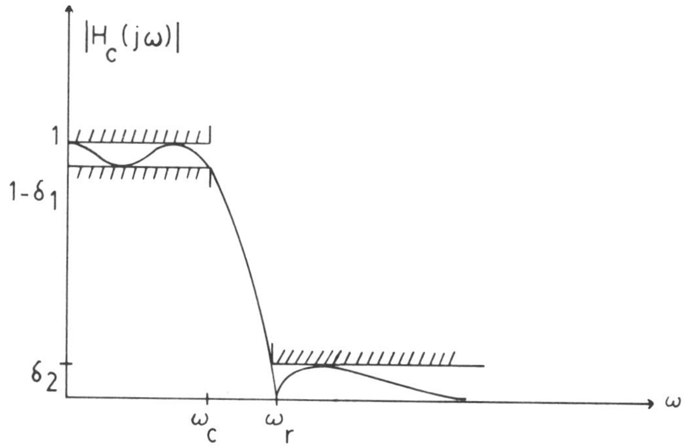

Figure 8.5. Sample elliptic LPF magnitude response.

The squared magnitude response for a Chebyshev-II design is of the form

$$|H_c(j\omega)|^2 = \frac{1}{1 + \varepsilon^2 \left[\dfrac{T_N^2(\omega_r/\omega_c)}{T_N^2(\omega_r/\omega)} \right]}. \tag{8.1.10}$$

It is readily observed from the response at cutoff that (8.1.8) still holds for ε^2, and since (8.1.6) and (8.1.10) are the same for $\omega = \omega_r$, (8.1.9) holds as well for N. The zeros of this design are obviously not at infinity, but rather lie along the $j\omega$ axis to produce the equiripple stopband characteristic. The pole and zero locations of the Chebyshev designs have been tabulated extensively [9], and we will assume access to such tabulations, or equivalent computer-aided-design (CAD) programs, in the sections to follow.

Elliptic Design

The sharpest transition from passband to stopband for given δ_1, δ_2, and N is achieved by the elliptic (or Cauer) design. In fact, the elliptic design is optimum in this sense. The magnitude response of an elliptic filter is equiripple in both the passband and stopband, as illustrated in figure 8.5.

The squared magnitude response is of the form

$$|H_c(j\omega)|^2 = \frac{1}{1 + \varepsilon^2 U_N^2(\omega/\omega_c)}, \tag{8.1.11}$$

where $U_N(\omega)$ is a Jacobian elliptic function. A discussion of elliptic functions is beyond the scope of this book, and we will rely on the availability of tabulations [9] or CAD programs [16] to obtain continuous-time elliptic filter designs. We can observe, however, from figure 8.5 that the zeros of $H_c(s)$ must lie along the $j\omega$ axis, as for the Chebyshev-II case.

It is reasonable at this point to ask why the elliptic design is not always chosen to meet specifications of the type in figure 8.1 since it does so with lower-order N than for any other filter design. One reason, as we have already noted, is that the Chebyshev designs are monotonic in one band, and the Butterworth design is monotonic in both. Another important reason is that the phase response of the elliptic design is more nonlinear in the passband, especially near cutoff, than the response of the other designs.

8.2 IMPULSE-INVARIANT TRANSFORMATION

As discussed in the introduction to this chapter, a number of different transformations can be employed to convert continuous-time filter designs into discrete-time designs. The most straightforward of these is probably the impulse-invariant transformation described in this section. The most generally useful is the bilinear transformation described in the next section.

Let $h_c(t)$ be the impulse response corresponding to $H_c(s)$, and define the continuous- to discrete-time transformation by setting

$$h(n) = h_c(nT). \tag{8.2.1}$$

Thus, in effect, we sample the continuous-time impulse response to produce that of the discrete-time filter. As we have seen in section 6.1, the frequency response $H'(\omega)$ corresponding to (8.2.1) is the Fourier transform of the continuous-time function

$$h_c^*(t) = \sum_{n=-\infty}^{\infty} h_c(nT)\delta(t - nT) \tag{8.2.2}$$

and hence

$$H'(\omega) = \frac{1}{T}\sum_{k=-\infty}^{\infty} H_c\left[j\left(\omega - k\frac{2\pi}{T}\right)\right]. \tag{8.2.3}$$

The system function corresponding to (8.2.1) is likewise

$$H(z)\big|_{z=e^{sT}} = \frac{1}{T}\sum_{k=-\infty}^{\infty} H_c\left(s - jk\frac{2\pi}{T}\right), \tag{8.2.4}$$

which is the many-to-one transformation from the s plane to the z plane that we also studied in section 6.2. Therefore, the impulse-invariant transformation does map the $j\omega$-axis and the left-half s plane into the unit circle and its interior, respectively, as desired.

Although the impulse response of the continuous-time filter is preserved in the sense of (8.2.1) by this transformation, the frequency response can be altered significantly, often to the point of being useless. In particular, (8.2.3) implies that $H'(\omega)$ is an aliased version of $H_c(j\omega)$, as illustrated in figure 8.6 for a lowpass design. Hence, the stopband characteristics are maintained adequately in the discrete-time frequency response only if the aliased tails of $H_c(j\omega)$ are sufficiently small. The passband is also affected, but this effect is usually less pronounced than for the stopband.

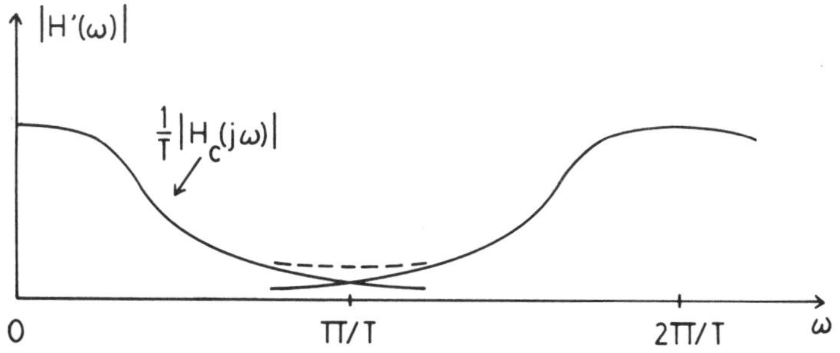

Figure 8.6. Magnitude response resulting from impulse-invariant transformation, showing aliasing.

It is clear from figure 8.6 that the Butterworth and Chebyshev-I lowpass designs are more appropriate for impulse-invariant transformation than are the Chebyshev-II and elliptic designs because the former are monotonic in the stopband, while the latter are not. In particular, the equiripple stopband characteristic of the Chebyshev-II and elliptic designs will, in general, be lost if the impulse-invariant transformation is used. Likewise, this transformation cannot be applied directly to highpass and bandstop designs.

The impulse-invariant transformation is not usually performed directly in the form of (8.2.1). Rather, $H_c(s)$ is first expanded in a partial fraction expansion to produce

$$H_c(s) = \sum_{k=1}^{N} \frac{A_k}{s - s_k}, \tag{8.2.5}$$

where we have assumed that there are no multiple poles. Then, since

$$h_c(t) = \sum_{k=1}^{N} A_k e^{s_k t} u(t), \tag{8.2.6}$$

we have from (8.2.1) that

$$h(n) = \sum_{k=1}^{N} A_k e^{s_k n T} u(n) \tag{8.2.7}$$

and thus

$$H(z) = \sum_{k=1}^{N} \frac{A_k}{1 - e^{s_k T} z^{-1}}. \tag{8.2.8}$$

Therefore, the parameters of $H(z)$ may be obtained directly from $H_c(s)$ without bothering to evaluate $h_c(t)$ or $h(n)$.

Note that the s-plane poles at s_k are mapped into z-plane poles at $p_k = e^{s_k T}$ by the impulse-invariant transformation. However, this transformation is not simply a mapping of $H_c(s)$ to $H(z)$ via the change of variables $z = e^{sT}$, as we can see from (8.2.4).

In particular, the N zeros of $H_c(s)$, including those at infinity, do not generally map in this manner. A transformation that does map the zeros, as well as the poles, directly by $z = e^{sT}$ can be defined and is, in fact, called the *matched z transform*. However, although perfectly usable filters can be designed in this way, no special time- or frequency-domain properties are preserved by this transformation, and it is not widely used.

A final observation concerning the impulse-invariant transformation is that for $T \ll 1$, (i.e., $\omega_s \gg 1$), the $1/T$ factor in (8.2.3) implies large gain in the resulting discrete-time filter. To keep the maximum of $|H'(\omega)|$ comparable to that of $|H_c(j\omega)|$, the transformation is often defined in the form

$$H(z) = \sum_{k=1}^{N} \frac{A_k T}{1 - e^{s_k T} z^{-1}}, \tag{8.2.9}$$

which implies that

$$h(n) = Th_c(nT). \tag{8.2.10}$$

Thus, the impulse response is attenuated by the factor T, and the frequency response is then

$$H'(\omega) = \sum_{k=-\infty}^{\infty} H_c\left[j\left(\omega - k\frac{2\pi}{T}\right)\right]. \tag{8.2.11}$$

EXAMPLE. Assume a second-order continuous-time design with the system function

$$H_c(s) = \frac{s + a}{(s + a)^2 + b^2}.$$

Expanding $H_c(s)$ in a partial-fraction expansion, we produce

$$H_c(s) = \frac{1/2}{s + a + jb} + \frac{1/2}{s + a - jb}.$$

Therefore, the impulse-variant transformation yields a discrete-time design with the system function

$$\begin{aligned} H(z) &= \frac{1/2}{1 - e^{(-a+jb)T} z^{-1}} + \frac{1/2}{1 - e^{(-a-jb)T} z^{-1}} \\ &= \frac{1 - e^{-aT}(\cos bT) z^{-1}}{1 - 2e^{-aT}(\cos bT) z^{-1} + e^{-2aT} z^{-2}}. \end{aligned}$$

Note that although we mapped the poles via $p_k = e^{s_k T}$, the zero at $s = -a$ was transformed into a zero at $z = e^{-aT}(\cos bT)$. Note also that the radius r of the z-plane poles

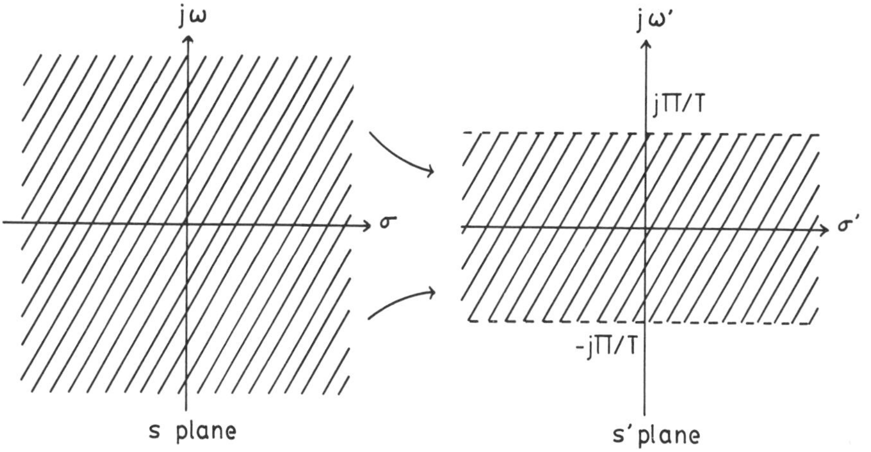

Figure 8.7. Desired mapping from s plane to s' plane to avoid aliasing.

is determined solely by the real part of s_k as $r = e^{-aT}$; while the angle θ of the z-plane poles is determined solely by the imaginary part of s_k as $\theta = bT$.

8.3 BILINEAR TRANSFORMATION

To avoid aliasing of the frequency response as encountered with the impulse-invariant transformation, we need a one-to-one mapping from the s plane to the z plane. The problem with the transformation $z = e^{sT}$ is that it is many-to-one, as we have seen in chapter 6. However, if we could first employ a one-to-one transformation from s to s', which compresses the entire s plane into the strip $-\pi/T \leq \text{Im}(s') \leq \pi/T$, then s' could be transformed to z via $z = e^{s'T}$ with no effect from aliasing. The desired mapping from the s plane to the s' plane is depicted in figure 8.7.

Such a one-to-one transformation from s to s' is given by

$$s' = \frac{2}{T} \tanh^{-1}\left(\frac{sT}{2}\right). \tag{8.3.1}$$

The nature of this transformation is seen most readily from its effect on the $j\omega$ axis. Substituting $s = j\omega$ and $s' = j\omega'$ into (8.3.1), we obtain

$$\omega' = \frac{2}{T} \tan^{-1}\left(\frac{\omega T}{2}\right) \tag{8.3.2}$$

which is plotted in figure 8.8. Thus, the entire ω axis is compressed into the interval $(-\pi/T, \pi/T)$ for ω' in a one-to-one manner, as desired. We also note that although the relationship between ω and ω' is nonlinear, it is approximately linear for small $\omega \approx \omega'$.

The desired transformation from s to z is now obtained by inverting (8.3.1) to produce

$$s = \frac{2}{T}\tanh\left(\frac{s'T}{2}\right)$$

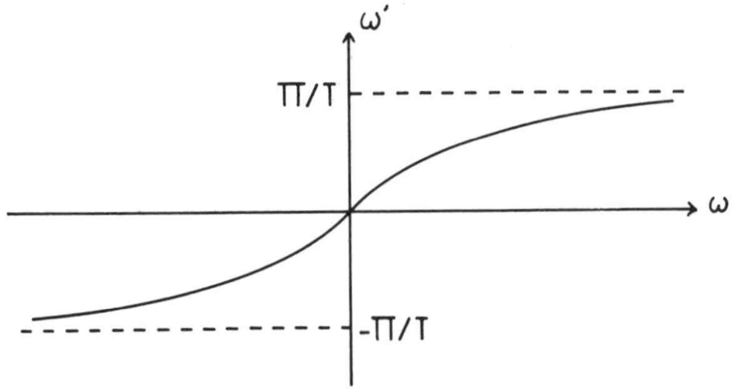

Figure 8.8. Mapping from ω to ω' via the bilinear transformation.

and setting $z = e^{s'T}$, or

$$s' = (1/T)\ln z,$$

which yields

$$s = \frac{2}{T}\tanh\left(\frac{\ln z}{2}\right).$$

This formidable-looking expression is actually straightforward since

$$\tanh x = \frac{e^{x} - e^{-x}}{e^{x} + e^{-x}} = \frac{1 - e^{-2x}}{1 + e^{-2x}},$$

from which we obtain

$$s = \frac{2}{T}\left(\frac{1 - z^{-1}}{1 + z^{-1}}\right). \tag{8.3.3}$$

Hence, the discrete-time filter design is obtained from the continuous-time design via the *bilinear transformation*

$$H(z) = H_c(s)\big|_{s=(2/T)(1-z^{-1})/(1+z^{-1})}. \tag{8.3.4}$$

Note that, unlike the impulse-invariant transformation, the bilinear transformation is algebraic, one-to-one, and thus invertible. The inverse transformation is derived from (8.3.3) as

$$z = \frac{1 + \frac{T}{2}s}{1 - \frac{T}{2}s}. \tag{8.3.5}$$

To check that the bilinear transformation has the desired properties stated in section 8.0, we note from (8.3.5) that for $\mathrm{Re}(s) = 0$, $|z| = 1$, and thus the $j\omega$ axis is indeed mapped into the unit circle. Also, $\mathrm{Re}(s) < 0$ implies $|z| < 1$, and thus the left-half s plane is mapped into the interior of the unit circle, as required for stability.

Because of the nonlinear relationship in (8.3.2), it is necessary to distinguish between the discrete-time frequency ω' and the continuous-time frequency ω. To see that ω' is indeed the variable corresponding to discrete-time frequency, we set $z = e^{j\omega' T}$ in (8.3.3) to produce

$$s = \frac{2}{T}\left(\frac{1 - e^{-j\omega' T}}{1 + e^{-j\omega' T}}\right) = \frac{2j \sin\dfrac{\omega' T}{2}}{T \cos\dfrac{\omega' T}{2}}$$

$$= \frac{2}{T} j \tan\frac{\omega' T}{2} = j\omega$$

and thus, from (8.3.2), $s = j\omega$ as required. Two frequencies of particular importance are $\omega = 0$, where $\omega' = 0$ and $z = 1$, and $\omega = \infty$, where $\omega' = \pi/T = \omega_s/2$ and $z = -1$. The latter case shows that poles or zeros at infinity in $H_c(s)$ are mapped to $z = -1$ in $H(z)$ by the bilinear transformation. Therefore, all-pole continuous-time lowpass designs (which have N zeros at infinity), such as the Butterworth and Chebyshev-I, will produce N zeros in $H(z)$ at $z = -1$ when the bilinear transformation is employed.

The nonlinear relationship between ω and ω' in (8.3.2) is known as *frequency warping*. The corresponding effect on $H'(\omega')$ relative to $H_c(j\omega)$ is obtained from (8.3.4) as

$$H'(\omega') = H_c(j\omega)\big|_{\omega=(2/T)\tan(\omega' T/2)}, \tag{8.3.6}$$

and is depicted in figure 8.9. Note that $H_c(j\omega)$ is compressed in frequency by this transformation, but otherwise, the characteristics of $H_c(j\omega)$ are preserved in $H'(\omega')$. In particular, if $|H_c(j\omega)|$ is equiripple in the passband or stopband, then $|H'(\omega')|$ is also. This property is the most important feature of the bilinear transformation and has made it the most widely employed transformation for classical designs.

To achieve a discrete-time design with specified values for ω'_c and ω'_r, these values must first be *prewarped* according to

$$\omega = \frac{2}{T}\tan\left(\frac{\omega' T}{2}\right) \tag{8.3.7}$$

in order to determine the appropriate values of ω_c and ω_r for the corresponding continuous-time design. And since the discrete-time *transition ratio* ω'_r/ω'_c is less than ω_r/ω_c due to the compression effect of warping, the specifications for the discrete-time filter are actually "tighter" than those for the prewarped continuous-time design. This effect can sometimes produce a lower-order discrete-time filter than would otherwise have been expected on the basis of continuous-time design theory alone.

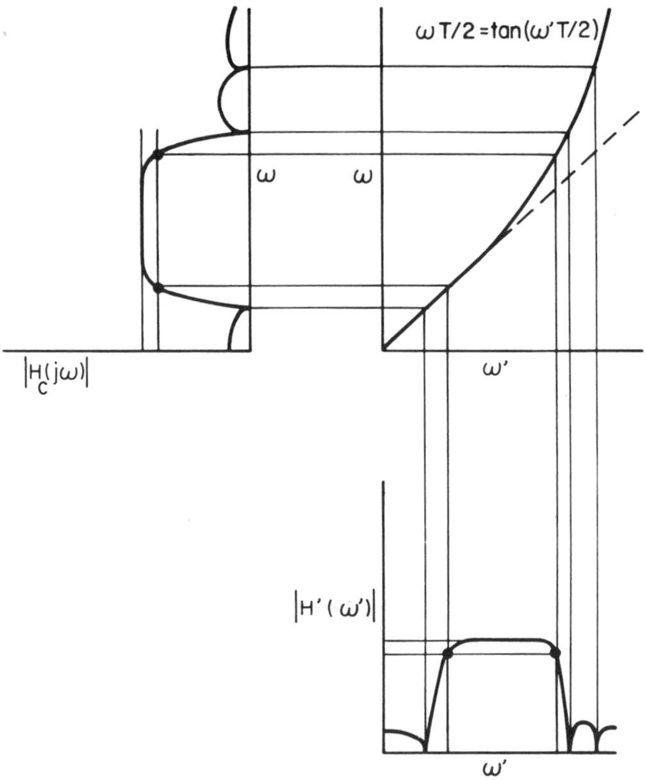

Figure 8.9. Illustration of the bilinear transformation of $H_c(j\omega)$ to $H'(\omega')$.

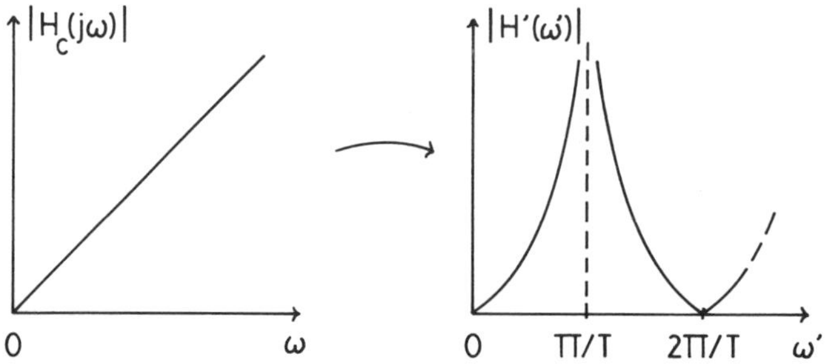

Figure 8.10. Effect of the bilinear transformation on a nonpiecewise-constant response.

As implied by the above discussion and figure 8.9, filter designs with piecewise-constant specifications are the most appropriate for bilinear transformation. Consider, for example, what happens in the bilinear transformation of the continuous-time differentiation characteristic of figure 8.10. The effect of warping is to change the linear magnitude response into a nonlinear (tangent) magnitude response, and an acceptable differentiator is achieved only for small ω'. This is the reason why the Bessel design is not amenable to the bilinear transformation, in general; i.e., its linear phase response is similarly distorted.

Although the bilinear transformation is defined by (8.3.4) as an algebraic substitution of (8.3.3) into $H_c(s)$, this is a tedious procedure in practice, and the transformation is not usually performed in this manner. Rather, since the entire s plane is mapped according to (8.3.3), the poles and zeros are themselves mapped in this way, and these plus a gain factor are all we need to specify $H(z)$. Specifically, writing $H_c(s)$ in terms of its poles and zeros as

$$H_c(s) = K \frac{\prod_{m=1}^{M_c} (s - \sigma_m)}{\prod_{k=1}^{N} (s - s_k)}, \tag{8.3.8}$$

the bilinear transformation must produce

$$H(z) = b_0 (1 + z^{-1})^{N-M_c} \frac{\prod_{m=1}^{M_c} (1 - z_m z^{-1})}{\prod_{k=1}^{N} (1 - p_k z^{-1})}, \tag{8.3.9}$$

where

$$z_m = \frac{1 + \frac{T}{2}\sigma_m}{1 - \frac{T}{2}\sigma_m} \quad \text{and} \quad p_k = \frac{1 + \frac{T}{2}s_k}{1 - \frac{T}{2}s_k}. \tag{8.3.10}$$

Note that we have included in (8.3.9) the $N - M_c$ zeros of $H_c(s)$ at infinity that are mapped to $z = -1$. The gain factor b_0 is obtained by equating the dc gains $H(1) = H_c(0)$, or the gains at any other convenient frequency. Therefore, by simply calculating the values of z_m and p_k according to (8.3.10), and obtaining b_0 from gain considerations, the system function $H(z)$ is completely specified.

The complete design procedure for a discrete-time filter using the bilinear transformation is summarized in table 8.1. The conversion of a lowpass design into another type of design (highpass, bandpass, or bandstop) or a different lowpass design can be accomplished by spectral transformation in either Step 2 or Step 6, as we shall see in the next section. In Step 2, for example, we may want to scale a lowpass design obtained from filter tables having $\omega_c = 1$ into another lowpass design with the given (prewarped) value for ω_c. In this simple case, we need only multiply all of the pole and zero locations from the filter tables by ω_c to obtain the desired σ_m and s_k.

Table 8.1 Design Procedure Using Bilinear Transformation

1.	Prewarp the specified values of ω_c' and ω_r' via equation (8.3.7) to determine ω_c and ω_r.
2.	Obtain the continuous-time design with specifications ω_c, ω_r, δ_1, δ_2, and thus N.
3.	Transform the zeros σ_m and poles s_k via equation (8.3.10) to produce z_m and p_k.
4.	Calculate b_0 via $H(1) = H_c(0)$, or equivalent.
5.	$H(z)$ is then given by equation (8.3.9).
6.	(Convert $H(z)$ to another discrete-time design via spectral transformation, if desired.)

The following relatively complex filter design was obtained by the above procedure using only tabulated data and a hand-held scientific calculator.

EXAMPLE. Assume that we wish to design a discrete-time elliptic lowpass filter having the following specifications:

$$\begin{aligned} \omega_c'/2\pi &= 1000\,\text{Hz} \\ \omega_r'/2\pi &= 1500\,\text{Hz} \\ 1/T &= 10{,}000\,\text{Hz} \\ 1-\delta_1 &= -0.25\,\text{dB} \\ \delta_2 &= -50\,\text{dB} \end{aligned}$$

Prewarping the specified values for ω_c' and ω_r' via (8.3.7), we obtain

$$\begin{aligned} \omega_c &= 6498.39 \\ \omega_r &= 10{,}190.51 \\ \omega_r/\omega_c &= 1.568 \end{aligned}$$

Tabulated data [9] for elliptic filters includes designs with $1-\delta_1 = -0.28$ dB, which is close enough to the specified value for most purposes. A satisfactory design having the following specifications is then found in the tables for an order of $N = 5$:

$$\begin{aligned} \omega_r/\omega_c &= 1.556 \\ 1-\delta_1 &= -0.28\,\text{dB} \\ \delta_2 &= -50.10\,\text{dB} \end{aligned}$$

The corresponding poles and zeros are given in the tables for $\omega_c = 1$, as follows:

$$\begin{aligned} s_1, s_2 &= -0.09699 \pm j1.0300 \\ s_3, s_4 &= -0.33390 \pm j0.7177 \\ s_5 &= -0.49519 \\ \sigma_1, \sigma_2 &= \pm j1.6170 \\ \sigma_3, \sigma_4 &= \pm j2.4377 \end{aligned}$$

The fifth zero is, of course, at infinity. The actual pole/zero locations for our value of ω_c are obtained simply by multiplying the above poles and zeros by $\omega_c = 6498.39$. From (8.3.10) the corresponding z-plane poles and zeros are found to be

$$\begin{aligned} p_1, p_2 &= 0.7542 \pm j0.5692 = 0.9449 \angle \pm 37.04^\circ \\ p_3, p_4 &= 0.7278 \pm j0.3635 = 0.8135 \angle \pm 26.54^\circ \\ p_5 &= 0.7228 \\ z_1, z_2 &= 0.5673 \pm j0.8235 = 1.0 \angle \pm 55.44^\circ \\ z_3, z_4 &= 0.2290 \pm j0.9734 = 1.0 \angle \pm 76.76^\circ \\ z_5 &= -1.0 \end{aligned}$$

The associated pole/zero diagram is shown in figure 8.11. The coefficients of the cascade form for the discrete-time filter are obtained from (5.3.4) and (5.3.5) as

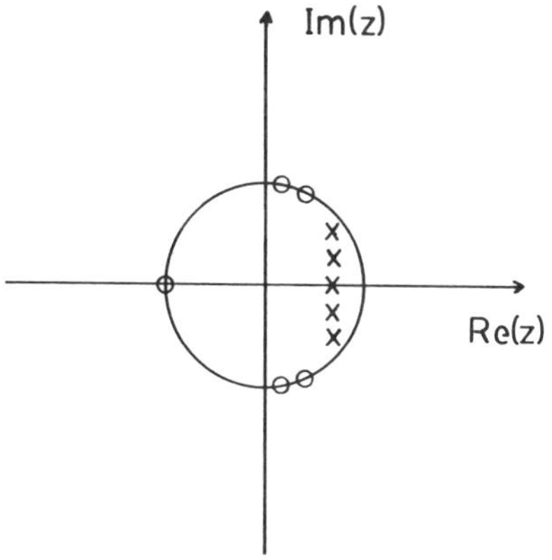

Figure 8.11. Pole/zero diagram for the fifth-order elliptic LPF example.

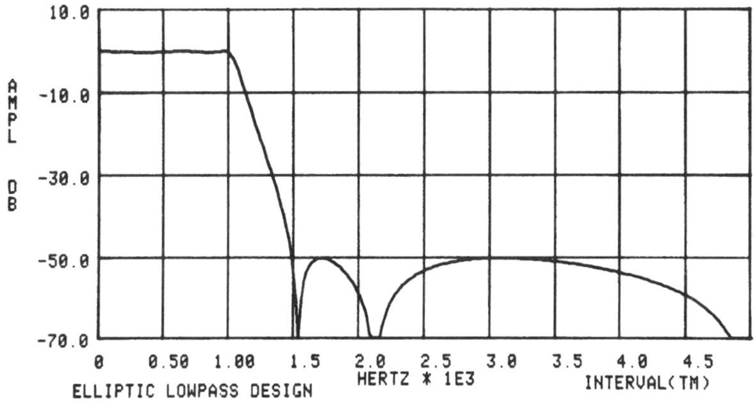

Figure 8.12. Magnitude response of the fifth-order elliptic LPF example.

$$\begin{aligned}
\alpha_{11} &= -1.5084 & \alpha_{21} &= 0.8928\\
\alpha_{12} &= -1.4556 & \alpha_{22} &= 0.6618\\
\alpha_{13} &= -0.7228 & \alpha_{23} &= 0.\\
\beta_{11} &= -1.1346 & \beta_{21} &= 1.0\\
\beta_{12} &= -0.4580 & \beta_{22} &= 1.0\\
\beta_{13} &= +1.0 & \beta_{23} &= 0.
\end{aligned}$$

Finally, the gain factor b_0 is determined by setting the dc gain $H(1) = 1$, which yields

$$b_0 = 1/121.47 = 0.008233.$$

A plot of the resulting magnitude response is given in figure 8.12.

For comparison, a Chebyshev type-I filter was also designed to the same specifications using (8.1.6) through (8.1.9). The required order of the filter from (8.1.9) is $N = 8$. $H_c(s)$

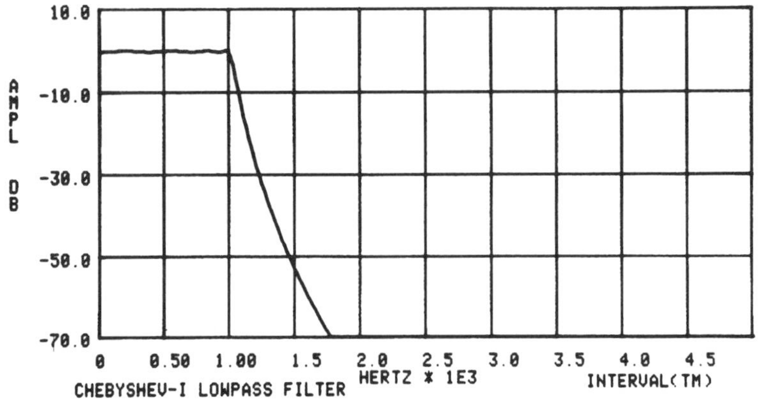

Figure 8.13. Magnitude responses of the eight-order Chebyshev-I LPF example.

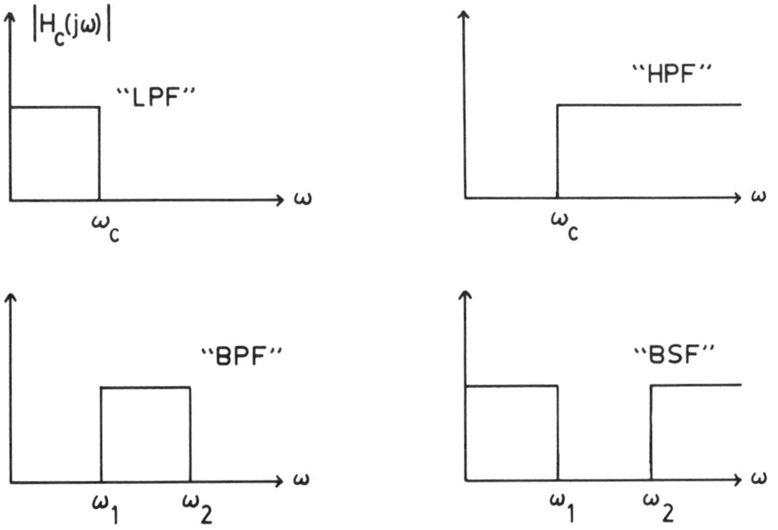

Figure 8.14. Ideal lowpass, highpass, bandpass, and bandstop magnitude responses.

was found from (8.1.6) for $\omega_c = 1$ and $s = j\omega$, and the poles were then scaled by $\omega_c = 6498.39$, as above. After bilinear transformation, the magnitude responses in figure 8.13 was computed for the resulting digital filter.

8.4 SPECTRAL TRANSFORMATIONS

Most classical filter-design tables or CAD programs yield lowpass designs that must be converted by spectral transformation into highpass, bandpass, or bandstop designs if the desired filter is one of the latter types. Of course, CAD programs often incorporate and perform the spectral transformations directly, when required. The ideal magnitude responses for these four filter types and the corresponding cutoff frequencies are illustrated in figure 8.14.

Table 8.2 Spectral Transformations for Continuous-Time Designs

1.	$H_c(s/\omega_c)$ is lowpass with cutoff ω_c.
2.	$H_c(\omega_c/s)$ is highpass with cutoff ω_c.
3.	$H_c\left[\dfrac{s^2 + \omega_1\omega_2}{s(\omega_2 - \omega_1)}\right]$ is bandpass with cutoffs ω_1 and ω_2.
4.	$H_c\left[\dfrac{s(\omega_2 - \omega_1)}{s^2 + \omega_1\omega_2}\right]$ is bandstop with cutoffs ω_1 and ω_2.

Spectral transformations may be performed on either continuous-time or discrete-time designs. If the bilinear transformation is employed for the continuous-to-discrete-time transformation, then spectral transformation may be performed either before or after the bilinear transformation, as convenient. If, however, the impulse-invariant transformation is used, spectral transformation is more generally applicable after the discrete-time design has been obtained since, as we have noted, highpass or bandstop continuous-time designs are not amenable to impulse-invariant transformation because of aliasing.

The spectral transformations for continuous-time designs are given in table 8.2, where it is assumed that $H_c(s)$ is a lowpass design with a cutoff frequency of one radian/second. The resulting filter designs have cutoff frequencies of ω_c or ω_1 and ω_2, as indicated. Note, for example, that a zero at $s = \infty$ in the lowpass design is transformed to a zero at $s = 0$ in the highpass design, to two zeros at $s = 0$ and ∞ in the bandpass design, or to two zeros at $s = \pm j\sqrt{\omega_1\omega_2}$ in the bandstop design.

The spectral transformations for discrete-time designs were first given by Constantionides [10] and have a very interesting form. Let $H(z)$ be the given lowpass design and $H_d(z)$ be the desired filter design. We want an algebraic transformation which replaces z in $H(z)$ by some function $g(z)$ to produce

$$H_d(z) = H[g(z)]. \tag{8.4.1}$$

To derive a rational function $H_d(z)$ from a rational $H(z)$, $g(z)$ must itself be rational. And to preserve stability and causality, $g(z)$ must map the unit circle and its interior into themselves. The latter point implies that

$$|g(e^{j\omega T})| = 1 \tag{8.4.2}$$

and thus $g(z)$ is an allpass function.

The appropriate allpass functions for the four spectral transformations of interest are given in table 8.3, where $H(z)$ is assumed to have a cutoff frequency of ω_c. Note that the parameters of the discrete-time spectral transformations are functions only of the z-plane angles $\omega T = 2\pi f/f_s$ associated with the various cutoff frequencies. These transformations may be performed either by direct substitution of $g(z)$ into (8.4.1) or by calculating the new pole locations P_k as the root(s) of the equation

$$p_k = g(P_k) \tag{8.4.3}$$

Table 8.3 Spectral Transformations for Discrete-Time Designs

$H_d(z)$	$g(z)$	Parameters
Lowpass with cutoff Ω_c	$\dfrac{z-\alpha}{1-\alpha z}$	$\alpha = \dfrac{\sin\left(\dfrac{\omega_c T - \Omega_c T}{2}\right)}{\sin\left(\dfrac{\omega_c T + \Omega_c T}{2}\right)}$
Highpass with cutoff Ω_c	$-\left(\dfrac{z-\alpha}{1-\alpha z}\right)$	$\alpha = \dfrac{\cos\left(\dfrac{\Omega_c T + \omega_c T}{2}\right)}{\cos\left(\dfrac{\Omega_c T - \omega_c T}{2}\right)}$
Bandpass with cutoffs Ω_1 and Ω_2	$-\left(\dfrac{z^2+\beta_1 z+\beta_2}{1+\beta_1 z+\beta_2 z^2}\right)$	$\beta_2 = \dfrac{K-1}{K+1}$ $\beta_1 = \dfrac{-2\alpha K}{K+1}$ $\alpha = \dfrac{\cos\left(\dfrac{\Omega_2 T + \Omega_1 T}{2}\right)}{\cos\left(\dfrac{\Omega_2 T - \Omega_1 T}{2}\right)}$ $K = \cot\left(\dfrac{\Omega_2 T - \Omega_1 T}{2}\right)\tan\dfrac{\omega_c T}{2}$
Bandstop with cutoffs Ω_1 and Ω_2	$\dfrac{z^2+\beta_1 z+\beta_2}{1+\beta_1 z+\beta_2 z^2}$	$\beta_2 = \dfrac{1-K}{1+K}$ $\beta_1 = \dfrac{-2\alpha}{K+1}$ $\alpha = \dfrac{\cos\left(\dfrac{\Omega_2 T + \Omega_1 T}{2}\right)}{\cos\left(\dfrac{\Omega_2 T - \Omega_1 T}{2}\right)}$ $K = \tan\left(\dfrac{\Omega_2 T - \Omega_1 T}{2}\right)\tan\dfrac{\omega_c T}{2}$

and likewise for the new zeros Z_k. The lowpass and highpass transformations will, of course, produce an Nth-order filter; while the bandpass and bandstop transformations result in a $2N$th-order filter.

There is a frequency-warping effect associated with these transformations, as one would expect. To see this, let ω be the frequency variable before spectral transformation and Ω be the frequency variable after transformation. Then, in the lowpass case, we have from $e^{j\omega T} = g(e^{j\Omega T})$ that

$$e^{j\omega T} = \frac{e^{j\Omega T} - \alpha}{1 - \alpha e^{j\Omega T}}$$

which can be solved for Ω to produce

$$\Omega = \frac{1}{T}\tan^{-1}\left[\frac{(1-\alpha^2)\sin\omega T}{2\alpha + (1+\alpha^2)\cos\omega T}\right]. \tag{8.4.4}$$

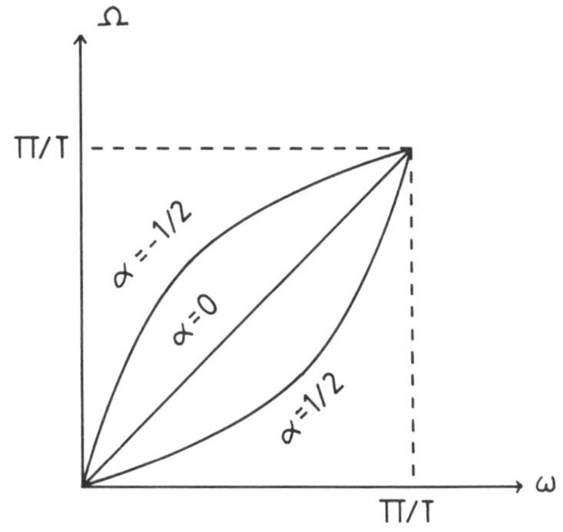

Figure 8.15. Warping effect of the lowpass-to-lowpass spectral transformation.

This relationship is plotted in figure 8.15 for several values of α. It is clear that because of this warping effect, as in the case of the bilinear transformation, spectral transformations are applicable primarily to filter designs with piecewise-constant specifications.

Replacing z by $g(z)$ is, of course, equivalent to replacing z^{-1} by $g^{-1}(z)$, and this suggests the possibility of replacing all delays in the filter implementation by the appropriate allpass networks to make a filter with tunable cutoff frequencies. This can indeed be done to a nonrecursive implementation of an FIR filter, although the resulting filter will be IIR with a recursive implementation. If the original implementation is recursive, however, this scheme produces an unrealizable network, in general, because delay-free feedback loops are created. A useful exception to this general rule can be produced in the bandpass case by forcing $K = 1$ and thus $\beta_2 = 0$. Then, z^{-1} is replaced by

$$g^{-1}(z) = -\frac{z^{-1}(z^{-1} - \alpha)}{1 - \alpha z^{-1}}, \tag{8.4.5}$$

which does not have a delay-free transmission path. The effect of setting $K = 1$ is to cause the bandwidth of the bandpass filter to equal that of the lowpass design, i.e.,

$$\Omega_2 - \Omega_1 = \omega_c.$$

For narrowband filters, α is given approximately by

$$\alpha \approx \cos \Omega_0 T,$$

where Ω_0 is the center frequency of the passband, i.e.,

$$\Omega_0 = \frac{\Omega_1 + \Omega_2}{2}.$$

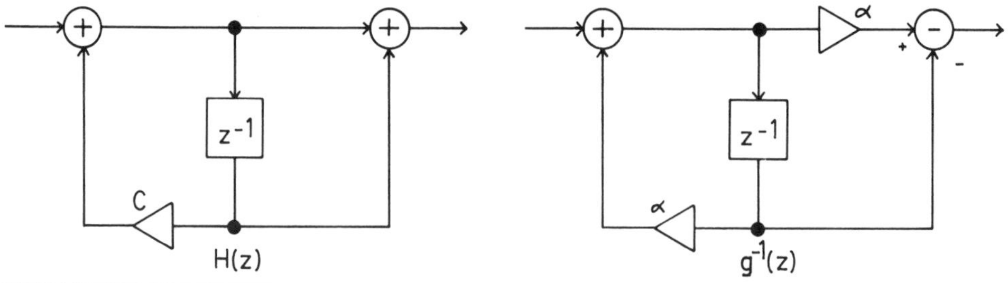

Figure 8.16. First-order lowpass filter $H(z)$, and $g^{-1}(z)$ for lowpass-to-highpass transformation.

EXAMPLE. Consider the spectral transformation of the lowpass function

$$H(z) = \frac{1 + z^{-1}}{1 - cz^{-1}}$$

into a highpass design by substituting

$$g^{-1}(z) = -\left(\frac{z^{-1} - \alpha}{1 - \alpha z^{-1}}\right)$$

in place of z^{-1}. Networks realizing $H(z)$ and $g^{-1}(z)$ are shown in figure 8.16, from which it is evident that this substitution cannot be made in the actual implementation without creating a delay-free loop. However, algebraic substitution of $g^{-1}(z)$ for z^{-1} in $H(z)$ yields

$$H_d(z) = \frac{1 - \left(\frac{z^{-1} - \alpha}{1 - \alpha z^{-1}}\right)}{1 + c\left(\frac{z^{-1} - \alpha}{1 - \alpha z^{-1}}\right)}$$

$$= \frac{(1 + \alpha)(1 - z^{-1})}{(1 - \alpha c) + (c - \alpha)z^{-1}}$$

$$= \left(\frac{1 + \alpha}{1 - \alpha c}\right) \frac{1 - z^{-1}}{1 - \left(\frac{\alpha - c}{1 - \alpha c}\right) z^{-1}},$$

which has an implementation similar to that of $H(z)$ itself. Figures 8.17(a) and (b) show pole/zero diagrams of $H(z)$ and $H_d(z)$, respectively and we note that the zero of $H(z)$ at $z = -1$ is mapped to $z = +1$ in $H_d(z)$, as required by the form of $g(z)$ for the highpass transformation. The pole at $z = c$ is mapped to one at $z = (\alpha - c)/(1 - \alpha c)$, which we could also have determined by solving the equation

$$c = g(P_1) = -\frac{P_1 - \alpha}{1 - \alpha P_1}.$$

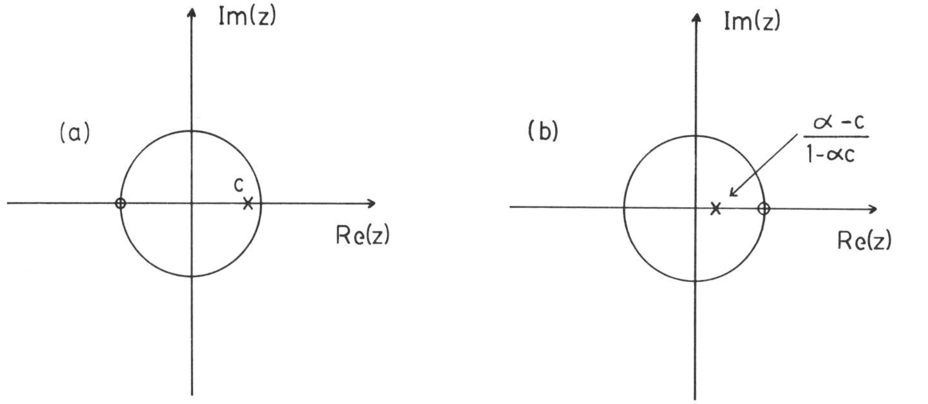

Figure 8.17. Pole/zero diagrams for (a) lowpass filter $H(z)$ and (b) highpass filter $H_d(z)$.

PROBLEMS

8.1 For a ninth-order Butterworth LPF with a 3-dB cutoff frequency of 1 kHz:
 a. Show the pole locations for $H_c(s)$ in the s plane.
 b. Sketch the magnitude response $|H_c(j\omega)|$ in dB versus $\log \omega$.
 c. Find the frequencies at which $|H_c(j\omega)| = -27$ dB, -54 dB, and -81 dB.

8.2 Design a second-order discrete-time Butterworth LPF by the impulse-invariant transformation for $\omega_c = \sqrt{2}a$.

8.3 Give $H_c(s)H_c(-s)$ for a third-order Chebyshev-I LPF with cutoff frequency $\omega_c = 1$ and passband ripple of 1 dB. (*Note*: Convert $1 - \delta_1 = -1$ dB to a magnitude value.)

8.4 Design a first-order LPF with a 3-dB cutoff frequency of 100 Hz and a sampling rate of 1 kHz via
 a. The impulse-invariant transformation.
 b. The bilinear transformation.
 Scale both designs for a peak gain of unity.

8.5 Design a first-order HPF with a 3-dB cutoff frequency of 100 Hz and a sampling rate of 1 kHz via
 a. The matched z transform.
 b. The bilinear transformation.
 Scale both designs for a peak gain of unity.

8.6 Sketch the magnitude responses for the discrete-time filters derived by bilinear and spectral transformation of the following Chebyshev-I, Chebyshev-II, and elliptic analog designs:
 a. 3rd-order LPF's b. 3rd-order HPF's
 c. 4th-order LPF's d. 4th-order HPF's
 e. 6th-order BPF's (*Note*: not 12th-order filters.)
 f. 8th-order BSF's (*Note*: not 16th-order filters.)

8.7 Design a 4th-order Butterworth BPF with cutoff frequencies of $\Omega_1 = 0.15\omega_s$ and $\Omega_2 = 0.35\omega_s$ by bilinear and spectral transformation.

8.8 Show that an Nth-order analog Butterworth LPF with cutoff frequency ω_c can be transformed to an Nth-order Butterworth HPF with the same cutoff frequency simply by adding N zeros at $s = 0$ and scaling the gain. What is the corresponding relationship for discrete-time Butterworth LPF's and HPF's designed via the bilinear transformation?

Sketch the HPF magnitude response in dB versus $\log \omega$, indicating the rolloff slope.

8.9 Show that $\cosh^{-1} y \approx \ln 2y$ for $y \geq 4$. Rewrite (8.1.9) for the order N of a Chebyshev filter when this condition is satisfied.

8.10 What order Butterworth and Chebyshev-II filters would be required to satisfy the following specifications on a discrete-time lowpass filter: (*Hint*: Use problem 8.9)

3-dB cutoff frequency: 200 Hz
Stopband: above 800 Hz
Sampling rate: 10 kHz
Minimum stopband attenuation: 72 dB

Sketch the magnitude response in both cases.

8.11 Given the continuous-time LPF $H_c(s) = 1/(s + 1)$:

a. Sketch the pole/zero diagrams in the s plane after the four continuous-time spectral transformations in table 8.2.

b. Sketch the pole/zero diagrams in the z plane after these four filters are transformed to discrete-time form via the bilinear transformation.

8.12 Given $H(z) = (1 + z^{-1})/(1 - cz^{-1})$ for a LPF:

a. Draw the network for a BPF with variable center frequency Ω_0 by incorporating a network that implements the spectral transformation in (8.4.5) into the LPF network.

b. Find $H_d(z)$ for the BPF and sketch its pole/zero diagram.

c. Specify the coefficients c and α for a narrowband BPF with 3-dB cutoff frequencies Ω_1 and Ω_2.

9. FIR FILTER DESIGN TECHNIQUES

9.0 INTRODUCTION

As we saw in section 5.3, an FIR filter is easily constrained to have one of two particularly useful properties: namely, linear-phase or linear-plus-90°-phase response corresponding to even or odd symmetry, respectively, in its impulse response. Therefore, techniques for the design of FIR filters are of considerable interest. The transformation techniques of the preceding chapter are not applicable to FIR design because, in general, they produce filters with poles, as well as zeros, and thus with infinite-duration impulse responses. The three general techniques most commonly employed for FIR design are described in this chapter.

The desired (ideal) frequency response for the FIR filter will be denoted by $H_d'(\omega)$ with corresponding impulse response $h_d(n)$. Since $H_d'(\omega)$ is simply the z transform of $h_d(n)$ evaluated on the unit circle, we have

$$H_d'(\omega) = \sum_{n=-\infty}^{\infty} h_d(n) e^{-j\omega nT} \tag{9.0.1}$$

and

$$h_d(n) = \frac{T}{2\pi} \int_{-\pi/T}^{\pi/T} H_d'(\omega) e^{j\omega nT} \, d\omega, \tag{9.0.2}$$

which has the form of an infinite Fourier series representation for $H_d'(\omega)$, with $h_d(n)$ as the Fourier coefficients. The series is, in fact, infinite if $H_d'(\omega)$ or any of its derivatives is discontinuous, which is the case for most filters of interest. On the other hand, the actual frequency response of the FIR filter is given by

$$H'(\omega) = \sum_{n=0}^{M} h(n)e^{-j\omega nT}, \tag{9.0.3}$$

and thus $H'(\omega)$ corresponds to a finite Fourier series approximation for $H'_d(\omega)$. This will be a useful way in which to view the FIR design problem.

9.1 WINDOW-FUNCTION TECHNIQUE

The simplest and most obvious way to design an FIR filter is just to truncate the ideal response $h_d(n)$ outside the interval $0 \leq n \leq M$ to produce $h(n)$, i.e.,

$$h(n) = \begin{cases} h_d(n), & n = 0, 1, \ldots, M \\ 0, & \text{otherwise.} \end{cases} \tag{9.1.1}$$

It is well known that the minimum mean-square error (MSE) in approximating a periodic function by a finite Fourier series is obtained by truncating the infinite Fourier series for that function. Hence the design specified by (9.1.1) corresponds, in fact, to the minimum MSE approximation of $H'_d(\omega)$ by $H'(\omega)$. However, truncation of the Fourier series also produces the familiar Gibbs phenomenon, which will be manifested in $H'(\omega)$, especially if $H'_d(\omega)$ is discontinuous. And since all frequency-selective filters are ideally discontinuous at their band edges (as illustrated in figure 8.14), simple truncation of the impulse response will often yield an unacceptable FIR design.

To show more precisely the effect of simple truncation, we rewrite (9.1.1) as

$$h(n) = w_R(n) \cdot h_d(n), \tag{9.1.2}$$

where $w_R(n)$ is the rectangular window function defined in section 7.3 as

$$w_R(n) = \begin{cases} 1, & n = 0, 1, 2, \ldots, M \\ 0, & \text{otherwise.} \end{cases} \tag{9.1.3}$$

From (3.4.6), we then have

$$H'(\omega) = W'_R(\omega) * H'_d(\omega), \tag{9.1.4}$$

where $W'_R(\omega)$ is the z transform of $w_R(n)$ evaluated for $z = e^{j\omega T}$. The sequence $w_R(n)$ has the same form as the impulse response of a simple averaging filter; hence, from the analysis of that filter in section 4.3, we obtain

$$|W'_R(\omega)| = \left| \frac{\sin\,(M+1)\dfrac{\omega T}{2}}{\sin\dfrac{\omega T}{2}} \right|, \tag{9.1.5}$$

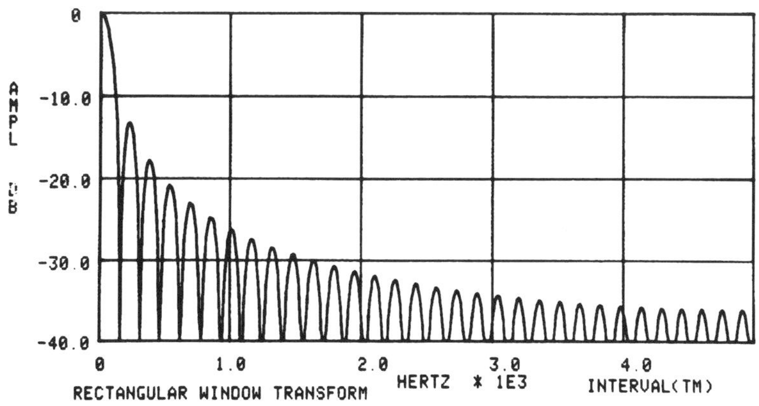

Figure 9.1. Rectangular window transform $|W_R'(\omega)|$ in dB.

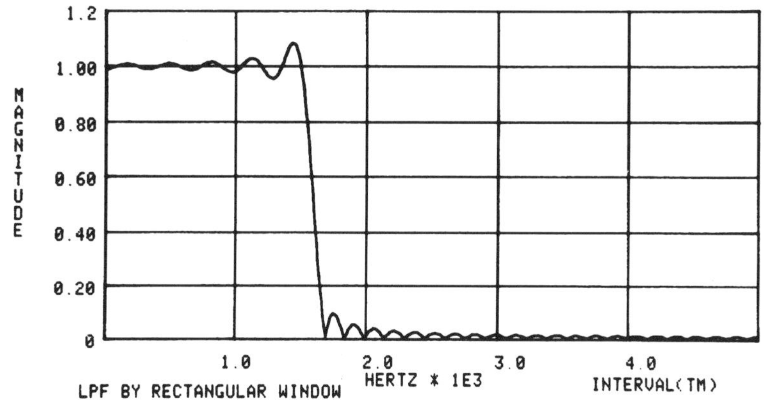

Figure 9.2. Magnitude response for LPF designed via rectangular window, showing Gibbs phenomenon.

which is plotted on a logarithmic (dB) scale in figure 9.1.

The convolution of $W_R'(\omega)$ with the desired magnitude response $H_d'(\omega)$ produces the Gibbs phenomenon in $H'(\omega)$, as illustrated in figure 9.2 for an ideal lowpass filter. Note the characteristic 9-percent *overshoot* associated with the Gibbs phenomenon in both the passband and stopband of $|H'(\omega)|$. This overshoot remains approximately constant no matter how large M is made, with only the width of the ripples becoming narrower as M is increased. Thus, the maximum passband ripple is 0.75 dB above the desired (0 dB) gain; while the attenuation associated with the maximum stopband ripple is only about 21 dB. $|H'(\omega)|$ is redrawn in logarithmic form in figure 9.3.

The basic problem with the rectangular window is the abruptness of the corresponding truncation of $h_d(n)$. As we shall see, this problem can also be viewed as the result of trying to obtain too narrow a transition band in $H'(\omega)$. What we need instead are window functions that are tapered at the ends and thereby produce smaller ripples in $H'(\omega)$ (at the expense of wider transition bands). Many window functions are available for this purpose,

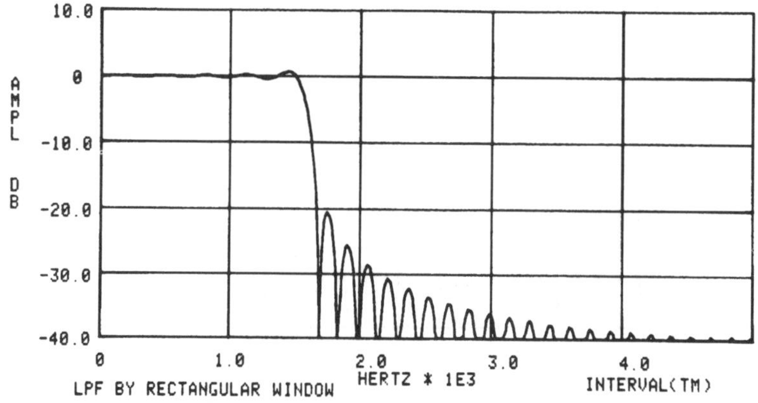

Figure 9.3. Magnitude response for LPF designed via rectangular window, in dB.

although most were originally developed for use in the spectrum analysis of finite-duration signals (discussed in section 7.3). In that context, the (shifted) window-function transform $W'(\omega - \omega_0)$ is itself the frequency response of the effective bandpass filters comprising the spectrum analyzer. Several of the most popular and useful of these window functions are described below.

The Hanning window corresponds to a simple raised cosine and is given by

$$w_N(n) = \frac{1}{2}\left[1 - \cos\frac{2\pi n}{M}\right] \tag{9.1.6}$$

$$= \sin^2\frac{\pi n}{M}, \qquad n = 0, 1, 2, \ldots, M.$$

Of course, this and all other window functions equal zero outside the interval $0 \leq n \leq M$. Note that the Hanning window has an actual length of $M - 1$, not $M + 1$, because the two end values defined by (9.1.6) are also zero. The first sidelobe of the window-function transform $W'_N(\omega)$ is down by 31 dB from the mainlobe amplitude, and the mainlobe width is $4\omega_s/M$ (including negative ω). (See figure 7.8.) In contrast, the first sidelobe of the rectangular window is down by 13 dB, and the mainlobe width is $2\omega_s/(M + 1)$. When $W'_N(\omega)$ is convolved with $H'_d(\omega)$ to produce $H'(\omega)$, the maximum stopband ripple of the resulting (frequency-selective) filter is down by about 44 dB from the passband gain, as shown in figure 9.4, compared with only 21 dB for the rectangular window. However, since the width of the transition band corresponds roughly to the mainlobe width $W'_N(\omega)$, it is more than twice that resulting from the rectangular window.

The Hamming window is given by

$$w_M(n) = 0.54 - 0.46\cos\frac{2\pi n}{M}, \qquad n = 0, 1, 2, \ldots, M, \tag{9.1.7}$$

which also corresponds to a raised cosine, but with different weights for the constant and cosine terms. Note that the Hamming window does not taper to end value of zero, but

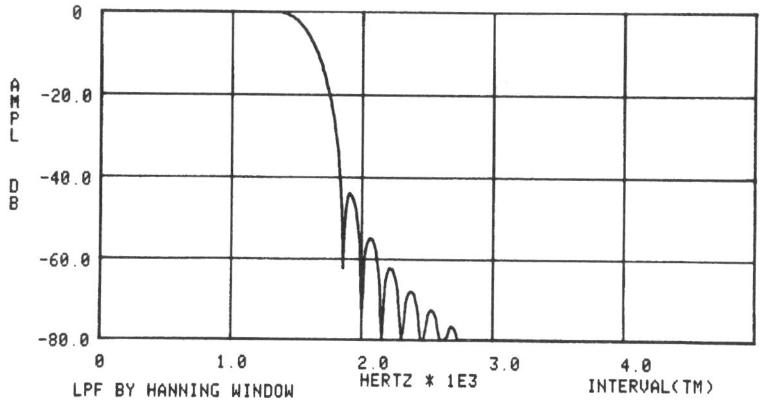

Figure 9.4. Magnitude response for LPF designed via Hanning window.

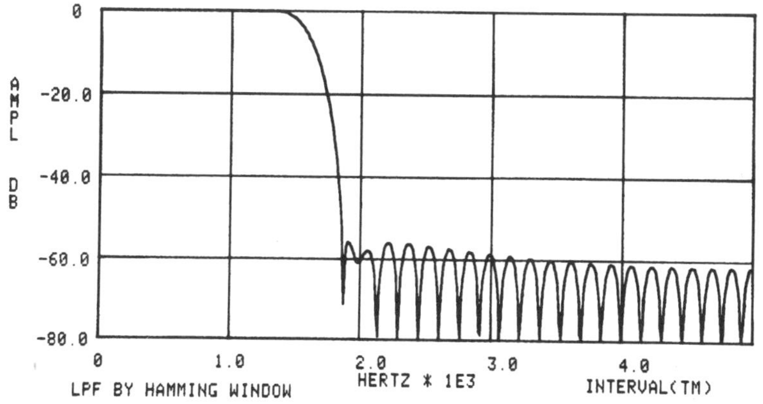

Figure 9.5. Magnitude response for LPF designed via Hamming window.

rather to 0.08. The transform $W'_M(\omega)$ is shown in figure 7.9, from which it is seen that the peak sidelobe is now the third, not the first, and it is down by about 41 dB from the mainlobe level. The mainlobe width, however, is about the same as for the Hanning window.

The maximum stopband ripple for a frequency-selective filter designed via the Hamming window is approximately 53 dB below the passband gain, as seen in figure 9.5. Thus, since the transition widths resulting from the Hamming and Hanning windows are about the same, the Hamming window is usually more appropriate for FIR filter design than is the Hanning. In section 7.3, we noted that the opposite is often the case in spectrum analysis because the sidelobes of $W'_N(\omega)$ fall off at a rate of 18 dB/octave; while $W'_M(\omega)$ falls off at only 6 dB/octave.

An even lower sidelobe level is provided by the Blackman window

$$w_B(n) = 0.42 - 0.5\cos\frac{2\pi n}{M} + 0.08\cos\frac{4\pi n}{M}, \qquad n = 0, 1, 2, \ldots, M. \tag{9.1.8}$$

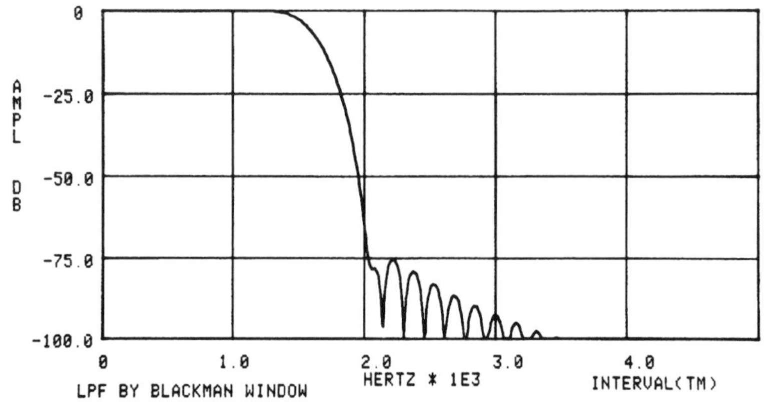

Figure 9.6. Magnitude response for LPF designed via Blackman window.

Table 9.1 Window Parameters, Including Peak Sidelobe Level of the Window Transform Itself, and Approximate Transition Bandwidth and Maximum Stopband Ripple (Minimum Attenuation) for the Corresponding Filter Designs

Window	Peak Sidelobe Level (dB)	Transition Bandwidth (ω_s/M)	Max. Stopband Ripple (dB)
Rectangular	−13	0.9	−21
Hanning	−31	3.1	−44
Hamming	−41	3.3	−53
Blackman	−57	5.5	−74

The maximum sidelobe for this window is down by 57 dB; while the maximum stopband ripple for the resulting FIR filter is down by 74 dB. The width of the mainlobe is $6\,\omega_s/M$, which is 50 percent larger than for the Hanning and Hamming windows. $W_B'(\omega)$ is shown in figure 7.10, and a resulting filter in figure 9.6. The salient features of the rectangular, Hanning, Hamming and Blackman windows are summarized in table 9.1, including more accurate estimates of the transition bandwidths for the resulting FIR filter designs.

A very flexible family of window functions has been developed by Kaiser. These windows are nearly optimum in the sense of having the largest energy in the mainlobe for a given peak sidelobe level. They are closely related to the *prolate spheroidal wavefunctions*, which are the optimum time-limited, continuous-time functions in a similar sense. The Kaiser windows are of the form

$$w_K(n) = \frac{I_0[\beta\sqrt{1-(1-2n/M)^2}]}{I_0[\beta]}, \qquad n = 0, 1, \ldots, M, \tag{9.1.9}$$

where $I_0[\ ]$ is the modified zeroth-order Bessel function of the first kind and β is a shape parameter determining the tradeoff between the mainlobe width and the peak sidelobe level. Typical values for β are in the range $4 < \beta < 9$. $I_0[\ ]$ is most easily computed from its power series expansion

Table 9.2 Parameter Values for the Kaiser Window Transform and Corresponding Filter Designs

β Parameter	Peak Sidelobe Level (dB)	Transition Bandwidth (ω_s/M)	Max. Stopband Ripple (dB)
2.0	−19	1.5	−29
3.0	−24	2.0	−37
4.0	−30	2.6	−45
5.0	−37	3.2	−54
6.0	−44	3.8	−63
7.0	−51	4.5	−72
8.0	−59	5.1	−81
9.0	−67	5.7	−90
10.0	−74	6.4	−99

$$I_0[x] = 1 + \sum_{m=1}^{\infty} \left[\frac{(x/2)^m}{m!} \right]^2 \tag{9.1.10}$$

with the first fifteen terms being sufficient for most applications. The effect of β on the various features of interest is illustrated in table 9.2, which is due to Kaiser [11].

Empirical design formulas have also been derived by Kaiser for the parameters β and M, given the desired values for the transition bandwidth and the peak sidelobe level. Referring to figure 8.1, let the normalized transition bandwidth be defined by

$$\Delta\omega = (\omega_r - \omega_c)/\omega_s \tag{9.1.11}$$

and the stopband attenuation be written as

$$A = -20 \log_{10} \delta_2. \tag{9.1.12}$$

(Of course, for filters designed by the window-function technique, $\delta_1 \approx \delta_2$, as defined in figure 9.13.) Then, the order M of the FIR filter is closely approximated by

$$M \approx \frac{A - 7.95}{14.36\Delta\omega}. \tag{9.1.13}$$

The shape parameter β can be determined from

$$\beta = \begin{cases} 0.1102(A - 8.7), & A \geq 50 \\ 0.5842(A - 21)^{0.4} + 0.07886(A - 21), & 21 < A < 50, \end{cases} \tag{9.1.14}$$

which is accurate to better than 1 percent over the useful range of A.

EXAMPLE. To demonstrate the window-function technique, we will derive two FIR counterparts to the example IIR design in section 8.3. The specifications for that lowpass filter were:

$$\begin{aligned} \omega_c/2\pi &= 1000\,\text{Hz} \\ \omega_r/2\pi &= 1500\,\text{Hz} \\ \omega_s/2\pi &= 10{,}000\,\text{Hz} \\ A &= 50\,\text{dB} \end{aligned}$$

Referring to table 9.1, we see that a Hamming window will provide about 53 dB stopband attenuation and thus will satisfy the specification on A. The transition bandwidth will be somewhat less than the mainlobe width of the window, i.e.,

$$\omega_r - \omega_c \approx \frac{3.3\omega_s}{M} \qquad \text{or} \qquad M \approx \frac{3.3}{\Delta\omega},$$

and hence we find that a value of $M = 66$ is sufficient for this design. The ideal lowpass response $H_d'(\omega)$ should have a cutoff frequency of $\omega_d/2\pi = 1250$ Hz because the transition band is centered on this frequency. Therefore, the ideal impulse response $h_d(n)$ is derived from (9.0.2) or the impulse-invariant transformation [see (8.2.10)] as

$$h_d(n) = \frac{\sin\left[\omega_d T(n - M/2)\right]}{\pi(n - M/2)} \tag{9.1.15}$$

where we have included a delay of $M/2$ to provide even symmetry about the point $n = M/2$, corresponding to that in $w_M(n)$. Multiplying $h_d(n)$ by $w_M(n)$, we obtain a causal, linear-phase FIR filter meeting the given specifications.

A lower-order FIR filter satisfying the same specifications can be derived using the Kaiser window. In particular, from (9.1.13) we find that a value of

$$M = \frac{50 - 7.95}{14.36(0.05)} \approx 59$$

is sufficient for this design. The shape parameter β for the Kaiser window is determined from (9.1.14) to be

$$\beta = 0.1102(50 - 8.7) = 4.55.$$

The same ideal response $h_d(n)$ as before is, of course, multiplied by $w_K(n)$ to obtain the filter design. The resulting frequency responses are shown in figures 9.7 and 9.8.

9.2 FREQUENCY-SAMPLING TECHNIQUE

For arbitrary, nonclassical specifications of $H_d'(\omega)$, the calculation of $h_d(n)$, $n = 0, 1, \ldots, M$, via (9.0.2) or an appropriate approximation can be a substantial computation task. In such situations, it may be preferable to employ a design technique that utilizes specified values of $H_d'(\omega)$ directly, without the necessity of determining $h_d(n)$. One simple technique of this type is based upon the specification of the samples of $H_d'(\omega)$ at $N = M + 1$ uniformly spaced frequencies ω_k in the interval $(0, \omega_s)$ and, as such, is an application of the discrete-Fourier transform (DFT). To simplify our derivation of this

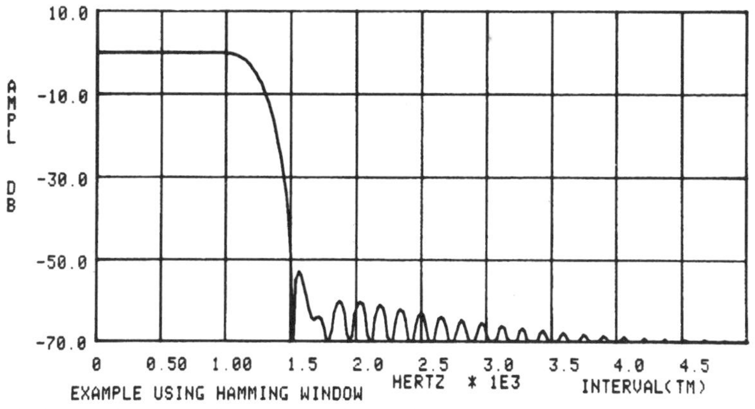

Figure 9.7. Magnitude response for LPF in the example, using the Hamming window with $M = 66$.

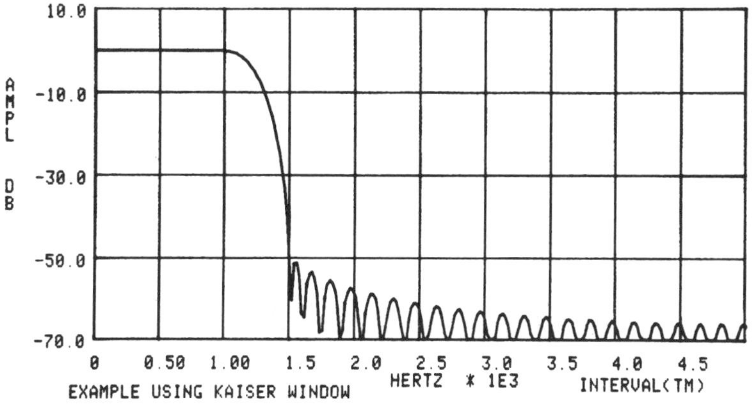

Figure 9.8. Magnitude response for LPF in the example, using the Kaiser window with $M = 59$ and $\beta = 4.55$.

technique, however, we will assume two appropriate forms for $h(n)$ and show that these do, in fact, lead to a frequency-sampling interpretation.

We wish to derive a linear-phase FIR filter with real nonzero $h(n)$, in general, for $n = 0, 1, \ldots, M$. Hence, the impulse response $h(n)$ must be symmetric about $n = M/2$. One such response is of the form

$$h(n) = A_0 + \sum_{k=1}^{[M/2]} 2A_k \cos\left(\frac{2\pi k(n + \frac{1}{2})}{M + 1}\right), \qquad n = 0, 1, \ldots, M, \tag{9.2.1}$$

where the A_k are real and $[M/2]$ denotes the integer part of $M/2$. We thus have a harmonic series expansion for $h(n)$ with fundamental angular frequency $2\pi/(M + 1)$ extending to (but not including) the Nyquist frequency. It is readily checked that (9.2.1) can be rewritten as

$$h(n) = \sum_{\substack{k=0 \\ k \neq N/2}}^{N-1} A_k e^{j\pi k/N} e^{j2\pi kn/N}, \qquad n = 0, 1, \ldots, N-1, \tag{9.2.2}$$

with $N = M + 1$ and $A_k = A_{N-k}$.

Therefore, we may write $h(n)$ as

$$h(n) = \sum_{\substack{k=0 \\ k \neq N/2}}^{N-1} h_k(n) \tag{9.2.3}$$

with corresponding transform

$$H(z) = \sum_{\substack{k=0 \\ k \neq N/2}}^{N-1} H_k(z), \tag{9.2.4}$$

where

$$h_k(n) = A_k e^{j\pi k/N} e^{j2\pi kn/N}, \qquad n = 0, 1, \ldots, N-1, \tag{9.2.5}$$

and

$$H_k(z) = \frac{A_k e^{j\pi k/N}(1 - z^{-N})}{1 - e^{j2\pi k/N} z^{-1}}.$$

Hence,

$$H_k'(\omega) = A_k e^{-j\omega T(N-1)/2} \frac{\sin \omega TN/2}{\sin [(\pi k/N - \omega T/2)]}, \tag{9.2.6}$$

which has a linear-phase factor, as expected, and a magnitude of

$$|H_k'(\omega)| = |A_k| \cdot \left| \frac{\sin \omega TN/2}{\sin [(\omega T/2 - \pi k/N)]} \right|. \tag{9.2.7}$$

The magnitude response $|H_k'(\omega)|$ has a maximum value of $N|A_k|$ at $\omega_k = k\omega_s/N$, as illustrated in figure 9.9. Since all other $|H_i'(\omega)|$, $i \neq k$, equal zero at $\omega = \omega_k$, it is clear from (9.2.6) that the only nonzero contribution to $H'(\omega)$ at $\omega = \omega_k$ is from $H_k'(\omega)$, and hence that

$$|H'(\omega_k)| = N|A_k|. \tag{9.2.8}$$

Therefore, by specifying the DFT samples of the desired magnitude response $|H_d'(\omega)|$ at the frequencies ω_k, and setting

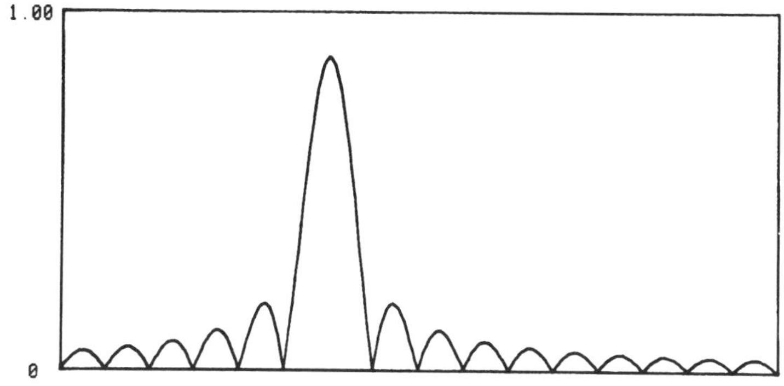

Figure 9.9. Magnitude response $|H_k'(\omega)|$ for a single frequency sampling component centered at ω_k.

$$A_k = \pm|H_d'(\omega_k)|/N, \tag{9.2.9}$$

we produce a filter design $H(z)$ from (9.2.1) for which

$$|H'(\omega_k)| = |H_d'(\omega_k)|.$$

That is, the desired and actual magnitude responses are equal at the N frequencies ω_k. In between these frequencies, $H'(\omega)$ is interpolated as the sum of the responses $H_k'(\omega)$, and its magnitude does not, in general, equal that of $H_d'(\omega)$. In this connection we note that, although the signs of the A_k are unspecified by (9.2.9), these signs should, in fact, alternate with increasing k in order to prevent large ripples in the passband(s) of $|H'(\omega)|$ between the frequencies ω_k. This effect is due to the phases associated with adjacent $H_k(z)$ in (9.2.4). (See problem 9.7.)

Frequency-sampling filters may also be designed using an $h(n)$ of the form (see problem 9.5)

$$h(n) = \sum_{k=1}^{[M/2]} 2A_k \sin\left(\frac{(2k-1)(n+\frac{1}{2})\pi}{M+1}\right), \qquad n = 0, 1, \ldots, M, \tag{9.2.10}$$

where the A_k are, as before, given by (9.2.9), but the corresponding ω_k are instead

$$\omega_k = \left(\frac{2k-1}{2N}\right)\omega_s, \tag{9.2.11}$$

which may, in some applications, fall at more convenient frequency values.

EXAMPLE. For an ideal lowpass filter,

$$|H_d'(\omega_k)| = \begin{cases} 1, & k = 0, 1, \ldots, P \\ 0, & k = P+1, \ldots, [M/2]. \end{cases}$$

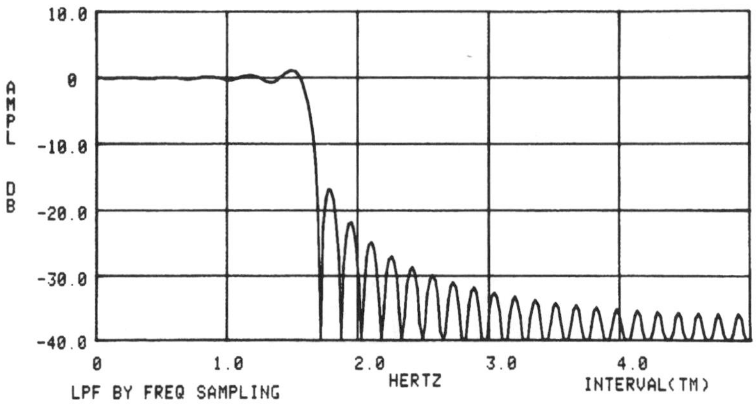

Figure 9.10. Frequency sampling LPF design with no transition sample.

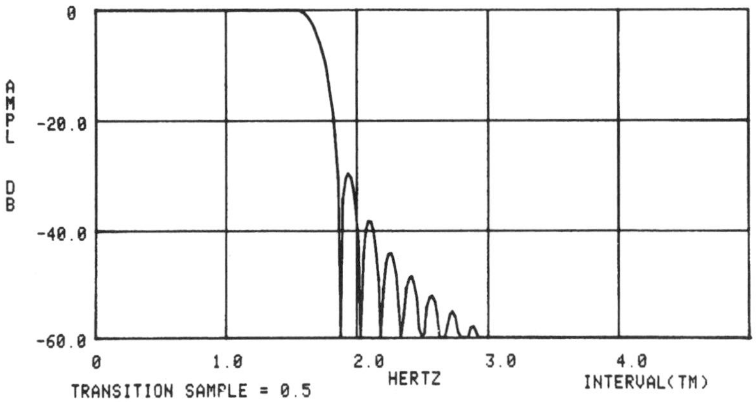

Figure 9.11. Frequency sampling LPF design with transition sample = 0.5.

Thus, from (9.2.9) and the associated discussion, we would choose

$$A_k = \begin{cases} (-1)^k/(M+1), & k = 0, 1, \ldots, P \\ 0, & k = P+1, \ldots, [M/2]. \end{cases}$$

The resulting magnitude response $|H'(\omega)|$ is shown in figure 9.10, from which we see that the frequency samples $|H'(\omega_k)|$ are indeed equal to the desired $|H_d'(\omega_k)|$. However, the response is very similar to that which would result from using the rectangular window, and the stopband attenuation is similarly disappointing (being here less than 20 dB). The problem, as before, is that we have tried to obtain too narrow a transition bandwidth. If, for example, we include a transition sample $A_p = 0.5(-1)^P/(M+1)$ with the other A_k, $k \neq P$, as above, the response becomes that of figure 9.11, which has a stopband attenuation of almost 30 dB. A simple trial-and-error search for the optimum value of the transition sample would quickly lead us to a value of approximately $A_P = 0.38(-1)^P/(M+1)$, which gives a stopband attenuation of about 40 dB, as shown in figure 9.12.

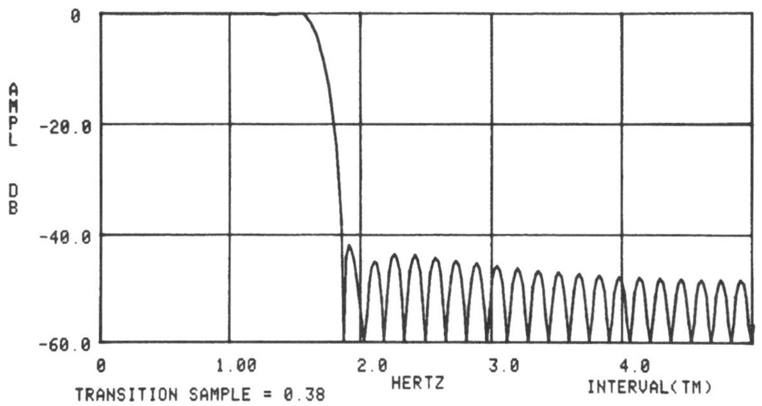

Figure 9.12. Frequency sampling LPF design with transition sample = 0.38.

Rabiner et al. [13] give a very complete discussion of frequency-sampling filters with tables of designs for one to four transition samples obtained by linear programming.

9.3 EQUIRIPPLE DESIGNS

We saw in section 8.1 that the IIR filter design of lowest order satisfying a classical frequency-selective specification is the elliptic design, which has ripples or *extrema* of uniform amplitude in the passband and stopband. It is reasonable to expect that this might also be true for FIR designs, and this is indeed the case. Such designs are called *equiripple*, optimum, or minimax FIR filters. Although a closed-form design technique is not available for these filters, an elegant and powerful iterative algorithm—the Remez exchange algorithm—does provide the basis for efficient design methods.

For convenience, the specification on the passband(s) is now changed to

$$1 + \delta_1 \geq |H'(\omega)| \geq 1 - \delta_1 \tag{9.3.1}$$

as illustrated for the lowpass case in figure 9.13, while the stopband specification remains $|H'(\omega)| \leq \delta_2$. Also, for convenience, $H'(\omega)$ is forced to be zero phase, and thus purely real, by making $h(n)$ symmetric about $n = 0$. Therefore, the filter order M is even, and

$$\begin{aligned} H'(\omega) &= \sum_{n=-M/2}^{M/2} h(n)e^{-j\omega nT} \\ &= h(0) + \sum_{n=1}^{M/2} h(n)\cos n\omega T \end{aligned} \tag{9.3.2}$$

since $h(n) = h(-n)$. The impulse response $h(n)$ thus corresponds to the coefficients of a finite Fourier cosine series for the periodic, real and even function $H'(\omega)$.

An important parameter in the Remez exchange algorithm is the total number of extrema in the (closed) Nyquist interval $[0, \omega_s/2]$. To determine this number, we note that the derivative of $H'(\omega)$ at each extremum must be zero except at the band edges (ω_c and ω_r). How many such points can there be for the function $H'(\omega)$ in (9.3.2)? Note first

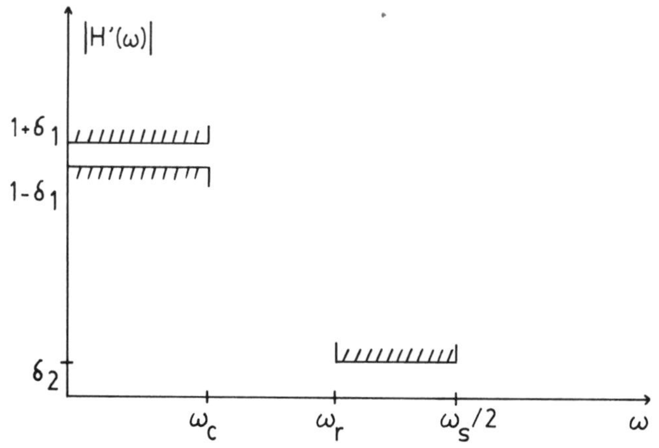

Figure 9.13. Modified magnitude response specification for FIR filter designs (LPF case).

that $\cos n\omega T$ can be written as a trigonometric polynomial in $\cos \omega T$ of order n, and hence that $H'(\omega)$ can also be written as such a polynomial of order $M/2$, i.e.,

$$H'(\omega) \quad = \quad \sum_{n=0}^{M/2} a_n (\cos \omega T)^n. \tag{9.3.3}$$

Taking the derivative and setting it to zero, we then have

$$\frac{d}{d\omega} H'(\omega) \quad = \quad 0 \quad = \quad -T(\sin \omega T) \sum_{n=1}^{M/2} n a_n (\cos \omega T)^{n-1}. \tag{9.3.4}$$

The factor $\sin \omega T$ implies solutions at 0 and $\omega_s/2$; while the remaining trigonometric polynomial is of order $M/2 - 1$ and thus has at most $M/2 - 1$ distinct roots. Hence, there are $M/2 - 1 + 2 = M/2 + 1$ possible extrema of zero slope. Adding the two extrema at the band edges ω_c and ω_r, we have a maximum total number of possible extrema of $M/2 + 3$.

A theorem from the theory of approximation states that the design is optimized for minimum ripple if, and only if, there are at least $M/2 + 2$ extrema of equal (weighted) amplitudes and alternating signs in the pass and stop bands. Such extrema are known as *alternations*. The additional alternation expected from the above calculation of $M/2 + 3$ possible extrema may occur (at 0 or $\omega_s/2$), in which case the design is *extraripple*; but this is not necessary for optimality. The three possible equiripple cases, including the extraripple case, are illustrated in figure 9.14. Note from the dips in the passband response and the zero crossings in the stopband that the zeros for this filter must be located in the z plane as depicted in figure 9.15. Hence, the order of this filter is 14, and there must be 9 or possibly 10 alternations, which checks with figure 9.14.

Since CAD software is readily available [14, 15] to design equiripple filters, we will only outline the general design formulations and the Remez exchange algorithm. For a given

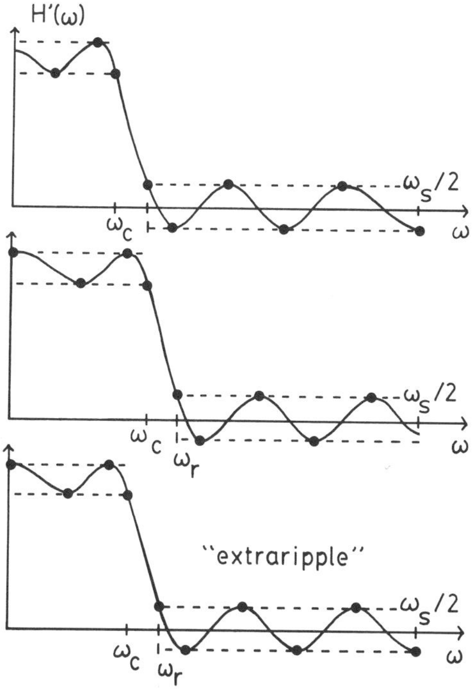

Figure 9.14. Sample equiripple designs, including the extraripple case, with dots indicating alternations.

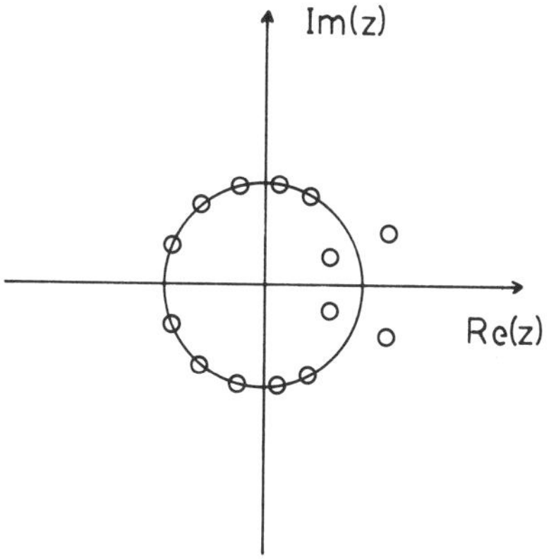

Figure 9.15. Zero locations for the equiripple designs in figure 9.14.

order M, we do not know a priori the minimum ripple amplitudes δ_1 and δ_2 that can be achieved (although this can be closely estimated). Two approaches have been taken to address this. Parks and McClellan [17] apply a weight K to the stopband specification to produce a single unknown $K\delta_2 = \delta_1 = \delta$, and δ is determined iteratively along with the filter coefficients. If the specifications are not met, the order is increased and the algorithm repeated. Hersey, Lewis, and Tufts [18], on the other hand, force the passband or stopband

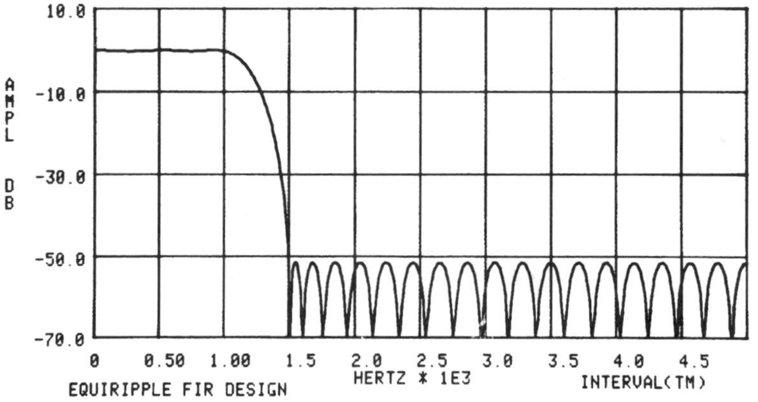

Figure 9.16. Magnitude response for equiripple LPF in example with $M = 46$.

constraint to be satisfied exactly and let $\delta = \delta_1$ or δ_2 for the other band be the single unknown to be determined iteratively.

To describe the Remez exchange algorithm, let $\omega_1^i, \omega_2^i, \ldots$ be the estimates at the ith iteration of the frequencies at which alternations of zero slope will occur, and let δ^i be the estimate of δ. In addition, we know that there are alternations at ω_c and ω_r and at either 0 or $\omega_s/2$ (let ω_0^i be our current estimate of which one). A trigonometric polynomial of the form in (9.3.3) is passed through the points $1 \pm \delta^i$ in the passband and $\pm \delta^i/K$ in the stopband at these frequencies with alternating signs such that $1 - \delta^i$ occurs at ω_c and $+\delta^i/K$ at ω_r. (Substitute $1 \pm \delta_1$ in the passband or $\pm \delta_2$ in the stopband for the method of Hersey, Lewis, and Tufts.) This new estimate of $H'(\omega)$ is then evaluated with high resolution to locate the frequencies $\omega_0^{i+1}, \omega_1^{i+1}, \ldots$ at which extrema actually occur (choosing ω_0^{i+1} by whether the extremum at 0 or $\omega_s/2$ is larger). These frequencies plus a new estimate δ^{i+1} form the inputs to the $(i + 1)$st iteration. (See problem 9.9.)

As an example, the specifications for the examples in sections 8.3 and 9.1 were used to design an equiripple lowpass filter by the method of Parks and McClellan. The resulting filter is of order $M = 46$ and is shown in figure 9.16. The substantial reduction in order over that using the Kaiser window ($M = 59$) is due primarily to the independent control of the passband and stopband ripple afforded by equiripple design methods, as opposed to either windowing or frequency-sampling techniques.

PROBLEMS

9.1 Find the maximum deviation of the passband gain from unity (0 dB) in percentage and in dB for frequency selective filters (e.g., lowpass or highpass) designed via the rectangular, Hanning, Hamming, and Blackman windows.

9.2 To be able to design FIR highpass, bandpass, and bandstop filters by the windowing method, we must be able to derive the corresponding ideal impulse responses $h_d(n)$. This is especially straightforward for M even because we can advance $h_d(n)$ by $M/2$ samples to make it symmetric about $n = 0$ so that $H_d'(\omega)$ is purely real.

a. Express an ideal HPF response $H_{\mathrm{HP}}'(\omega)$ in terms of an ideal LPF response $H_{\mathrm{LP}}'(\omega)$. You may wish to refer to figure 8.14.

b. Express an ideal BPF response $H'_{BP}(\omega)$ in terms of ideal LPF responses $H'_{LP1}(\omega)$ and $H'_{LP2}(\omega)$.
c. Express an ideal BSF response $H'_{BS}(\omega)$ in terms of an ideal BPF response $H'_{BP}(\omega)$.
d. Given (9.1.15) for the (causal) impulse response $h_{LP}(n)$ of an ideal LPF, express $h_{HP}(n)$, $h_{BP}(n)$, and $h_{BS}(n)$ for M even from your results in (a)–(c).

9.3 Specify the indicated designs (i.e., the order M or N, ideal cutoff ω_d, and other parameters) for a highpass filter with the following specifications:

Passband cutoff: 2 kHz
Stopband: 0 – 1 kHz
Stopband attenuation: 72 dB
Sampling rate: 10 kHz

a. FIR filter by Blackman windowing.
b. FIR filter by Kaiser windowing.
c. IIR Butterworth filter (for comparison).

9.4 An FIR bandpass filter is desired with a passband from 1.5 kHz to 2.5 kHz, stopbands from 0 to 0.5 kHz and from 3.5 to 5 kHz, and a sampling rate of 10 kHz. Stopband attenuation of at least 40 dB is specified.
a. What order (M) is required if the filter is designed via Hanning windowing?
b. What order for Kaiser windowing?
c. What order for the frequency-sampling method?
d. Give $h_d(n)$ and the cutoff frequencies ω_{d1} and ω_{d2} for the ideal filter required in the windowing method.
e. Specify the A_k for the frequency-sampling design.

9.5 Show that the expression for $h(n)$ in (9.2.10) corresponds to a frequency sampling design with samples at the frequencies

$$\omega_k = \frac{2k-1}{2N}\,\omega_s.$$

9.6 An FIR lowpass filter with 40-dB attenuation is to be designed by frequency sampling with a passband from 0 to 750 Hz, a stopband from 1750 Hz to 5 kHz, and a sampling rate of 10 kHz.
a. Specify the filter coefficients for the sine expansion in (9.2.10).
b. Repeat for the cosine expansion in (9.2.1).
c. Compare the filter orders M in (a) and (b).

9.7 Show that the signs of the A_k in the frequency-sampling designs should alternate with increasing k to prevent large notches between samples in the resulting passbands. (*Hint:* Sketch two adjacent responses $H'_k(\omega)$ and $H'_{k+1}(\omega)$ from (9.2.6) after removing the linear-phase (pure-delay) factor.)

9.8 Assume that the pole/zero diagram in figure 5.12 is for an equiripple FIR lowpass filter.
a. What is the order M of the filter?
b. How many alternations are there if the filter is not extraripple?
c. Sketch $|H'(\omega)|$ and check it with (b).

9.9 At the ith step of the Remez exchange algorithm, the $M/2$ extrema in $H_i'(\omega)$ give new frequency estimates ω_k^{i+1}, $k = 0, 1, \ldots, M/2 - 1$, (in addition to ω_c and ω_r) of the ultimate alternation frequencies. Therefore, the new estimates must satisfy

$$H_d'(\omega_k^{i+1}) - H_{i+1}'(\omega_k^{i+1}) = \pm\,\delta_1^{i+1} \quad \text{or} \quad \pm\,\delta_2^{i+1}.$$

Using $K\delta_2 = \delta_1 = \delta$ and equation (9.3.2) for $H_{i+1}'(\omega)$, write $M/2 + 2$ equations that can be solved for $h_{i+1}(n)$, $n = 0, 1, \ldots, M/2$, and δ^{i+1}. (In practice, only δ^{i+1} is calculated from these equations, and (9.3.3) is used otherwise.)

10. FILTER DESIGN BY MODELING

10.0 INTRODUCTION

The two preceding chapters presented the traditional design methods for IIR and FIR digital filters. These methods are grounded in the mathematical theory of approximation, where upper and lower constraints are placed on the deviation of $|H'(\omega)|$ from the ideal response. Another class of design methods is based on the statistical modeling of spectra; and in these methods, the least-squares error (LSE) criterion is employed. These methods are especially appropriate when the filter design or model is derived from experimental data having random fluctuations. However, they can also be useful for deterministic designs when the least-squares criterion is indicated, when nonclassical specifications are given for IIR filters, when poles and zeros of unequal number ($N \neq M$) are desired, and so forth. Complete coverage of this field, which is sometimes termed *modern spectrum analysis*, is beyond the scope of this book, and we will concentrate on deterministic filter design. However, the basics of statistical data modeling and spectrum analysis will also be included in section 10.5.

10.1 AUTOREGRESSIVE (ALL-POLE) FILTERS

The techniques of *autoregressive* (AR) or all-pole filter modeling underlie all of the methods in this field. Consider the causal AR filter model

$$H(z) = \frac{1}{A(z)} \tag{10.1.1}$$

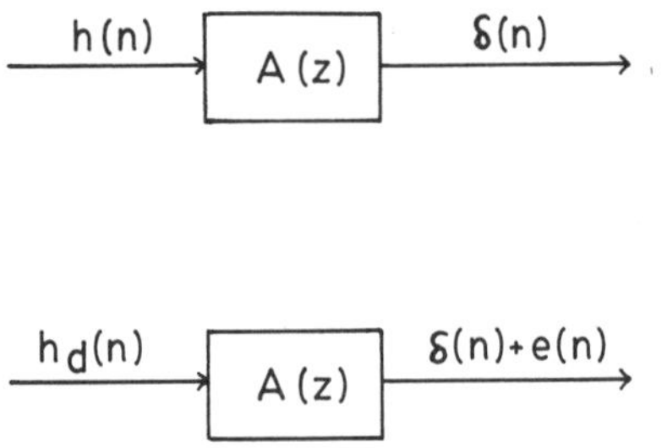

Figure 10.1. Inverse filter $A(z)$ applied to (a) model response $h(n)$ and (b) desired response or data $h_d(n)$.

where

$$A(z) = a_0 + a_1 z^{-1} + \cdots + a_N z^{-N}.$$

Then,

$$H(z)A(z) = 1 \tag{10.1.2}$$

or, in the time domain, using the alternate notation $a(n) = a_n$,

$$h(n) * a(n) = \delta(n). \tag{10.1.3}$$

The FIR filter $A(z)$ is thus the *inverse filter* for $H(z)$, which *whitens* $h(n)$ to produce $\delta(n)$, as depicted in figure 10.1(a). Using the $2N + 1$ consecutive values for $h(n)$ beginning at $n = -N$, (10.1.3) provides $N + 1$ linear equations from which to solve for $a(n)$. That is, given $h(0)$ through $h(N)$ plus the fact that $h(n) = 0$, $n < 0$, the convolution may be written in matrix form to produce the following $N + 1$ equations for $a(n)$:

$$\begin{bmatrix} h_0 & 0 & \cdots & 0 \\ h_1 & h_0 & \cdots & 0 \\ \vdots & & \ddots & \vdots \\ h_N & \cdots & \cdots & h_0 \end{bmatrix} \begin{bmatrix} a_0 \\ a_1 \\ \vdots \\ a_N \end{bmatrix} = \begin{bmatrix} 1 \\ 0 \\ \vdots \\ 0 \end{bmatrix} \tag{10.1.4}$$

However, given a desired or measured impulse response $h_d(n)$ that is not exactly AR and/or of greater than Nth order, we can only approximate it as in (10.1.1). Therefore, (10.1.3) becomes

$$h_d(n) * a(n) = \delta(n) + e(n), \tag{10.1.5}$$

where $e(n)$ is the approximation error, and $A(z)$ is only an approximate inverse filter as illustrated in figure 10.1(b). Writing (10.1.5) in matrix form, we have

$$\mathbf{H}_d \underline{a} = \underline{\delta} + \underline{e} \tag{10.1.6}$$

where

$$\mathbf{H}_d = \begin{bmatrix} h_{d0} & 0 & \cdots & 0 \\ h_{d1} & h_{d0} & & 0 \\ \vdots & & \ddots & \vdots \\ h_{dN} & h_{d,N-1} & \cdots & h_{d0} \\ \vdots & & & \vdots \\ h_{dL} & \cdots & \cdots & h_{d,L-N} \end{bmatrix}$$

$$\underline{a} = [a_0, a_1, \ldots, a_N]^t$$

$$\underline{\delta} = [1, 0, 0, \ldots, 0]^t$$

$$\underline{e} = [e_0, e_1, \ldots, e_L]^t.$$

Note that we have assumed more data samples $h_d(n)$, $n = 0, 1, \ldots, L$, than the minimum required (i.e., $L > N$) in order to approximate $h_d(n)$ closely (hopefully) over more than the minimum interval. Applying the least-squares error criterion, we seek to minimize

$$E_N = \underline{e}^t\underline{e} = \sum_{n=0}^{L} e^2(n). \tag{10.1.7}$$

This is a standard problem in least-squares estimation with overdetermined equations [19]. The minimizing solution $\underline{a}$ is obtained by multiplying through (10.1.6) by $\mathbf{H}_d^t$ to produce

$$\mathbf{H}_d^t\mathbf{H}_d\underline{a} = \mathbf{H}_d^t\underline{\delta} + \mathbf{H}_d^t\underline{e}. \tag{10.1.8}$$

It is then found that

$$\mathbf{H}_d^t\underline{e} = \underline{0}. \tag{10.1.9}$$

Note also that $\mathbf{H}_d^t\underline{\delta} = h_{d0}\underline{\delta}$, where the dimension of $\underline{\delta}$ varies as required. Equation (10.1.9) is an example of the *orthogonality principle* of least-squares estimation, which states that for the minimizing solution $\underline{a}$, the error will be orthogonal to the data [20].

In more compact notation, therefore, we have from (10.1.8) and (10.1.9) the *normal* equation

$$\mathbf{\Phi}_N\underline{a} = h_{d0}\underline{\delta} \tag{10.1.10}$$

where $\mathbf{\Phi}_N$ is the $(N + 1) \times (N + 1)$ symmetric *covariance* matrix

$$\mathbf{\Phi}_N = \mathbf{H}_d^t \mathbf{H}_d \tag{10.1.11}$$

and thus

$$\underline{a} = h_{d0} \mathbf{\Phi}_N^{-1} \underline{\delta}. \tag{10.1.12}$$

That is, $\underline{a}$ is proportional to the first column of $\mathbf{\Phi}_N^{-1}$. $\mathbf{\Phi}_N$ is a positive-definite matrix and thus invertible because the columns of $\mathbf{H}_d$ are linearly independent. This general approach to least-squares estimation of AR models (having a finite-duration squared-error norm without windowing, as in (10.1.7)) is known as the *covariance method of linear prediction* [21]. Efficient inversion of $\mathbf{\Phi}_N$ is achieved by Cholesky factorization.

As $L \to \infty$, the elements ϕ_{ij} of $\mathbf{\Phi}_N$ approach the autocorrelation values $r(i - j)$ where

$$r(m) = \sum_{n=0}^{\infty} h_d(n) h_d(n + m), \tag{10.1.13}$$

and since $h_d(n)$ is real, $r(-m) = r(m)$. Therefore, $\mathbf{\Phi}_N$ approaches the symmetric Toeplitz autocorrelation matrix

$$\mathbf{R}_N = \begin{bmatrix} r_0 & r_1 & \cdots & r_N \\ r_1 & r_0 & & \vdots \\ \vdots & & \ddots & \vdots \\ r_N & \cdots & \cdots & r_0 \end{bmatrix}. \tag{10.1.14}$$

The limiting form of (10.1.10) with $\mathbf{\Phi}_N = \mathbf{R}_N$ is known as the *Yule-Walker* equation. Additional insight may be gained in this limiting case by transforming (10.1.5) to obtain

$$E(z) = H_d(z) A(z) - 1.$$

By Parseval's Theorem, therefore, the error norm becomes

$$E_N = \sum_{n=0}^{\infty} e^2(n) = \frac{1}{\omega_s} \int_0^{\omega_s} |H_d'(\omega) A'(\omega) - 1|^2 \, d\omega. \tag{10.1.15}$$

This shows in the frequency domain the sense in which $A'(\omega)$ whitens $H_d'(\omega)$. Clearly, from (10.1.4), $E_N \to 0$ as the order $N \to \infty$. Hence, any response $H_d'(\omega)$, whether AR or not, may be approximated arbitrarily closely by $H'(\omega) = 1/A'(\omega)$ in the sense of (10.1.15) if N is chosen to be sufficiently large.

Given only a finite data block $h_d(n)$, $n = 0, 1, \ldots, L$, $\mathbf{R}_N$ may be approximated by zero padding the data to produce

$$\hat{\mathbf{R}}_N = \hat{\mathbf{H}}_d^t \hat{\mathbf{H}}_d = \begin{bmatrix} \hat{r}_0 & \hat{r}_1 & \cdots & \hat{r}_N \\ \hat{r}_1 & \hat{r}_0 & & \vdots \\ \vdots & & \ddots & \vdots \\ \hat{r}_N & \cdots & \cdots & \hat{r}_0 \end{bmatrix}, \tag{10.1.16}$$

where

$$\hat{\mathbf{H}}_d = \begin{bmatrix} h_{d0} & 0 & \cdots & 0 \\ h_{d1} & h_{d0} & & 0 \\ \vdots & & \ddots & \vdots \\ & & & h_{d0} \\ h_{dL} & \ddots & & \vdots \\ 0 & & \ddots & \vdots \\ \vdots & & & \vdots \\ 0 & \cdots & \cdots & h_{dL} \end{bmatrix}$$

and thus

$$\hat{r}(m) = \sum_{n=0}^{L-m} h_d(n)\, h_d(n + m) \tag{10.1.17}$$

for $0 \leq m \leq N$. Substitution of $\hat{\mathbf{R}}_N$ in place of $\mathbf{\Phi}_N$ in (10.1.10) and (10.1.12) constitutes the *autocorrelation method of linear prediction* [21]. That is, with the normalization of $\underline{a}$ given by

$$\underline{\hat{a}} = \underline{a}/a_0 = [1, \hat{a}_1, \ldots, \hat{a}_N]^t,$$

the autocorrelation method corresponds to the normal equation

$$\hat{\mathbf{R}}_N \underline{\hat{a}} = (h_{d0}/a_0)\underline{\delta} \tag{10.1.18}$$

or

$$\underline{\hat{a}} = (h_{d0}/a_0)\hat{\mathbf{R}}_N^{-1}\underline{\delta}. \tag{10.1.19}$$

Since the columns of $\hat{\mathbf{H}}_d$ are linearly independent, $\hat{\mathbf{R}}_N$ cannot be singular.

The autocorrelation method, in effect, windows the infinite sequence $h_d(n)$ by a rectangular window $w_R(n)$ of length $L + 1$, and hence the desired frequency response $H_d'(\omega)$ is convolved with the window transform $W_R'(\omega)$. The resulting AR filter is thus an approximation to $H_d'(\omega) * W_R'(\omega)$, not $H_d'(\omega)$ itself. In return for this disadvantage, however, there are several advantages to this method. One is that $A(z)$ is guaranteed to be minimum phase (all roots inside the unit circle), and hence $H(z)$ is guaranteed to be stable. This is not true for the covariance method, in general. In fact, if for some reason one wishes to model an unstable $h_d(n)$, the covariance method is the appropriate choice. However, for a stable $h_d(n)$, the covariance method will usually also produce a stable model. The proof of stability for the autocorrelation method will be given shortly.

EXAMPLE. As a simple example of the covariance and autocorrelation methods, we will obtain first-order models of the desired response $h_d(n) = c^n$, for $n = 0, 1, \ldots, L$, by these two methods. For the covariance method, we form the data matrix

$$\mathbf{H}_d = \begin{bmatrix} 1 & 0 \\ c & 1 \\ c^2 & c \\ \cdot & \cdot \\ \cdot & \cdot \\ \cdot & \cdot \\ c^L & c^{L-1} \end{bmatrix},$$

from which the covariance matrix for $N = 1$ is seen to be

$$\mathbf{\Phi}_1 = \mathbf{H}_d^t \mathbf{H}_d = \begin{bmatrix} \sum_{n=0}^{L} c^{2n} & \sum_{n=0}^{L-1} c^{2n+1} \\ \sum_{n=0}^{L-1} c^{2n+1} & \sum_{n=0}^{L-1} c^{2n} \end{bmatrix}.$$

Since $h_{d0} = 1$, the normal equation for $\underline{a}$ is of the form

$$\mathbf{\Phi}_1 \underline{a} = \underline{\delta}.$$

The two simultaneous linear equations comprising this vector equation yield the expected solution

$$\underline{a} = \begin{bmatrix} 1 \\ -c \end{bmatrix}.$$

Therefore, the filter model is just

$$H(z) = \frac{1}{1 - cz^{-1}}$$

or $h(n) = c^n u(n)$, which exactly matches the given $h_d(n)$ in this simple case, whether stable ($|c| < 1$) or not.

For the autocorrelation method, we zero-pad the given data to produce

$$\hat{\mathbf{H}}_d = \begin{bmatrix} 1 & 0 \\ c & 1 \\ c^2 & c \\ \cdot & c^2 \\ \cdot & \cdot \\ \cdot & \cdot \\ c^L & \cdot \\ & \cdot \\ 0 & c^L \end{bmatrix}$$

and thus

$$\hat{\mathbf{R}}_1 = \hat{\mathbf{H}}_d^t \hat{\mathbf{H}}_d = \begin{bmatrix} \sum_{n=0}^{L} c^{2n} & \sum_{n=0}^{L-1} c^{2n+1} \\ \sum_{n=0}^{L-1} c^{2n+1} & \sum_{n=0}^{L} c^{2n} \end{bmatrix}.$$

The normal equation is now

$$\hat{\mathbf{R}}_1 \underline{\hat{a}} = (1/a_0)\underline{\delta}.$$

Employing the second of the two equations comprising this vector equation to solve for $\hat{a}_1$, we have

$$\sum_{n=0}^{L-1} c^{2n+1} + \hat{a}_1 \sum_{n=0}^{L} c^{2n} = 0,$$

from which

$$\hat{a}_1 = -\left(\frac{\sum_{n=0}^{L-1} c^{2n+1}}{\sum_{n=0}^{L} c^{2n}}\right).$$

One can show that $|\hat{a}_1| < 1$ and hence that the model is stable for all values of c. Note, in particular, the following cases:

$$\hat{a}_1 \approx -c, \qquad |c| \ll 1$$

$$\hat{a}_1 \approx -1/c, \qquad |c| \gg 1$$

$$\hat{a}_1 = -L/(L+1), \qquad c = 1.$$

A second advantage of the autocorrelation method in addition to guaranteed stability is that a very efficient iterative algorithm exists for the solution of $\underline{\hat{a}}$. This is the Levinson-Durbin algorithm, which solves for $\underline{\hat{a}}$ recursively from first to Nth order. Let the Pth-order solution be $\underline{\hat{a}}_p = \underline{a}_P/a_{0P}$, i.e.,

$$\hat{\mathbf{R}}_P \underline{\hat{a}}_P = (h_{d0}/a_{0P})\underline{\delta}. \tag{10.1.20}$$

Assume the recursive relationship

$$\begin{aligned} \underline{\hat{a}}_{P+1} &= \begin{bmatrix} 1 \\ \hat{a}_{1P} \\ \cdot \\ \cdot \\ \cdot \\ \hat{a}_{PP} \\ 0 \end{bmatrix} + K_{P+1} \begin{bmatrix} 0 \\ \hat{a}_{PP} \\ \cdot \\ \cdot \\ \cdot \\ \hat{a}_{1P} \\ 1 \end{bmatrix} \\ &= \underline{\hat{a}}_P^f + K_{P+1} \underline{\hat{a}}_P^b \\ &= \underline{\hat{a}}_P^f + K_{P+1} \mathbf{J} \underline{\hat{a}}_P^f \end{aligned} \tag{10.1.21}$$

where K_{P+1} is a scalar and $\mathbf{J}$ is the permutation matrix

$$\mathbf{J} = \begin{bmatrix} 0 & \cdots\cdots & 0 & 1 \\ 0 & \cdots\cdots & 1 & 0 \\ \cdot & & \cdot & \cdot \\ \cdot & \cdot & & \cdot \\ \cdot & \cdot & & \cdot \\ 1 & 0 & \cdots\cdots & 0 \end{bmatrix}$$

The gain K_{P+1} for the backward component $\underline{\hat{a}}_P^b = \mathbf{J}\underline{\hat{a}}_P^f$ is called the *reflection coefficient*. Then, $\underline{\hat{a}}_{P+1}$ must satisfy

$$\hat{\mathbf{R}}_{P+1} \underline{\hat{a}}_{P+1} = (h_{d0}/a_{0,P+1})\underline{\delta} \tag{10.1.22}$$

or, substituting from (10.1.21),

$$\hat{\mathbf{R}}_{P+1} \underline{\hat{a}}_P^f + K_{P+1} \hat{\mathbf{R}}_{P+1} \mathbf{J} \underline{\hat{a}}_P^f = (h_{d0}/a_{0,P+1})\underline{\delta}. \tag{10.1.23}$$

Therefore, by the definition of $\underline{\hat{a}}_P^f$,

$$\begin{bmatrix} \hat{\mathbf{R}}_P \\ \hline \hat{r}_{P+1} \cdots\cdots \hat{r}_1 \end{bmatrix} \underline{\hat{a}}_P + K_{P+1} \begin{bmatrix} \hat{r}_1 \cdots\cdots \hat{r}_{P+1} \\ \hline \hat{\mathbf{R}}_P \end{bmatrix} \mathbf{J} \underline{\hat{a}}_P = \begin{bmatrix} h_{d0}/a_{0,P+1} \\ 0 \\ \cdot \\ \cdot \\ \cdot \\ 0 \end{bmatrix} \tag{10.1.24}$$

The second through the next-to-last equations in (10.1.24) are satisfied from (10.1.20) because $\hat{\mathbf{R}}_P$ is symmetric and Toeplitz. Letting

$$\Delta_{P+1} = [\hat{r}_{P+1}, \ldots, \hat{r}_1]\underline{\hat{a}}_P, \tag{10.1.25}$$

the last equation in (10.1.24) gives the reflection coefficient as

$$K_{P+1} = -\Delta_{P+1}/(h_{d0}/a_{0P}). \tag{10.1.26}$$

The first equation, on the other hand, yields from (10.1.20), (10.1.25), and (10.1.26)

$$\begin{aligned}(h_{d0}/a_{0,P+1}) &= (h_{d0}/a_{0P}) + K_{P+1}\Delta_{P+1} \\ &= (h_{d0}/a_{0P}) + K_{P+1}[-K_{P+1}(h_{d0}/a_{0P})] \\ &= (h_{d0}/a_{0P})[1 - K_{P+1}^2].\end{aligned} \tag{10.1.27}$$

Therefore,

$$h_{d0}^2 \le (h_{d0}/a_{0,P+1}) \le (h_{d0}/a_{0P}) \tag{10.1.28}$$

because, in the limit as $P \to \infty$, the impulse response is matched exactly by the AR model and thus $a_{0\infty} = 1/h_{d0}$. Note, however, that $a_{0N} \neq 1/h_{d0}$ unless $K_P = 0$ for all $P > N$, corresponding to an exact Nth-order AR model.

Another important observation may be made concerning the reflection coefficients. From (10.1.28), $(h_{d0}/a_{0P}) > 0$, and thus from (10.1.27)

$$|K_P| < 1, \qquad \text{for all } P. \tag{10.1.29}$$

This condition on the reflection coefficients is both necessary and sufficient to ensure the stability of $H(z) = 1/A(z)$, which may be proved by applying Rouche's Theorem [22] to (10.1.21). Rouche's Theorem states that if $F(z)$ and $G(z)$ are analytic functions inside and on a closed contour C (e.g., the unit circle), with $|G(z)| < |F(z)|$ on C, then $F(z)$ and $F(z) + G(z)$ have no zeros on C and the same number of zeros inside C. Let

$$F(z) = z^{P+1}A_P^f(z) = z^{P+1} + \hat{a}_{1P}z^P + \cdots + \hat{a}_{PP}z,$$

which is analytic everywhere except at $z = \infty$. The $P + 1$ zeros of $F(z)$ are those of $A_P(z)$ plus one at $z = 0$. Assume that these zeros are all inside the unit circle. Similarly, let

$$\begin{aligned}G(z) &= K_{P+1}z^{P+1}A_P^b(z) = K_{P+1}[\hat{a}_{PP}z^P + \cdots + \hat{a}_{1P}z + 1] \\ &= K_{P+1}A_P^f(z^{-1}).\end{aligned}$$

Hence, the zeros of $G(z)$ are all outside the unit circle (i.e., $G(z)$ is *maximum phase*); but since $|K_{P+1}| < 1$, $|G(z)| < |F(z)|$ for $z = e^{j\omega T}$, that is, on the unit circle. Therefore, by

(10.1.21) and Rouche's Theorem, all $P + 1$ zeros of $A_{P+1}(z) = z^{-(P+1)}[F(z) + G(z)]$ are inside the unit circle. The sufficiency of the condition $|K_P| < 1$ for stability is thus proved by induction with an initial polynomial $A_0(z) = 1$. The proof of its necessity is similar.

Summary

We will summarize the Levinson-Durbin algorithm by detailing several iterations.

Iteration 0

The zeroth-order solution is simply the scalar

$$\underline{\hat{a}}_0 = 1.$$

The normal equation (10.1.20) is $\hat{r}_0\underline{\hat{a}}_0 = (h_{d0}/a_{00})$, from which

$$(h_{d0}/a_{00}) = \hat{r}_0.$$

Iteration 1

The first-order solution is then, from (10.1.21) with $P = 0$,

$$\underline{\hat{a}}_1 = \begin{bmatrix} 1 \\ 0 \end{bmatrix} + K_1 \begin{bmatrix} 0 \\ 1 \end{bmatrix} = \begin{bmatrix} 1 \\ K_1 \end{bmatrix} = \begin{bmatrix} 1 \\ \hat{a}_{11} \end{bmatrix}.$$

To determine K_1, we calculate Δ_1 from (10.1.25) as

$$\Delta_1 = [\hat{r}_1]\underline{\hat{a}}_0 = \hat{r}_1$$

and hence from (10.1.26)

$$K_1 = -\Delta_1/(h_{d0}/a_{00}) = -\hat{r}_1/\hat{r}_0.$$

Note that $|K_1| < 1$, as expected. Finally, from (10.1.27),

$$\begin{aligned}(h_{d0}/a_{01}) &= (h_{d0}/a_{00})(1 - K_1^2) \\ &= \hat{r}_0(1 - \hat{r}_1^2/\hat{r}_0^2).\end{aligned}$$

Iteration 2

The second-order solution is then

$$\underline{\hat{a}}_2 = \begin{bmatrix} 1 \\ \hat{a}_{11} \\ 0 \end{bmatrix} + K_2 \begin{bmatrix} 0 \\ \hat{a}_{11} \\ 1 \end{bmatrix} = \begin{bmatrix} 1 \\ K_1(1 + K_2) \\ K_2 \end{bmatrix} = \begin{bmatrix} 1 \\ \hat{a}_{12} \\ \hat{a}_{22} \end{bmatrix}$$

where

$$\begin{aligned}
\Delta_2 &= [\hat{r}_2, \hat{r}_1]\underline{\hat{a}}_1 = \hat{r}_2 + \hat{a}_{11}\hat{r}_1 \\
K_2 &= -\Delta_2/(h_{d0}/a_{01}) \\
(h_{d0}/a_{02}) &= (h_{d0}/a_{01})(1 - K_2^2).
\end{aligned}$$

The recursion proceeds in this manner through the Nth iteration where $P + 1 = N$, or until $K_{P+1} = 0$ which indicates that the AR model exactly matches the given autocorrelation values $\hat{r}(n)$ for $n = 0, 1, \ldots, P$.

A significant feature of the Levinson-Durbin algorithm is the associated reduction in computation required to solve the normal equations for $\underline{\hat{a}}$. The solution of a general set of N linear equations by, for example, Gaussian elimination requires $O(N^3)$ operations and $O(N^2)$ memory locations. Cholesky decomposition requires fewer computations and less memory, but these are still $O(N^3)$ and $O(N^2)$, respectively. In contrast, the Levinson-Durbin algorithm requires only $O(N^2)$ operations and $O(N)$ memory locations. However, in estimating computational complexity, we must also keep in mind that $O(NL)$ computations are needed to calculate the correlation sequence $\hat{r}(m)$, and for $L \gg N$, this dominates the other computations.

The reflection coefficient sequence $\{K_1, K_2, \ldots, K_N\}$ is equivalent to the vector $\underline{\hat{a}}$ in the sense that one can be obtained directly from the other. The *step-up* algorithm to convert $\{K_P\}$ to $\underline{\hat{a}}$ is simply the recursive application of (10.1.21) with $\underline{\hat{a}}_0^f = [1 \quad 0]^t$ to obtain $\underline{\hat{a}}_P$, $P = 1, 2, \ldots, N$. The *step-down* algorithm converts $\underline{\hat{a}}$ to $\{K_P\}$ and is also derived from (10.1.21). Those equations may be grouped into pairs of the form

$$\begin{aligned}
\hat{a}_{nP} &= \hat{a}_{n,P-1} + K_P\hat{a}_{P-n,P-1} \\
\hat{a}_{P-n,P} &= \hat{a}_{P-n,P-1} + K_P\hat{a}_{n,P-1},
\end{aligned} \tag{10.1.30}$$

which may be solved for $\hat{a}_{n,P-1}$ to produce

$$\hat{a}_{n,P-1} = \frac{\hat{a}_{nP} - K_P\hat{a}_{P-n,P}}{1 - K_P^2}. \tag{10.1.31}$$

Also, from the last equation in (10.1.21), we have $K_P = \hat{a}_{PP}$. Hence, $\underline{\hat{a}}_{P-1}$ may be derived from $\underline{\hat{a}}_P$ recursively for $P = N, N - 1, \ldots, 2$ to produce $\{K_P\}$. Note that for the autocorrelation method, $|K_P| < 1$, and thus the denominator $1 - K_P^2$ in (10.1.31) cannot be zero. However, in other applications of the step-down algorithm, we can have $|K_P| = 1$, in which case (10.1.31) cannot be used.

An additional application of the step-down algorithm is that, given an arbitrary $\underline{\hat{a}}$, it constitutes a simple stability test for the corresponding $H(z) = 1/A(z)$. That is, if $|K_P| \geq 1$ for any P, then $H(z)$ is not stable. For example, although the covariance method does not guarantee stability, the resulting model can be easily checked for stability by means of the step-down algorithm.

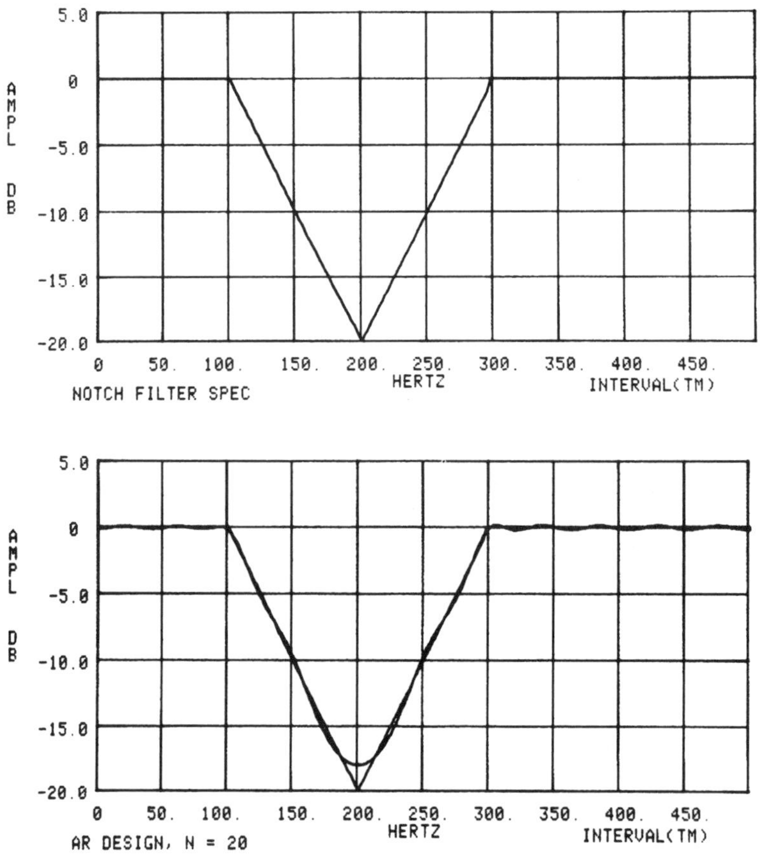

Figure 10.2. (a) Desired magnitude response and (b) 20th-order AR design.

In many applications, one is not given a causal sequence $h_d(n)$ to approximate, but rather $|H_d'(\omega)|$, from which $\hat{r}(m)$ must be estimated. Sampling $|H_d'(\omega)|$ on a dense set of uniformly spaced frequencies $\omega_k = k\omega_s/K$, we produce the DFT $|\hat{H}_d(k)|$ which may be inverted to yield $\hat{h}_d(n)$. However, $|H_d'(\omega)|$ is zero-phase, and thus $\hat{h}_d(n)$ is noncausal and symmetric about $n = 0$. Therefore, let

$$\hat{r}(m) = \sum_{n=-L}^{L-m} \hat{h}_d(n)\hat{h}_d(n + m) \tag{10.1.33}$$

for $m \geq 0$ and $L < K/2$.

As an example, we apply the autocorrelation method to approximate the frequency response $|H_d'(\omega)|$ shown in figure 10.2(a) with $K = 512$, $L = 50$, and $N = 20$. The resulting AR design $|H'(\omega)| = 1/|A'(\omega)|$ is shown in figure 10.2(b). Note that the greatest deviation from the desired response occurs at the bottom of the V-shaped notch. This is to be expected since this is an all-pole design and we have no zeros with which to produce a sharp notch. The passbands are almost equiripple, but this is not a general property of

AR designs. In fact, in the vicinity of abrupt transitions, large ripples will be produced; and hence such abrupt changes are to be avoided in the specification of $|H_d'(\omega)|$.

The linear normal equations for the coefficient vector $\underline{a}$ in this section have resulted from the choice of the inverse-filter error $e(n) = h_d(n) * a(n) - \delta(n)$ in (10.1.5) to which to apply the least-squares error criterion. It may be argued [23] that this is not the error we really wish to minimize and hence that the resulting estimate $\underline{a}$ is not the best least-squares estimate. Rather, by these arguments,

$$e(n) = h_d(n) - h(n) \tag{10.1.34}$$

is the preferred error measure. The least-squares norm E_N then becomes

$$E_N = \sum_{n=0}^{L} [h_d(n) - h(n)]^2 \tag{10.1.35}$$

or, in the limiting case as $L \to \infty$,

$$\begin{aligned} E_N &= \sum_{n=0}^{\infty} [h_d(n) - h(n)]^2 \\ &= \frac{1}{\omega_s} \int_0^{\omega_s} |H_d'(\omega) - H'(\omega)|^2 \, d\omega. \end{aligned} \tag{10.1.36}$$

This error norm, however, leads to nonlinear equations for $\underline{a}$, which must be solved iteratively. Steiglitz and McBride [24] have proposed an efficient iterative algorithm to estimate both the denominator and numerator coefficients for $H(z)$ in the more general ARMA case, and we will adapt their algorithm here to the AR-only case.

We note first that

$$\begin{aligned} h_d(n) &= h_d(n) * \delta(n) = h_d(n) * [h(n) * a(n)] \\ &= h(n) * h_d(n) * a(n). \end{aligned}$$

Therefore, defining

$$\mathbf{H} = \begin{bmatrix} h_0 & 0 & \cdots & 0 \\ h_1 & h_0 & & 0 \\ \vdots & & \ddots & \vdots \\ h_L & \cdots & \cdots & h_0 \end{bmatrix} \qquad \underline{h}_d = \begin{bmatrix} h_{d0} \\ h_{d1} \\ \vdots \\ h_{dL} \end{bmatrix} \qquad \underline{h} = \begin{bmatrix} h_0 \\ h_1 \\ \vdots \\ h_L \end{bmatrix}$$

we have in vector form from (10.1.34)

$$\underline{h}_d = \mathbf{H}\mathbf{H}_d\underline{a} = \underline{h} + \underline{e}. \tag{10.1.37}$$

The minimizing solution $\underline{a}$ is approximately

$$\underline{a} = (\mathbf{H}\mathbf{H}_d)^{\#}\underline{h} \tag{10.1.38}$$
$$= (\mathbf{H}_d^t\mathbf{H}^t\mathbf{H}\mathbf{H}_d)^{-1}(\mathbf{H}\mathbf{H}_d)^t\underline{h},$$

where $(\ )^{\#}$ denotes the *pseudo-inverse* matrix. This solution is only approximate (although quite close to the exact solution) because we have neglected the fact that the model $h(n)$ and thus $\mathbf{H}$ and $\underline{h}$ are themselves functions of $\underline{a}$. The matrix product $\mathbf{HH}_d$ corresponds to the convolution $h(n) * h_d(n)$. Hence, (10.1.37) tells us to prefilter the data $h_d(n)$ by the model $h(n)$ before solving for $\underline{a}$, which we can only do iteratively. That is, letting the ith estimate be $\underline{a}_i$ with corresponding $\mathbf{H}_i$ and $\underline{h}_i$ we solve for $\underline{a}_{i+1}$ from (10.1.38) by

$$\underline{a}_{i+1} = (\mathbf{H}_i\mathbf{H}_d)^{\#}\underline{h}_i. \tag{10.1.39}$$

The convergence of this iterative algorithm has been found to be rapid (on the order of 5 to 10 iterations) [23]. Either the covariance or autocorrelation method can be used to provide the initial vector $\underline{a}_0$. A potential difficulty is that the algorithm may not converge if the order (N) of the model $h(n)$ is chosen to be larger than that of the data $h_d(n)$. Hence, it is important not to overestimate the order of the model.

10.2 MOVING-AVERAGE (ALL-ZERO) FILTERS

Moving-average (MA) filters are another name for FIR filters. From chapter 9, we already have three techniques for designing linear-phase FIR filters with arbitrary magnitude responses. Another class of techniques that generally provides minimum-phase filter designs is available from the field of statistical modeling. Two of these techniques are described in this section.

The first technique is Durbin's method [25], which is straightforward and highly effective if an MA model is appropriate to the data. The MA filter is of the form $H(z) = B(z)$, where

$$B(z) = b_0 + b_1 z^{-1} + \cdots + b_M z^{-M}. \tag{10.2.1}$$

Since an Lth-order AR model can approximate any desired filter with arbitrary accuracy if L is made sufficiently large (as discussed in the previous section), let $1/A_L(z)$ be a high-order AR model of $h_d(n)$ or $|H_d'(\omega)|$ with $L \gg M$. As noted previously, the AR filter cannot directly model sharp notches in $|H_d'(\omega)|$, but for sufficiently large L, the "information" concerning such notches is nevertheless contained in $A_L(z)$ and can be extracted in the design of $B(z)$.

Therefore, we desire

$$B(z) \approx 1/A_L(z) \tag{10.2.2}$$

or, in the time domain,

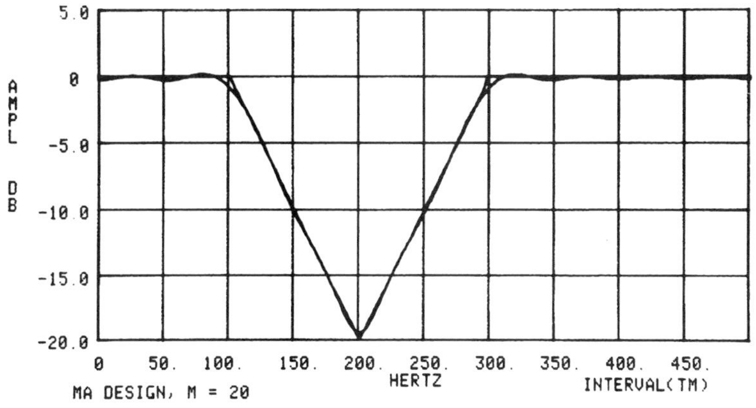

Figure 10.3. 20th-order MA design from Durbin's method for specification in figure 10.2.

$$a_L(n) * b(n) = \delta(n) + e(n), \tag{10.2.3}$$

where $e(n)$ is the approximation error. This is exactly analogous to (10.1.5), and thus we estimate $b(n)$ such that $1/B(z)$ is an AR model of the FIR filter $A_L(z)$. Since $A_L(z)$ is FIR, the autocorrelation method may be directly applied to the "data" sequence $a_L(n)$ (without windowing) to compute $b(n)$. Therefore, $B(z)$ is guaranteed to be minimum phase, whether the polynomial $A_L(z)$ is or not.

As an example, an MA model with $M = 20$ was made of the ideal filter in figure 10.2(a) using an intermediate AR model of order $L = 128$. Such a high-order AR model is not difficult to compute via the autocorrelation method with the Levinson-Durbin algorithm. The resulting magnitude response is shown in figure 10.3. Comparing this with the AR model having $N = 20$ in figure 10.2(b), we note that the notch for the MA design is deeper, as expected; but the accuracy near the band edges is not as good as in the AR case because we now have no poles with which to produce sharp peaks or edges in $|H'(\omega)|$. The ripple in the passbands of both designs is about the same.

Durbin's method has several other useful applications besides direct estimation of MA models. Given a polynomial $B(z)$ that is not minimum phase (such as the denominator $A(z)$ of an unstable AR filter), a minimum-phase equivalent $\tilde{B}(z)$ with $|\tilde{B}'(\omega)| \approx |B'(\omega)|$ can be obtained by factoring the zero-phase polynomial $G(z) = B(z)B(z^{-1})$. That is, forming the autocorrelation function

$$g(n) = b(n) * b(-n), \tag{10.2.4}$$

the Yule-Walker equation is solved to produce an AR model $A_L(z)$ with $L \gg M$ (noting that $g(n) = 0$ for $|n| > M$). Alternatively, we may be given $G(z)$ itself to factor. If $G(z)$ is factorable, then Durbin's method provides a close approximation to the minimum-phase factor. If, however, $G(z)$ is not factorable (corresponding to single zeros on the unit circle and thus a nonpositive power spectrum $G'(\omega)$), the estimate $\tilde{B}(z)$ cannot be a true factor. Nevertheless, it can be a useful approximate factor in the sense that $\tilde{G}(z) = \tilde{B}(z)\tilde{B}(z^{-1})$

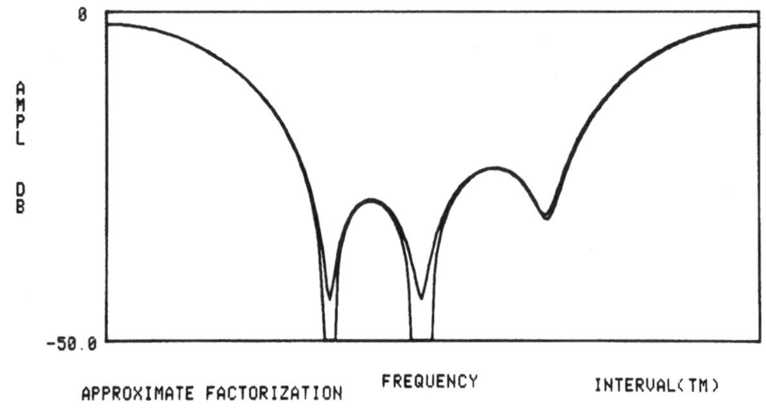

Figure 10.4. Unfactorable $G'(\omega)$ versus factorable $\tilde{G}'(\omega)$ from Durbin's method.

is now factorable and $\tilde{G}'(\omega)$ is positive. For example, figure 10.4 shows an MA power spectrum $G'(\omega)$ with negative values (off the dB scale) for two intervals on the ω axis, along with $\tilde{G}'(\omega)$ from Durbin's method. Note that $\tilde{G}'(\omega)$ closely approximates $G'(\omega)$ everywhere except in those intervals where $G'(\omega)$ is negative.

Each of the two steps in Durbin's method employs a whitening criterion to design the inverse filters $A_L(z)$ and $B(z)$. A single, overall whitening criterion may be applied by requiring that $H_d(z)/B(z)$ be approximately white. That is, defining the inverse filter

$$H_b(z) = 1/B(z), \tag{10.2.5}$$

the equivalent of (10.1.5) in the MA case is

$$h_d(n) * h_b(n) = \delta(n) + e(n). \tag{10.2.6}$$

Minimization of the least-squares norm of $e(n)$, however, now leads to nonlinear equations for $b(n)$. An iterative algorithm, similar to that of Steiglitz and McBride, has been proposed by Judell [26] to estimate $b(n)$. Hannan [80] has proposed a similar method in terms of DFT's. Noting that

$$h_b(n) * h_b(n) * h_d(n) * b(n) = \delta(n) + e(n) \tag{10.2.7}$$

and letting

$$\mathbf{H}_b = \begin{bmatrix} h_{b0} & 0 & \cdots & 0 \\ h_{b1} & h_{b0} & & 0 \\ \vdots & & \ddots & \vdots \\ h_{bL} & \cdots & \cdots & h_{b0} \end{bmatrix}, \tag{10.2.8}$$

we have

$$\mathbf{H}_b \mathbf{H}_b \mathbf{H}_d \underline{b} = \underline{\delta} + \underline{e}. \tag{10.2.9}$$

The minimizing solution $\underline{b}$ is thus approximately

$$\underline{b} = (\mathbf{H}_b \mathbf{H}_b \mathbf{H}_d)^{\#} \underline{\delta} \tag{10.2.10}$$

and the implied iteration is

$$\underline{b}_{i+1} = (\mathbf{H}_{bi} \mathbf{H}_{bi} \mathbf{H}_d)^{\#} \underline{\delta}. \tag{10.2.11}$$

The solution from Durbin's method makes an excellent starting vector $\underline{b}_0$ for the iteration. An unexpected difficulty is that the iteration given by (10.2.11) goes in exactly the wrong direction. That is, letting $\Delta \underline{b}_{i+1} = \underline{b}_{i+1} - \underline{b}_i$, the elements of $\Delta \underline{b}_{i+1}$ have appropriate magnitudes, but incorrect signs [26]. Therefore, we define a modified iteration by

$$\begin{aligned} \underline{\hat{b}}_{i+1} &= \underline{\hat{b}}_i - \Delta \underline{b}_{i+1} \\ &= 2\underline{\hat{b}}_i - \underline{b}_{i+1}. \end{aligned} \tag{10.2.12}$$

The modified iteration converges rapidly (3 to 5 iterations).

As a final note, we observe that $e(n) = h_d(n) - h(n)$, with $h(n) = b(n)$, is not usually an appropriate error measure for the least-squares criterion in the MA case, unlike the AR and ARMA cases, because the minimizing solution is then just the rectangularly windowed response $b(n) = h_d(n) w_R(n)$.

10.3 ARMA (POLE/ZERO) FILTERS

The general ARMA case is, as might be expected, more complicated than the separate AR and MA cases previously investigated. We will present several ARMA methods beginning, as before, with the simplest and moving to those with stronger error criteria. The simplest are based on *equation error* measures having no directly meaningful interpretations as do the inverse-filter error and the *fitting error* $h_d(n) - h(n)$. Since all sorts of ideal equations can be written, minimization of equation error may or may not lead to useful modeling methods; one must simply try each method to find out. Methods will be given for the case where the ideal response $h_d(n)$ is given and also for the case where the autocorrelation function $\hat{r}(n)$ is given or estimated as in (10.1.33).

We seek the general ARMA model

$$H(z) = \frac{B(z)}{A(z)} \tag{10.3.1}$$

where, since arbitrary gain is not needed in both $B(z)$ and $A(z)$, we constrain $a_0 = 1$. Ideally, therefore, $H_d(z) \approx H(z)$ or

$$H_d(z)A(z) \approx B(z). \tag{10.3.2}$$

Linear equations may be produced for $a(n)$ and $b(n)$ by writing (10.3.2) in the time domain as

$$h_d(n) * a(n) = b(n) + e(n), \tag{10.3.3}$$

where $e(n)$ is now simply equation error. Forcing $e(n) = 0$ for $n \le M$, we may write (10.3.3) in the vector form [27]

$$\mathbf{H}_d \underline{a} = \begin{bmatrix} \mathbf{H}_{d1} \\ --- \\ \mathbf{H}_{d2} \end{bmatrix} \underline{a} = \begin{bmatrix} \underline{b} \\ --- \\ \underline{e}_2 \end{bmatrix} = \underline{b}_L. \tag{10.3.4}$$

Minimization of $\underline{e}_2^t \underline{e}_2$ via the lower partition of (10.3.4) and the covariance method provides the estimate of $\underline{a}$. In particular, partitioning $\mathbf{H}_{d2}$ as $[\underline{h}_{d2} \mid \mathbf{H}'_{d2}]$ with

$$\underline{h}_{d2} = \begin{bmatrix} h_{d,M+1} \\ \vdots \\ h_{dL} \end{bmatrix} \qquad \mathbf{H}'_{d2} = \begin{bmatrix} h_{dM} & \cdots & h_{d,M-N+1} \\ \vdots & & \vdots \\ h_{d,L-1} & \cdots & h_{d,L-N} \end{bmatrix}$$

and defining $\underline{a}'$ by

$$\underline{a} = \begin{bmatrix} 1 \\ --- \\ \underline{a}' \end{bmatrix},$$

the lower partition of (10.3.4) becomes

$$\mathbf{H}'_{d2}\, \underline{a}' = -\underline{h}_{d2} + \underline{e}_2 \tag{10.3.5}$$

and thus

$$\begin{aligned} \underline{a}' &= -(\mathbf{H}'_{d2})^{\#}\, \underline{h}_{d2} \\ &= -(\mathbf{H}'^{t}_{d2}\mathbf{H}'_{d2})^{-1}(\mathbf{H}'^{t}_{d2}\underline{h}_{d2}). \end{aligned} \tag{10.3.6}$$

Because this estimate of $\underline{a}'$ provides a least-squares prediction of $-\underline{h}_{d2}$ from $\mathbf{H}'_{d2}$ by (10.3.5), this method may be interpreted as *linear prediction* on the *tail* of $h_d(n)$.

Given $\underline{a}$, there are several ways to estimate $\underline{b}$. The upper partition of (10.3.4) yields

$$\underline{b} = \mathbf{H}_{d1}\, \underline{a}, \tag{10.3.7}$$

but this corresponds to simple truncation or windowing of $\underline{b}_L = \mathbf{H}_d \underline{a}$ and is known to give poor results, in general. It is much better to utilize all of the information in $\underline{b}_L$ by, for example, applying Durbin's method to $\underline{b}_L$ to estimate $\underline{b}$. More simply, and almost equivalently, one may simply step down the Lth-order estimate $\underline{b}_L$ to the Nth-order estimate $\underline{b}$ via the step-down algorithm given in section 10.1.

If we are given $|H_d'(\omega)|$ and thus $\hat{r}(n)$ instead of $h_d(n)$, the above mentioned method must be modified somewhat. Since $r(n) = h_d(n) * h_d(-n)$, the power spectrum (transform of $r(n)$) is given by

$$S(z) = H_d(z)\, H_d(z^{-1}) \approx \frac{B(z)B(z^{-1})}{A(z)A(z^{-1})}. \tag{10.3.8}$$

Therefore,

$$S(z)A(z) \approx \frac{B(z)B(z^{-1})}{A(z^{-1})} = \frac{G(z)}{A(z^{-1})} \tag{10.3.9}$$

or, in the time domain, using the estimated autocorrelation $\hat{r}(n)$,

$$\hat{r}(n) * a(n) = g(n) * h_a(-n) + e(n), \tag{10.3.10}$$

where $H_a(z) = 1/A(z)$ and $e(n)$ is the equation error. By analogy with the lower partition of (10.3.4) since $g(n) * h_a(-n) = 0$ for $n > M$, we may write

$$\mathbf{R}_2 \underline{a} = \underline{e}_2$$

or, as in (10.3.5),

$$\mathbf{R}_2' \underline{a}' = -\underline{r}_2 + \underline{e}_2, \tag{10.3.11}$$

where

$$\underline{r}_2 = \begin{bmatrix} \hat{r}_{M+1} \\ \vdots \\ \hat{r}_L \end{bmatrix} \qquad \mathbf{R}_2' = \begin{bmatrix} \hat{r}_M & \cdots & \hat{r}_{M-N+1} \\ \vdots & & \vdots \\ \hat{r}_{L-1} & \cdots & \hat{r}_{L-N} \end{bmatrix}.$$

The equation-error norm $\underline{e}_2^t \underline{e}_2$ is thus minimized by

$$\begin{aligned} \underline{a}' &= -(\mathbf{R}_2')^{\#}\, \underline{r}_2 \\ &= -(\mathbf{R}_2'^t \mathbf{R}_2')^{-1} (\mathbf{R}_2'^t \underline{r}_2). \end{aligned} \tag{10.3.12}$$

The matrix product $\mathbf{R}_2'^t \mathbf{R}_2'$ corresponds to a *correlation of correlation*, and (10.3.11) may be interpreted as linear prediction on the tail of $\hat{r}(n)$ [28, 29].

Given $\underline{a}$, there are several ways, as before, to estimate $\underline{b}$. A simple and effective way is to apply Durbin's method to the residual MA autocorrelation function

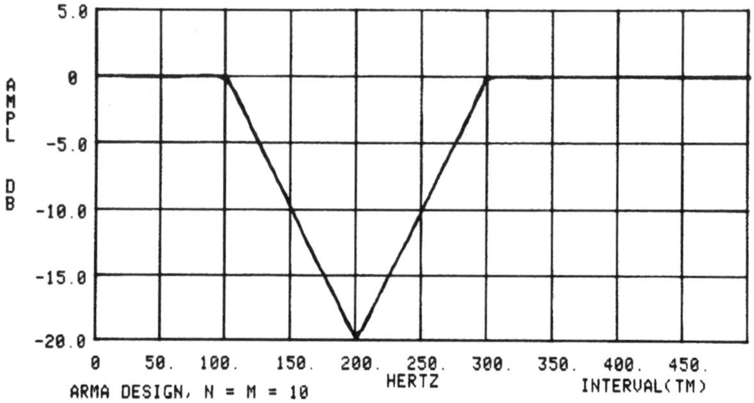

Figure 10.5. Tenth-order ARMA design for specification in figure 10.2.

$$\hat{g}(n) = \hat{r}(n) * a(n) * a(-n) \tag{10.3.13}$$

for which, given $2L + 1$ values $\hat{r}(n)$, $n = -L, \ldots, L$, we can calculate the $2(L - N) + 1$ values $\hat{g}(n)$, $n = -L + N, \ldots, L - N$, from (10.3.13).

Durbin's method may also be extended to provide a complete ARMA modeling technique [30, 31]. The first step is still to obtain a high-order AR model $1/A_L(z)$ with $L \gg N, M$ by an AR method which is appropriate to the data. Then,

$$\frac{1}{A_L(z)} \approx \frac{B(z)}{A(z)} \tag{10.3.14}$$

and thus, in the time domain

$$a_L(n) * b(n) = a(n) + e(n). \tag{10.3.15}$$

Note that (10.3.15) is of the form of (10.3.3) except that $a(n)$ and $b(n)$ are interchanged because $A_L(z)$ is the approximate inverse of $H_d(z)$. The error $e(n)$ is again simply equation error. Therefore, $b(n)$ may be estimated from the tail of $a_L(n)$ by the covariance method of linear prediction, as in (10.3.4) through (10.3.6). Noting that the residual sequence

$$a_{L+M}(n) = a_L(n) * b(n) \tag{10.3.16}$$

is of finite duration $L + M + 1$, we may estimate $a(n)$ by simply stepping down the $(L + M)$th-order AR estimate $a_{L+M}(n)$ to Nth order.

As an example, this extension of Durbin's method was applied to the example of sections 10.1 and 10.2 with $L = 128$ and $M = N = 10$. The resulting $|H_d'(\omega)|$ is shown in figure 10.3.1. Note that the sharp band edges and the V-shaped notch are all closely modeled, and the passband ripple is much less than in the separate AR and MA cases (which have the same total number (20) of degrees of freedom). The other equation-error method of (10.3.8) through (10.3.13) produces a similar ARMA model.

Judell's ARMA method minimizes the least-squares norm of the inverse-filter error, not equation error, and is an approximation to the maximum-likelihood estimator [26]. The inverse-filter error is defined by

$$h_b(n) * h_d(n) * a(n) \quad = \quad \delta(n) + e(n). \tag{10.3.17}$$

Writing this in matrix form, we have

$$\mathbf{H}_b \mathbf{H}_d \underline{a} \quad = \quad \underline{\delta} + \underline{e} \tag{10.3.18}$$

from which results the iterative solution

$$\underline{a}_{i+1} \quad = \quad (\mathbf{H}_{bi} \mathbf{H}_d)^{\#} \, \underline{\delta}. \tag{10.3.19}$$

This is just the covariance method with $h_d(n)$ prefiltered by $h_{bi}(n)$. Likewise,

$$\mathbf{A} \mathbf{H}_b \mathbf{H}_b \mathbf{H}_d \underline{b} \quad = \quad \underline{\delta} + \underline{e}, \tag{10.3.20}$$

where

$$\mathbf{A} \quad = \quad \begin{bmatrix} a_0 & 0 & \cdots & 0 \\ \vdots & \ddots & & \vdots \\ a_N & & \ddots & \vdots \\ \vdots & \ddots & & \vdots \\ 0 & \cdots \; a_N & \cdots & a_0 \end{bmatrix},$$

from which

$$\underline{b}_{i+1} \quad = \quad (\mathbf{A}_{i+1} \mathbf{H}_{bi} \mathbf{H}_{bi} \mathbf{H}_d)^{\#} \, \underline{\delta}. \tag{10.3.21}$$

This is again the covariance method, with $h_{bi}(n) * h_d(n)$ further prefiltered by $a_{i+1}(n) * h_{bi}(n)$. As noted in section 10.2, the iteration defined by (10.3.21) takes us in exactly the wrong direction and, therefore, is modified to produce

$$\underline{\hat{b}}_{i+1} \quad = \quad 2\underline{\hat{b}}_i - \underline{b}_{i+1}. \tag{10.3.22}$$

Durbin's extended method provides a good initial ARMA model for the iteration since all we need is $\underline{b}_0$. Convergence is rapid.

The Steiglitz-McBride algorithm [23, 24] minimizes instead the least-squares norm of the fitting error $e(n) = h_d(n) - h(n)$, as discussed in section 10.1. The matrix version of the inverse filter $H_a(z) = 1/A(z)$ is

$$\mathbf{H}_a = \begin{bmatrix} h_{a0} & 0 & \cdots & 0 \\ h_{a1} & h_{a0} & & 0 \\ \vdots & & \ddots & \vdots \\ h_{aL} & \cdots & \cdots & h_{a0} \end{bmatrix}, \tag{10.3.23}$$

and the partition consisting of the first $M + 1$ columns is

$$\mathbf{H}'_a = \begin{bmatrix} h_{a0} & \cdots & 0 \\ \vdots & \ddots & \vdots \\ h_{aM} & \cdots & h_{a0} \\ \vdots & & \vdots \\ h_{aL} & \cdots & h_{a,L-M} \end{bmatrix}$$

We can then write, as in (10.1.37),

$$\underline{h}_d = \mathbf{H}_a \mathbf{H}_d \underline{a} = \underline{h} + \underline{e} = \mathbf{H}'_a \underline{b} + \underline{e}, \tag{10.3.24}$$

from which

$$\underline{a} = (\mathbf{H}_a \mathbf{H}_d)^{\#} \mathbf{H}'_a \underline{b}. \tag{10.3.25}$$

On the other hand,

$$\underline{h} = \mathbf{H}'_a \underline{b} = \underline{h}_d - \underline{e}, \tag{10.3.26}$$

from which

$$\underline{b} = (\mathbf{H}'_a)^{\#} \underline{h}_d. \tag{10.3.27}$$

Substituting (10.3.27) into (10.3.25), we eliminate the explicit dependence of $\underline{a}$ on $\underline{b}$ and define the iteration as

$$\underline{a}_{i+1} = (\mathbf{H}_{ai} \mathbf{H}_d)^{\#} \mathbf{H}'_{ai} (\mathbf{H}'_{ai})^{\#} \underline{h}_d. \tag{10.3.28}$$

A good initial estimate $\underline{a}_0$ is provided by the covariance method on the tail of $h_d(n)$. At convergence, $\underline{b}$ is obtained from (10.3.27). As discussed earlier, it is important not to overestimate the number (N) of poles in order to ensure convergence [23], which is again rapid.

In the important special case of $M = N - 1$, corresponding to an arbitrary sum of N exponentials, Evans and Fischl [32] and Kumaresan et al. [33] have developed a modification of the covariance method that also minimizes the fitting-error norm. From (10.3.26) and (10.3.27),

$$\underline{e} = \underline{h}_d - \mathbf{H}'_a(\mathbf{H}'_a)^{\#}\,\underline{h}_d$$
$$= (\mathbf{I} - \mathbf{H}'_a(\mathbf{H}'^t_a\mathbf{H}'_a)^{-1}\mathbf{H}'^t_a)\,\underline{h}_d. \tag{10.3.29}$$

The matrix $\mathbf{H}'_a(\mathbf{H}'^t_a\mathbf{H}'_a)^{-1}\mathbf{H}'^t_a$ is a projection operator that projects $\underline{h}_d$ onto the subspace spanned by the columns of $\mathbf{H}'_a$. Hence, $\underline{e}$ is the difference between this projection of $\underline{h}_d$ and $\underline{h}_d$ itself. The matrix $(\mathbf{I} - \mathbf{H}'_a(\mathbf{H}'^t_a\mathbf{H}'_a)^{-1}\mathbf{H}'^t_a)$ is *idempotent* (i.e., it is unchanged when multiplied by itself) and also symmetric. Therefore, the least-squares error norm is given by

$$\underline{e}^t\underline{e} = \underline{h}^t_d(\mathbf{I} - \mathbf{H}'_a(\mathbf{H}'^t_a\mathbf{H}'_a)^{-1}\mathbf{H}'^t_a)\,\underline{h}_d. \tag{10.3.30}$$

In the case of $M = N - 1$, the matrix $\mathbf{H}'_a$ is $(L + 1) \times N$, and it can be shown that

$$\mathbf{I} - \mathbf{H}'_a(\mathbf{H}'^t_a\mathbf{H}'_a)^{-1}\mathbf{H}'^t_a = \mathbf{A}'^t(\mathbf{A}'\mathbf{A}'^t)^{-1}\mathbf{A}' \tag{10.3.31}$$

where $\mathbf{A}'$ is the lower $(L + 1 - N) \times (L + 1)$ partition of $\mathbf{A}$ given by

$$\mathbf{A}' = \begin{bmatrix} a_N & \cdots & a_0 & 0 & \cdots & 0 \\ 0 & a_N & \cdots & a_0 & \cdots & 0 \\ \vdots & & \ddots & & \ddots & \vdots \\ 0 & \cdots & 0 & a_N & \cdots & a_0 \end{bmatrix}$$

For equation (10.3.31) to hold, it must be true that $\mathbf{A}'\mathbf{H}'_a = 0$, which can be seen from

$$\mathbf{A}\mathbf{H}_a = \begin{bmatrix} \mathbf{A}'' \\ \hline \mathbf{A}' \end{bmatrix} [\mathbf{H}'_a \,|\, \mathbf{H}''_a] = \mathbf{I} = \begin{bmatrix} \mathbf{I} & | & 0 \\ \hline 0 & | & \mathbf{I} \end{bmatrix}.$$

Therefore, the error norm in (10.3.30) becomes

$$\underline{e}^t\underline{e} = \underline{h}^t_d\mathbf{A}'^t(\mathbf{A}'\mathbf{A}'^t)^{-1}\mathbf{A}'\underline{h}_d. \tag{10.3.32}$$

But $\mathbf{A}'\underline{h}_d$ corresponds to the convolution $a(n) * h_d(n)$ for $n = M + 1, \ldots, L$, and may thus be written also as $\mathbf{H}_{d2}\underline{a}$, where $\mathbf{H}_{d2}$ was defined in (10.3.4) as

$$\mathbf{H}_{d2} = \begin{bmatrix} h_{dN} & \cdots & h_{d0} \\ \vdots & & \vdots \\ h_{dL} & \cdots & h_{d,L-N} \end{bmatrix}$$

Hence, (10.3.32) may be written instead as

$$\underline{e}^t\underline{e} = \underline{a}^t\mathbf{H}^t_{d2}(\mathbf{A}'\mathbf{A}'^t)^{-1}\mathbf{H}_{d2}\underline{a}. \tag{10.3.33}$$

Note, for comparison, from (10.3.4) that the least-squares norm for the covariance method on the tail of $h_d(n)$ is just

$$\underline{e}_2^t \underline{e}_2 = \underline{a}^t \mathbf{H}_{d2}^t \mathbf{H}_{d2} \underline{a} = \underline{a}^t \mathbf{\Phi}_2 \underline{a}. \tag{10.3.34}$$

Therefore, the difference between minimization of the fitting-error norm $\underline{e}^t\underline{e}$ and the equation-error norm $\underline{e}_2^t\underline{e}_2$ is the inclusion of the *matrix prefilter* $(\mathbf{A}^t\mathbf{A}^{\prime\prime})^{-1}$ in the modified covariance matrix

$$\hat{\mathbf{\Phi}}_2 = \mathbf{H}_{d2}^t (\mathbf{A}^t \mathbf{A}^{\prime t})^{-1} \mathbf{H}_{d2}. \tag{10.3.35}$$

Writing $\mathbf{\Phi}_2$ in the partitioned form

$$\mathbf{\Phi}_2 = [\underline{h}_{d2} \,|\, \mathbf{H}_{d2}^{\prime}]^t [\underline{h}_{d2} \,|\, \mathbf{H}_{d2}^{\prime}] = \left[\begin{array}{c|c} \phi_0 & \underline{\phi}_2^t \\ \hline \underline{\phi}_2 & \mathbf{\Phi}_2^{\prime} \end{array}\right] \tag{10.3.36}$$

we note that the equation-error solution in (10.3.6) was just

$$\underline{a}^{\prime} = -(\mathbf{\Phi}_2^{\prime})^{-1} \underline{\phi}_2. \tag{10.3.37}$$

Therefore, partitioning the modified covariance matrix $\hat{\mathbf{\Phi}}_2$ as in (10.3.36), the minimizing solution for the fitting-error norm becomes

$$\hat{\underline{a}}^{\prime} = -(\hat{\mathbf{\Phi}}_2^{\prime})^{-1} \hat{\underline{\phi}}_2. \tag{10.3.38}$$

Kumaresan et al. [33] have given an iterative algorithm to determine $\hat{\underline{a}}^{\prime}$ based upon an efficient inversion of the (possibly large) matrix prefilter $(\mathbf{A}^t\mathbf{A}^{\prime\prime})^{-1}$.

10.4 LATTICE STRUCTURES

The Levinson-Durbin algorithm implies a *lattice* structure for FIR (MA) filters that can be manipulated to produce IIR lattice structures as well (both AR and ARMA). The MA lattice is particularly useful for minimum-phase FIR filters, and both it and the AR lattice have found widespread application in speech analysis/synthesis systems and in adaptive systems. Allpass filters are also nicely implemented in lattice form.

Recall from section 10.1 that the inverse (FIR) filter $A(z)$ can be generated recursively from the reflection coefficients K_p via the step-up algorithm

$$\underline{a}_P = \underline{a}_{P-1}^f + K_P \underline{a}_{P-1}^b \tag{10.4.1}$$

$$= \underline{a}_{P-1}^f + K_P \mathbf{J} \underline{a}_{P-1}^f, \qquad P = 1, 2, \ldots, N,$$

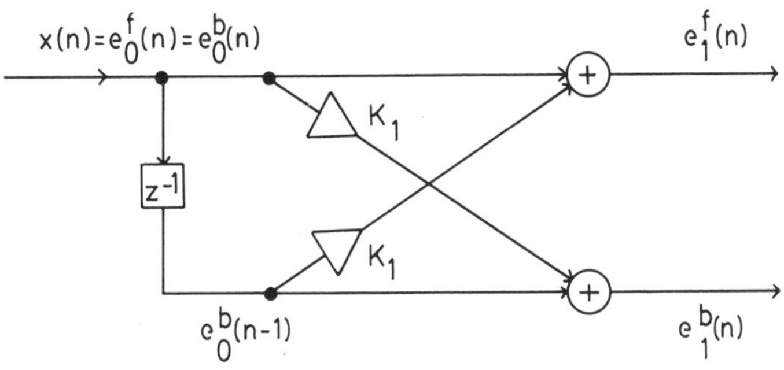

Figure 10.6. First section in FIR lattice.

where $\underline{a}_0^f = [1\ 0]^t$ and the final coefficient vector is $\underline{a} = \underline{a}_N$. This filter is scaled such that the first coefficient a_{0P} equals unity for all P. Therefore, the initial *extended* forward and backward filters are given by

$$\underline{a}_1^f = \begin{bmatrix} \underline{a}_1 \\ --- \\ 0 \end{bmatrix} = \begin{bmatrix} 1 \\ K_1 \\ 0 \end{bmatrix}$$

and

$$\underline{a}_1^b = \begin{bmatrix} 0 \\ --- \\ \mathbf{J}\underline{a}_1 \end{bmatrix} = \begin{bmatrix} 0 \\ K_1 \\ 1 \end{bmatrix}$$

Let the *unextended* forward and backward filters $\underline{a}_1$ and $\mathbf{J}\underline{a}_1$ have outputs $e_1^f(n)$ and $e_1^b(n)$, respectively. The first-order filters $\underline{a}_1$ and $\mathbf{J}\underline{a}_1$ may be implemented as shown in figure 10.6.

The next unextended forward and backward filters $\underline{a}_2$ and $\mathbf{J}\underline{a}_2$ are given by

$$\underline{a}_2 = \underline{a}_1^f + K_2\underline{a}_1^b$$

and thus

$$\mathbf{J}\underline{a}_2 = \underline{a}_1^b + K_2\underline{a}_1^f$$

with outputs $e_2^f(n)$ and $e_2^b(n)$, respectively. Note that the output of the extended (delayed) backward filter $\underline{a}_1^b$ is $e_1^b(n-1)$. Hence,

$$e_2^f(n) = e_1^f(n) + K_2 e_1^b(n-1)$$

and (10.4.2)

$$e_2^b(n) = e_1^b(n-1) + K_2 e_1^f(n).$$

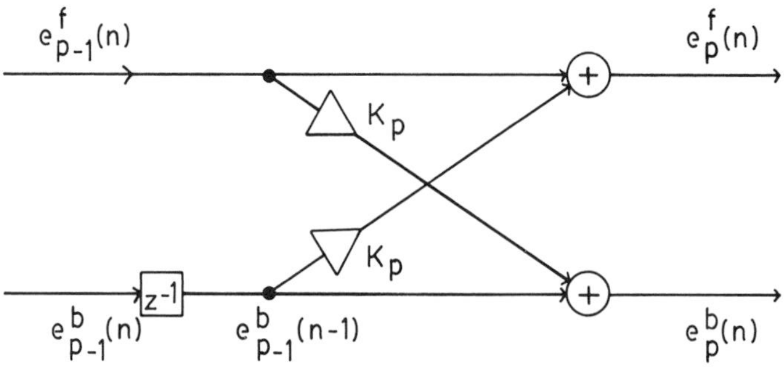

Figure 10.7. General FIR lattice section.

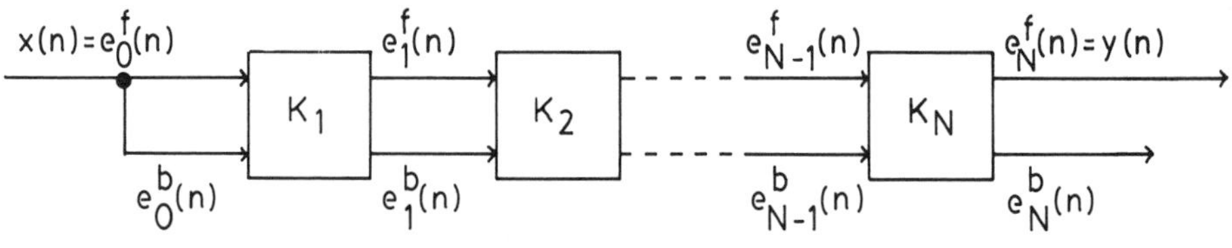

Figure 10.8. Interconnection of FIR lattice sections.

The filters $\underline{a}_2$ and $\mathbf{J}\underline{a}_2$ are thus produced by cascading the lattice section in figure 10.7 (for $P = 2$) with the first section in figure 10.6 (which is of exactly the same form). Subsequent filters $\underline{a}_P$ and $\mathbf{J}\underline{a}_P$ are generated by cascading additional lattice sections of the form in figure 10.7, corresponding to the general difference equations

$$\begin{aligned} e^f_P(n) &= e^f_{P-1}(n) + K_P e^b_{P-1}(n-1) \\ e^b_P(n) &= e^b_{P-1}(n-1) + K_P e^f_{P-1}(n). \end{aligned} \tag{10.4.3}$$

The final forward output $e^f_N(n)$ constitutes the output of the FIR filter $A(z)$, which is minimum-phase if $|K_P| < 1$ for all P. Alternatively, $e^b_N(n)$ is the output of the *backward* FIR filter $z^{-N}A(z^{-1})$ which is maximum-phase if $|K_P| < 1$ for all P. The interconnection of the FIR lattice sections, each having the form in figure 10.7, is illustrated in figure 10.8.

A lattice structure for the AR filter $H(z) = 1/A(z)$ is generated by rewriting (10.4.3) as

$$\begin{aligned} e^f_{P-1}(n) &= e^f_P(n) - K_P e^b_{P-1}(n-1) \\ e^b_P(n) &= e^b_{P-1}(n-1) + K_P e^f_{P-1}(n). \end{aligned} \tag{10.4.4}$$

Clearly, if the input $x(n) = e^f_0(n) = e^b_0(n)$ to the FIR filter $A(z)$ produces the output $y(n) = e^f_N(n)$, then the input $x(n) = e^f_N(n)$ to the inverse filter $H(z) = 1/A(z)$ will produce the output $y(n) = e^f_0(n) = e^b_0(n)$. Letting $P = 1$ in (10.4.4), we thus have

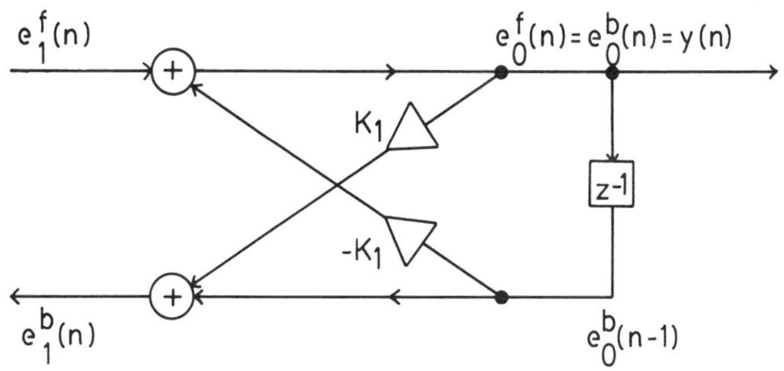

Figure 10.9. Last section in recursive (AR) lattice.

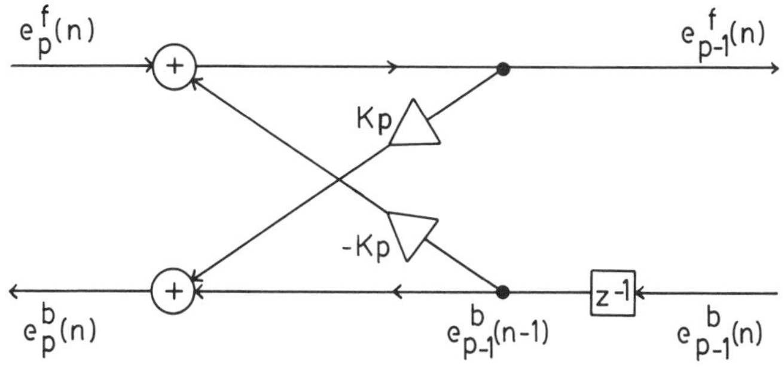

Figure 10.10. General recursive lattice section.

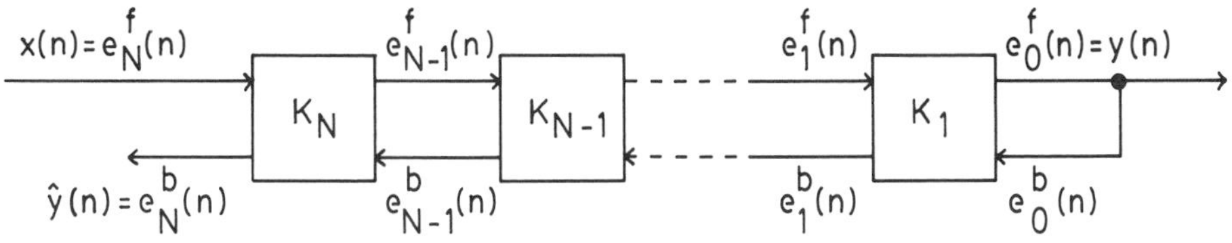

Figure 10.11. Interconnection of recursive (AR) lattice sections.

$$y(n) = e_0^f(n) = e_1^f(n) - K_1 e_0^b(n-1)$$

$$e_1^b(n) = e_0^b(n-1) + K_1 e_0^f(n)$$

which can be implemented by the *recursive-lattice* section shown in figure 10.9. The rest of the lattice is produced by cascading sections of the form of figure 10.10 (in front of the first section) for $P = 2, \ldots, N$, corresponding to the difference equations in (10.4.4). The input $x(n)$ to the filter is $e_N^f(n)$, as illustrated in figure 10.11. The filter is stable if, and only if, $|K_P| < 1$ for all P. Note that while the system function from the input $x(n) = e_N^f(n)$ to the output $y(n) = e_0^f(n)$ is $1/A(z)$, that from $e_0^f(n)$ to the auxillary output $\hat{y}(n) = e_N^b(n)$ is still $z^{-N}A(z^{-1})$ because the original difference equations in (10.4.3) are still satisfied.

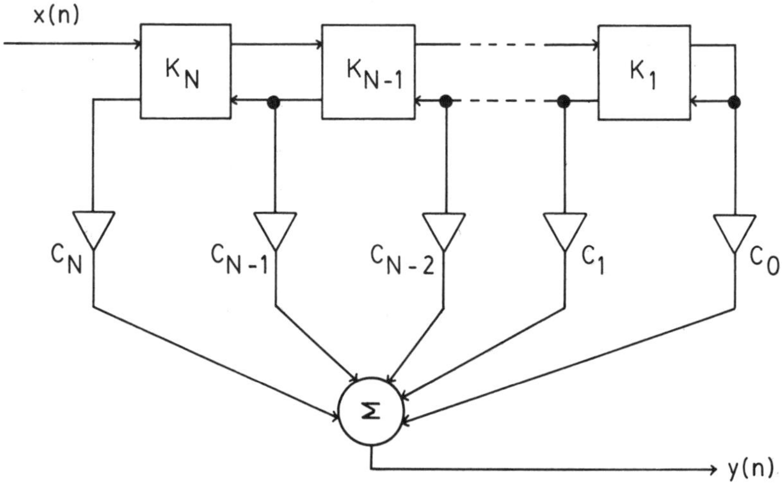

Figure 10.12. General IIR (ARMA) lattice implementation.

Therefore, the system function $\hat{H}(z)$ from $x(n)$ to the auxillary output $\hat{y}(n)$ is the cascade of these two system functions, i.e.,

$$\hat{H}(z) = \frac{z^{-N}A(z^{-1})}{A(z)}, \tag{10.4.5}$$

which is allpass. The recursive lattice thus provides an alternative implementation for allpass filters.

The ease with which filter coefficients can be constrained to ensure linear phase, minimum phase, or stability is closely related to the sensitivity of the system function to variation (quantization) of the coefficients, as discussed in the next chapter. Indeed, as thus expected, the lattice structures have excellent sensitivity properties since the only constraint required for minimum phase or stability is $|K_P| < 1$ for all P. We also note that linear phase in the FIR lattice requires simply that $|K_N| = 1$ (see problem 10.10).

The recursive lattice may be modified to implement on ARMA filter by the addition of *taps* as shown in figure 10.12 [34]. Note that the system function to the c_P tap is just

$$\hat{H}_P(z) = \frac{z^{-P}A_P(z^{-1})}{A(z)}, \tag{10.4.6}$$

where the $A_P(z)$ are the z transforms corresponding to the intermediate solutions $\underline{a}_P$ from the Levinson-Durbin algorithm. That is, the $\underline{a}_P$ are obtained by stepping down the coefficient vector $\underline{a}$. The overall system function of the tapped recursive lattice is thus

$$H(z) = \frac{1}{A(z)}\sum_{P=0}^{N} c_P z^{-P}A_P(z^{-1}) = \frac{B(z)}{A(z)}, \tag{10.4.7}$$

from which we determine that

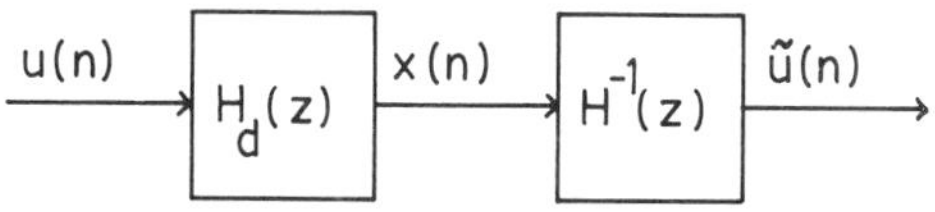

Figure 10.13. Inverse filter approach to stochastic process modeling.

$$b_k = \sum_{P=k}^{N} c_P a_{P-k,P}. \tag{10.4.8}$$

Given $\underline{b}$ and the $\underline{a}_P$, (10.4.8) may be solved for c_P, $P = N, \ldots, 1, 0$. Of course, an ARMA filter may also be realized by cascading an FIR structure with the recursive lattice, but this requires N more delays.

10.5 SPECTRUM ANALYSIS BY MODELING

The techniques of the preceding sections can be used to model stochastic signals as well as impulse responses. In so doing, we may be interested in the model itself for purposes of system identification, data reduction, or signal synthesis. On the other hand, the model may be only a means by which to estimate the signal spectrum. The advantages of this approach include smoother spectral estimates than with the periodogram and higher spectral resolution for short data records. This section is a brief introduction to this area, which is sometimes termed *modern spectrum analysis* [35, 36].

Our problem is to choose an appropriate model for the (wide-sense) stationary signal $x(n)$ and to estimate the parameters of the model, given only a finite number (K) of samples. Let the model for the signal be that white noise $u(n)$ of variance σ^2 was filtered by $H_d(z)$ with $h_d(0) = 1$ to produce $x(n)$. As before, the model of $H_d(z)$ is denoted by $H(z)$. Then, if $x(n)$ were inverse-filtered by $H^{-1}(z)$, as depicted in figure 10.13, we would recover an estimate $\tilde{u}(n)$ of the input signal consisting of $u(n)$ plus an estimation error $e(n)$. If $e(n)$ is also white and uncorrelated with $u(n)$, then $\tilde{u}(n)$ is white, as well. Therefore, to estimate the coefficients of the model $H(z)$, we will seek to whiten the output $\tilde{u}(n)$ of the inverse filter.

If an AR model $H(z) = 1/A(z)$ with $a_0 = 1$ is chosen, then the inverse filter $H^{-1}(z) = A(z)$ is FIR, and

$$x(n) * a(n) = \tilde{u}(n). \tag{10.5.1}$$

As in section 10.1, the covariance method or the autocorrelation method will be derived, depending upon how we handle the finite duration of the data $x(n)$. Assume first that the data are explicitly windowed by a rectangular window $w_R(n)$ of length K, leading to the autocorrelation estimate

$$\begin{aligned}\tilde{r}(m) &= \frac{1}{K} \sum_{n=-\infty}^{\infty} [w_R(n)x(n)][w_R(n+m)x(n+m)] \\ &= \frac{1}{K} \sum_{n=0}^{K-1-m} x(n)x(n+m) \end{aligned} \tag{10.5.2}$$

for $m = 0, 1, \ldots, L$. As shown in section 7.5, the estimate $\tilde{r}(m)$ is biased by the windowing, i.e.,

$$E[\tilde{r}(m)] = w_T(m) r(m), \tag{10.5.3}$$

where $w_T(m)$ is a triangular window of length $2K - 1$, centered about $n = 0$. By our model, the power spectrum of $x(n)$ is

$$S(z) = \sigma^2 H_d(z) H_d(z^{-1}) \approx \frac{\tilde{\sigma}^2}{A(z)A(z^{-1})}. \tag{10.5.4}$$

Thus, in the time domain using the estimated autocorrelation function,

$$\tilde{r}(n) * a(n) = \tilde{\sigma}^2 h_a(-n) + e(n), \tag{10.5.5}$$

where $H_a(z) = 1/A(z)$ and $e(n)$ is the equation error. Noting that $h_a(0) = 1$ and $h_a(-n) = 0$ for $n > 0$, we write (10.5.5) in matrix form for $n = 0, 1, \ldots, N$, forcing $e(n) = 0$ over this interval, to produce the normal equation

$$\tilde{\mathbf{R}}_N \underline{a} = \tilde{\sigma}^2 \underline{\delta}, \tag{10.5.6}$$

where $\tilde{\mathbf{R}}_N$ is the Toeplitz matrix

$$\tilde{\mathbf{R}}_N = \begin{bmatrix} \tilde{r}_0 & \cdots & \tilde{r}_N \\ \vdots & \ddots & \vdots \\ \tilde{r}_N & \cdots & \tilde{r}_0 \end{bmatrix}. \tag{10.5.7}$$

Comparing (10.5.6) with (10.1.18) and (10.5.2) with (10.1.17), we see that this is simply the autocorrelation method with $\tilde{\sigma}^2$ in place of (h_{d0}/a_0) and $\hat{r}(m)$ scaled by $1/K$. Hence, the Levinson-Durbin algorithm is applicable, and the resulting $A(z)$ is guaranteed to be minimum-phase. The implied power spectral estimate is, from (10.5.4),

$$\tilde{S}'(\omega) = \frac{\tilde{\sigma}^2}{|A'(\omega)|^2} \tag{10.5.8}$$

but because of the windowing, this is an estimate of $S'(\omega) * W_T'(\omega)$, not $S'(\omega)$ itself.

Given more than the minimum number of autocorrelation values (i.e., $L > N$), we can instead write (10.5.5) as

$$\tilde{\mathbf{R}}_L' \underline{a} = \tilde{\sigma}^2 \underline{\delta} + \underline{e}, \tag{10.5.9}$$

where $\tilde{\mathbf{R}}_L'$ consists of the first $N + 1$ columns of $\tilde{\mathbf{R}}_L$, and then minimize the equation-error norm $\underline{e}^t \underline{e}$ by

$$\underline{a} = \tilde{\sigma}^2 (\tilde{\mathbf{R}}'_L)^{\#} \underline{\delta}. \tag{10.5.10}$$

Improved results are also possible using the unbiased autocorrelation estimates

$$\tilde{r}(m) = \frac{1}{K - m} \sum_{n=0}^{K-1-m} x(n)x(n + m) \tag{10.5.11}$$

for $0 \leq m \leq N$ or L (to remove the effect of windowing), but then $A(z)$ is no longer guaranteed to be minimum phase. In modeling, this is generally unacceptable because the model $H(z)$ is not guaranteed to be stable. However, for spectrum analysis via (10.5.8), minimum phase is not required.

Although not obvious at first, whitening the estimated input signal $\tilde{u}(n)$ is closely related to minimizing its variance $\tilde{\sigma}^2$. To show this, we derive the covariance method in two different ways, each based upon one of these criteria. In section 10.1, we derived the autocorrelation method from the covariance method by approximating the ideal correlation matrix $\mathbf{R}_N$ first by the covariance matrix $\mathbf{\Phi}_N$ and then by the windowed $\hat{\mathbf{R}}_N$. Going backwards, we now replace $\tilde{\mathbf{R}}_N$ by $\tilde{\mathbf{\Phi}}_N$, where

$$\tilde{\mathbf{\Phi}}_N = \frac{1}{K - N} \mathbf{X}^t \mathbf{X} \tag{10.5.12}$$

and

$$\mathbf{X} = \begin{bmatrix} x_N & \cdots & x_0 \\ \vdots & & \vdots \\ x_{K-1} & \cdots & x_{K-N-1} \end{bmatrix}.$$

Each element $\tilde{\phi}_{ij}$ of $\tilde{\mathbf{\Phi}}_N$ is an (unwindowed) estimate of $r(i - j)$ of the form

$$\tilde{\phi}_{ij} = \frac{1}{K - N} \sum_{n=N}^{K-1} x(n - i)x(n - j). \tag{10.5.13}$$

The normal equation for the covariance method is thus, from (10.5.6),

$$\tilde{\mathbf{\Phi}}_N \underline{a} = \tilde{\sigma}^2 \underline{\delta} \tag{10.5.14}$$

Since $x(n)$ is not windowed to produce $\tilde{\mathbf{\Phi}}_N$, the resulting spectral estimate $\tilde{S}'(\omega)$ is not "smeared" by convolution with $W'_T(\omega)$. Neither, however, is $A(z)$ guaranteed to be minimum phase.

The second approach is simply to minimize the estimated input variance $\tilde{\sigma}^2$. Writing (10.5.1) in matrix form (without running off the ends of our data block), we have

$$\mathbf{X}\underline{a} = \underline{\tilde{u}}, \tag{10.5.15}$$

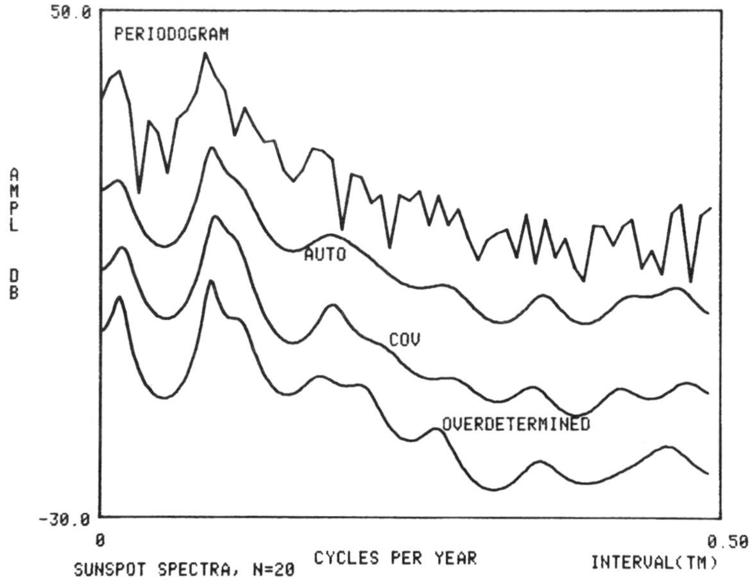

Figure 10.14. Estimated power spectra for sunspot data by periodogram, autocorrelation method, covariance method, and overdetermined normal equations.

where

$$\underline{\tilde{u}} = [\tilde{u}_N, \ldots, \tilde{u}_{K-1}]^t.$$

The obvious estimate of σ^2 is then

$$\tilde{\sigma}^2 = \frac{1}{K-N}\underline{\tilde{u}}'\underline{\tilde{u}} = \underline{a}'\tilde{\mathbf{\Phi}}_N\underline{a}. \tag{10.5.16}$$

Setting $\partial\tilde{\sigma}^2/\partial a_i = 0$, $i = 1, 2, \ldots, N$, we produce the normal equation for the covariance method in (10.5.14). (See problem 10.14.) Hence, whitening the spectrum of $\tilde{u}(n)$ and minimizing its variance are closely (but not exactly) related.

As an example of these spectral analysis methods, the power spectrum for the sequence of yearly sunspot activity levels [37] for 210 years has been estimated using the periodogram, the autocorrelation method, the covariance method, and the overdetermined equations with unbiased correlation estimates in (10.5.9) through (10.5.11). The corresponding spectral estimates are shown in figure 10.14, with $N = 20$ in the latter three cases and $L = 50$ in the last case. Note that the spectral definition (resolution) of features of potential interest seems to increase as we go from top to bottom in the figure, as expected. The largest spectral peak corresponds to the well-known 11-year sunspot cycle.

MA models may be produced from high-order AR models $\underline{a}_L(n)$ by Durbin's method, as described in section 10.2. Likewise, ARMA models may be obtained from $\tilde{r}(n)$ or $a_L(n)$ by the equation-error methods in (10.3.8) through (10.3.16).

Note that in deriving the autocorrelation method, and thus the covariance method, we minimized (set to 0) only the equation error in (10.5.5), not a direct measure of the whiteness of $u(n)$. The equation-error approach led to linear equations for $\underline{a}$, and the associated methods have had widespread and highly productive application [35–37]. However, the bias and variance of the estimated parameters can be reduced by invoking stronger error criteria at the expense of producing nonlinear equations and iterative algorithms. To conclude this chapter, we give a brief overview of several of these approaches.

A direct measure of the whiteness of the inverse filter output $\tilde{u}(n)$ is provided by $E(z)$, defined from (10.5.4) by

$$A(z^{-1})\tilde{S}(z)A(z) = \tilde{\sigma}^2 + E(z) \tag{10.5.17}$$

or, in the time domain,

$$a(-n) * \tilde{r}(n) * a(n) = \tilde{\sigma}^2\delta(n) + e(n). \tag{10.5.18}$$

Writing (10.5.18) in matrix form without running off the end of the available autocorrelation estimates, we have

$$\mathbf{A}''\tilde{\mathbf{R}}'_L\underline{a} = \tilde{\sigma}^2\underline{\delta} + \underline{e} \tag{10.5.19}$$

where

$$\mathbf{A}'' = \begin{bmatrix} a_0 & \cdots & a_N & 0 & \cdots & 0 \\ \vdots & \ddots & & \ddots & & \vdots \\ 0 & & \ddots & & \ddots & 0 \\ \vdots & & & \ddots & & \vdots \\ 0 & \cdots & 0 & a_0 & \cdots & a_N \end{bmatrix}$$

Approximate minimization of $\underline{e}'\underline{e}$ implies the iterative algorithm [38]

$$\underline{a}_{i+1} = \tilde{\sigma}^2_{i+1}(\mathbf{A}''_i\tilde{\mathbf{R}}'_L)^{\#}\underline{\delta}. \tag{10.5.20}$$

The estimated variance $\tilde{\sigma}^2_{i+1}$ is determined such that $a_{0,i+1} = 1$. An intuitive feel for (10.5.20) is provided by the realization that it closely approximates the covariance method applied to the estimated impulse response

$$\tilde{h}_{di}(n) = a_i(-n) * \tilde{r}(n)/\tilde{\sigma}^2.$$

The fitting-error approach, on the other hand, suggests minimization of a least-squares norm of the autocorrelation error

$$e(n) = \tilde{r}(n) - w_T(n)r_a(n), \tag{10.5.21}$$

where $r_a(n)$ is the autocorrelation function from the model $1/A(z)$ and specific account is taken of the bias in $\tilde{r}(n)$ due to windowing. The model autocorrelation $r_a(n)$ can be determined from $\underline{a}$ by stepping $\underline{a}$ down to $\{K_P\}$ and then solving for $\{r_a(P)\}$ from $\{K_P\}$ via the Levinson-Durbin algorithm (noting that $K_P = 0$ for $P > N$). Rewriting (10.5.21) as

$$\underline{\tilde{r}} = \mathbf{W}_T \underline{r}_a + \underline{e}, \tag{10.5.22}$$

where $\mathbf{W}_T$ is a diagonal matrix with diagonal elements $w_T(n)$, we need an operator $\mathcal{O}(\underline{\tilde{r}})$ such that

$$\mathcal{O}(\underline{\tilde{r}})\underline{a} = \underline{\tilde{r}} \tag{10.5.23}$$

in order to make the minimization of $\underline{e}^t\underline{e}$ an explicit function of $\underline{a}$. The Steiglitz-McBride algorithm provides such an operator of the form

$$\mathbf{H}_a \tilde{\mathbf{R}}_0' \underline{a} = \underline{\tilde{r}} \tag{10.5.24}$$

where

$$\tilde{\mathbf{R}}_0' = \begin{bmatrix} \tilde{r}_0 & 0 & \cdots & 0 \\ \vdots & \ddots & & \vdots \\ \tilde{r}_N & & & \tilde{r}_0 \\ \vdots & & & \vdots \\ \tilde{r}_L & \cdots & & \tilde{r}_{L-N} \end{bmatrix}, \tag{10.5.25}$$

corresponding to the prefiltering of the causal sequence $\tilde{r}(n)$, $n = 0, 1, \ldots, L$, by $h_a(n)$.

Therefore, combining (10.5.22) and (10.5.24), we have

$$\mathbf{H}_a \tilde{\mathbf{R}}_0' \underline{a} = \mathbf{W}_T \underline{r}_a + \underline{e}, \tag{10.5.26}$$

which implies the iterative algorithm

$$\underline{a}_{i+1} = (\mathbf{H}_{ai} \tilde{\mathbf{R}}_0')^{\#} \mathbf{W}_T \underline{r}_{ai}. \tag{10.5.27}$$

PROBLEMS

10.1 Identify the elements of the covariance matrix $\mathbf{\Phi}_N = \mathbf{H}_d^t \mathbf{H}_d$ in terms of the quantities

$$\phi_{ij} = \sum_{n=0}^{L} h_d(n-i)h_d(n-j).$$

Determine a simple recursive relationship to compute $\phi_{i-1,j-1}$ from ϕ_{ij}.

10.2 Given $h_d(0) = 1$, $h_d(1) = 0.9$, and $h_d(2) = 0.8$, find first-order AR models $H(z)$ for this data via the covariance and autocorrelation methods, and compare the results by finding the model response $h(n)$, $n = 0, 1, 2$, in each case.

10.3 Repeat problem 10.2 using the Steiglitz-McBride algorithm with an initial vector $\underline{a}_0 = [1, -0.8]^t$.

10.4 Given $\hat{r}(n) = \varrho^{|n|}$, find an AR model of appropriate order via the Levinson-Durbin algorithm.

10.5 Given samples of the unstable filter response $h_d(n) = 2^n u(n)$ for $n = 0, 1, 2, 3$, find first-order AR models $H(z)$ via the covariance and autocorrelation methods, and compare the results with the true filter $H_d(z)$.

10.6 Convert the coefficient vector $\underline{a} = [1, -2r\cos\theta, r^2]^t$ to a reflection coefficient sequence $\{K_P\}$ via the step-down algorithm. Draw the corresponding FIR lattice filter.

10.7 Test the stability of the filter

$$H(z) = 1/(1 - 1.6z^{-1} - 0.2z^{-2} + 0.5z^{-3})$$

via the step-down algorithm.

10.8 Show that an FIR filter $A_N(z)$ having $|K_N| > 1$, but $|K_P| < 1$, $P < N$, is *maximum-phase*, i.e., all its zeros lie outside the unit circle. (*Hint*: Consider a filter with $K'_N = 1/K_N$.)

10.9 A second-order FIR notch filter $H_d(z)$ having two zeros on the unit circle is modeled by a high-order AR filter $H_a(z) = 1/A_L(z)$.
   a. Using the fact that $A_L(z) \approx 1/H_d(z)$, sketch $a_L(n)$.
   b. Discuss the relative advantages of the covariance and autocorrelation methods in estimating the second-order MA model $h(n) = b(n)$ from $a_L(n)$.

10.10 Show that the condition $|K_N| = 1$ implies that the corresponding FIR lattice filter has either linear phase or linear-plus-90° phase. (We note that if $|K_P| < 1$, $P < N$, and $|K_N| = 1$, all the filter zeros lie exactly on the unit circle, but not otherwise.)

10.11 Derive and draw the recursive lattice implementation of the discrete-time oscillator $H(z) = 1/(1 - 2\cos\omega_0 Tz^{-1} + z^{-2})$.

10.12 Draw the recursive lattice implementation of the allpass filter

$$H(z) = \frac{0.7 + 0.17z^{-1} - 0.34z^{-2} + z^{-3}}{1 - 0.34z^{-1} + 0.17z^{-2} + 0.7z^{-3}}$$

10.13 Given the desired response $h_d(n) = 0.5, 1.0, 0.9, 0.8, 0.7$ for $n = 0, 1, 2, 3, 4$, respectively, find a first-order ARMA model $H(z)$ for this data, and compare $h(n)$ with $h_d(n)$.

10.14 Derive the normal equation (10.5.14) for the covariance method in the case of stationary data by minimizing $\tilde{\sigma}^2$ in (10.5.16), i.e., set

$$\frac{\partial \tilde{\sigma}^2}{\partial a_i} = 0, \qquad i = 1, 2, \ldots, N.$$

10.15 Given the unbiased autocorrelation estimate $\tilde{r}(n) = \cos \omega_0 nT$ for a stationary process, find an AR model $H(z)$ and the input power $\tilde{\sigma}^2$ via the Yule-Walker equation in (10.5.6) and the Levinson-Durbin algorithm. Interpret your results.

10.16 Given $H(z) = 1/A(z)$ and $\tilde{\sigma}^2$, it is not always simple computationally to find the corresponding autocorrelation function $\tilde{r}(n)$ as the inverse transform of $\tilde{S}(z) = \tilde{\sigma}^2 H(z) H(z^{-1})$. An alternate approach is to solve the Yule-Walker equation in (10.5.6) "backwards" for $\tilde{r}(n)$ given $\underline{a}$. Use this approach to determine $\tilde{r}(n)$ for $n = 0, 1, 2$, given that $\underline{a} = [1, -0.95, 0.9]'$ and $\tilde{\sigma}^2 = 1$. Then extrapolate $\tilde{r}(n)$ for $n = 3, 4, 5$ using the fact that ideally $\tilde{r}(n) * a(n) = 0$ for $n > 0$.

11. QUANTIZATION EFFECTS

11.0 INTRODUCTION

Up to now, we have really been studying *discrete-time* filters and signal processing, not *digital* filters and signal processing, because the signals were discrete in time but not in amplitude. Such discrete-time systems can be realized by, for example, switched-capacitor circuits where the amplitudes are represented by voltage or charge and are thus continuously variable. However, the vast majority of discrete-time systems are realized by digital computers or hardware, and the amplitudes of both signals and coefficients are also discrete variables in these systems. Strictly speaking, such systems are nonlinear, but fortunately, as we shall see, these effects can usually be modeled as linear perturbations or additive quantization noise.

There is an initial temptation to lump all quantization effects together into a single error analysis, but this is not a useful approach. After all, the filter coefficients are only quantized once—in the design process—and those values remain constant in the filter implementation. Hence, the effect of coefficient quantization is to perturb $h(n)$, $H(z)$ and $H'(\omega)$ from their ideal (designed) forms in a deterministic manner, and the system is still linear. We are free to check the quantized design. If it no longer meets the specifications, we can optimize it, redesign it, restructure it, and/or allocate more bits to satisfy the specifications. As we will see, the structure of the digital filter network has a dramatic effect on its sensitivity to coefficient quantization.

Signal quantization, on the other hand, due to the rounding or truncation of products within the filter, occurs in the operation of the filter and is usually best viewed as a random process. In most cases, each quantization operation can be modeled as producing white noise that is uncorrelated from other quantization noise sources in the filter; this greatly simplifies the analysis and/or synthesis of the filter. An important exception to this approach is the zero or constant input case where autonomous *limit cycle* oscillations may be produced. Even for this highly nonlinear phenomenon, linear analysis can be quite helpful, as we will show. Again, as for coefficient quantization, the network structure is the key design choice affecting signal-quantization noise levels and spectra.

It is especially important to consider quantization effects when designing filters for implementation in fixed-point arithmetic, such as with microprocessors or special-purpose digital hardware. Larger computers using floating-point arithmetic of 32 bits or more are usually sufficiently accurate that these effects can be safely neglected. Even there, the structural considerations for coefficient quantization can be significant for narrowband filters. Signal quantization, on the other hand, is intimately related to scaling and dynamic range considerations, and the self-scaling of floating-point arithmetic largely eliminates quantization noise problems. Therefore, we will restrict our study of quantization effects to the fixed-point case.

11.1 COEFFICIENT QUANTIZATION

Much insight can be gained into the problem of coefficient quantization by studying its effect on pole/zero locations in the z plane. Later, we will define formal coefficient sensitivity measures and derive bounds on and from them. Consider first the simple case of a first-order IIR filter $H(z) = 1/(1 - az^{-1})$. The coefficient a equals the pole location, and thus quantization of a directly defines the possible pole locations along the real axis. Fixed-point implementation with b fractional bits for a implies that we can implement any of the discrete values $a = L2^{-b}$ for integer L. Hence, the possible pole locations are uniformly spaced by 2^{-b} along the real axis in the z plane. This is a straightforward situation, and we must simply allocate enough bits (b) to satisfy our design criteria.

The situation becomes more interesting and potentially troublesome for second-order IIR filters. Let $H(z) = 1/(1 + a_1 z^{-1} + a_2 z^{-2})$ and recall that for complex poles p and p^*, $a_1 = -2\,\mathrm{Re}\,(p)$ and $a_2 = |p|^2$. Hence, quantization of a_1 quantizes the real part of the poles, as in the first-order case, but quantization of a_2 quantizes the radius of the poles with nonuniform spacing. This is illustrated in figure 11.1, where the possible pole locations for a stable filter are given by the intersection of vertical lines and concentric circles (for $b = 3$). That is, $a_1 = L_1 2^{-b}$ defines uniformly spaced vertical lines and $a_2 = L_2 2^{-b}$ defines circles with radii $r = \sqrt{a_2}$, at whose intersections the poles may be placed. Note the sparseness of possible pole locations in the vicinity of $z = \pm 1$ compared with $z = \pm j$. Hence, for a given value of b, we can place poles near $z = \pm 1$ with much less accuracy than elsewhere in the z plane.

Since second-order filters are the building blocks of the popular cascade and parallel forms, this simple analysis leads to some very important conclusions. First, narrowband lowpass (or highpass) filters will be the most sensitive to coefficient quantization because their poles fall near $z = 1$ (or $z = -1$). Hence, we can expect such filters to require high coefficient accuracy, or alternatively, we can look for other structures that reduce this sensitivity near $z = \pm 1$. Second, oversampling by substantially more than the minimum Nyquist rate leads to increased coefficient sensitivity and is thus to be avoided, in general,

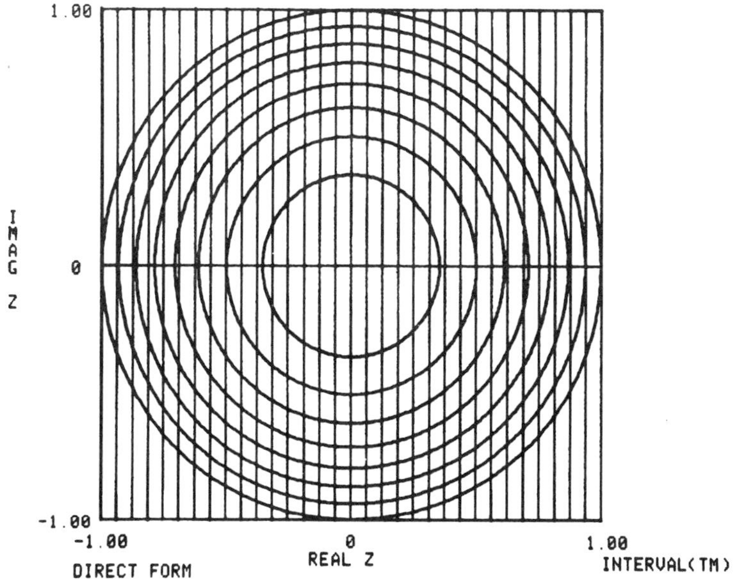

Figure 11.1. Possible pole locations for second-order direct form with $b = 3$ fractional bits.

because the poles are pushed closer to $z = 1$. This is counterintuitive at first because, as the sampling period T is decreased, we should be able to approximate the underlying continuous-time filter more closely. However, although aliasing is indeed reduced by oversampling, the sensitivity of the filter coefficients is increased and, of course, so is the required speed of the hardware, and these are generally the more important effects.

The analysis becomes more complicated as we increase the order of a filter in direct form beyond second order, but general conclusions can be anticipated. We have already seen that the coefficient sensitivity increased as we went from first to second order, and it is reasonable to expect this to get worse as we go above second order. It is indeed a well-known result in numerical analysis that the sensitivity of the roots of a polynomial to the accuracy of its coefficients increases with the order of the polynomial. In particular, although it is a simple matter to ensure that the quantized coefficients a_1 and a_2 for a second-order filter remain within the stability triangle (figure 5.10), it is not straightforward to ensure the stability of a direct-form filter of more than second order after coefficient quantization. (It is, however, easy to check the stability of the filter by the step-down procedure of chapter 10.) Even if the filter does remain stable, the poles can be displaced significantly from their design locations by coefficient quantization, and the specifications can be easily violated. Therefore, for IIR filters implemented with fixed-point arithmetic, the direct form is generally avoided, and the cascade or parallel forms are preferred because they are comprised of less sensitive first- and second-order sections.

Having studied the poles of an IIR filter, we next consider the effect of coefficient quantization on the filter zeros. Consider the case of an FIR filter in direct form, where the magnitude response is produced entirely by zeros. Such filters would seem to have the same severe sensitivity problem as IIR filters in direct form because $H(z)$ is a single

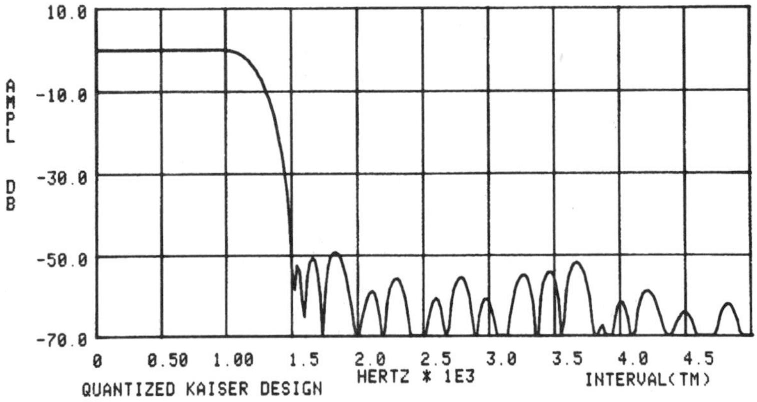

Figure 11.2. FIR lowpass design via Kaiser window from figure 9.8 with coefficient quantization to $b = 9$ bits.

higher-order polynomial. However, there is a very important difference: namely, most FIR designs are for linear-phase filters (with possibly an additional 90° phase shift), and the coefficients are thus constrained to satisfy the symmetry condition

$$b_m = \pm b_{M-m} \tag{11.1.1}$$

as shown in section 5.3. Hence, quantizing both sides of (11.1.1) to the same perturbed value for each m implies that the filter is still linear phase after quantization (plus 90° with the minus sign), and it is only the magnitude response that is perturbed. In particular, the zeros lying on the unit circle must remain on the unit circle unless they move enough to come together and split into reciprocal pairs. This constraint greatly reduces the sensitivity of most direct-form FIR filters and, unlike the IIR case, the direct form is widely employed. The parallel form is, of course, not applicable to FIR filters, and the cascade form, though available, is not widely used. (Note that for zeros not located on the unit circle, we would require a minimum order of four for the cascade-form sections to satisfy the symmetry condition in (11.1.1) in order to have the zeros remain in both conjugate and reciprocal pairs.)

As an example, consider the FIR lowpass filter designed with the Kaiser window and shown in figure 9.8. After quantization of the coefficients to $b = 9$ bits, the perturbed design is still useful, as seen in figure 11.2, and the minimum attenuation has dropped by only 2 dB to a level of 49 dB. Note, in particular, that the *zeros of transmission* in the stopband of $H'(\omega)$—due to the zeros of $H(z)$ on the unit circle—have been retained although their locations have been perturbed.

We now turn to the effect of coefficient quantization on the zeros of IIR filters. For the classical designs, all of which have zeros only on the unit circle, the numerator polynomial of the direct form will be linear phase and will thus have the same low sensitivity as a linear-phase FIR filter. However, we have already rejected direct-form IIR filters, in general, because of the sensitivity of the poles. Hence, we will consider only the parallel- and cascade-form cases.

As noted above, stopband attenuation is most dependent upon having zeros of transmission in $H'(\omega)$ (zeros on the unit circle). However, the zeros of the parallel form are not realized by individual first- or second-order terms, but by the way all of the parallel section transforms add together (see equation 5.2.4). That is, to determine the zeros, we must generate a common denominator for all of the terms in $H(z)$ and add them together to produce the higher-order numerator polynomial whose roots are the zeros. Hence, every coefficient α_{ji} or γ_{ji} in the parallel form affects every zero! This is a highly sensitive situation, and the zeros are in no way constrained to lie on the unit circle after quantization. Therefore, although the parallel form can be quite adequate for filters without stringent attenuation requirements, it should be avoided (or at least checked after coefficient quantization) for filters with demanding specifications.

It is clear from the above argument that the cascade form is more robust under coefficient quantization than the parallel form. Its poles, of course, have the same low sensitivity as the parallel form, but its zeros are also the roots of only first- or second-order polynomials and are thus less sensitive than the roots of higher-order polynomials. In addition, in the common case of zeros on the unit circle, the coefficients β_{2i} in the numerator terms $(\beta_{2i}z^{-2} + \beta_{1i}z^{-1} + 1)$ equal ± 1, as discussed in section 5.3, and hence are not changed by quantization. Therefore, although these zeros move along the unit circle with quantization of β_{1i}, they do not move off the unit circle (unless the zeros become real), and the stopband attenuation specifications are more readily satisfied.

As an example, consider the 6th-order bandstop filter design in figure 11.3(a), having a specified minimum attenuation over the stopband of 25 dB. Quantization of the cascade- and parallel-form coefficients to 6 fractional bits produces the perturbed magnitude responses shown in figure 11.3(b) and (c), respectively. Note that although the stopband ripples are no longer of equal amplitude in the cascade-form case, the minimum attenuation is still 24 dB; while the zeros of transmission are lost with the parallel form, and the minimum attenuation over the original stopband drops to only about 16 dB.

There remains the question of reducing the sensitivity of the poles of the cascade and parallel forms in the vicinity of $z = \pm 1$. This tends to be a problem only for very narrowband lowpass (or highpass) filters, as we have discussed. The problem, we saw, was that the poles fell at the intersections of vertical lines and concentric circles in the z plane, which intersect at very oblique angles near $z = \pm 1$. Consider instead the second-order state-space structure with an **A** matrix of the normal (or "coupled") form

$$\mathbf{A} = \begin{pmatrix} a_1 & a_2 \\ -a_2 & a_1 \end{pmatrix}. \tag{11.1.2}$$

This structure is depicted in figure 11.4. As shown in section 4.4, the poles of a filter incorporating this recursive structure are at $a_1 \pm ja_2$. Hence, by quantizing the coefficients a_1 and a_2, we directly quantize the real and imaginary parts of the poles, and the possible pole locations form a uniform grid of vertical and horizontal lines, depicted in figure 11.5 for $b = 3$. We observe that the density of possible pole locations near $z = \pm 1$ is now the

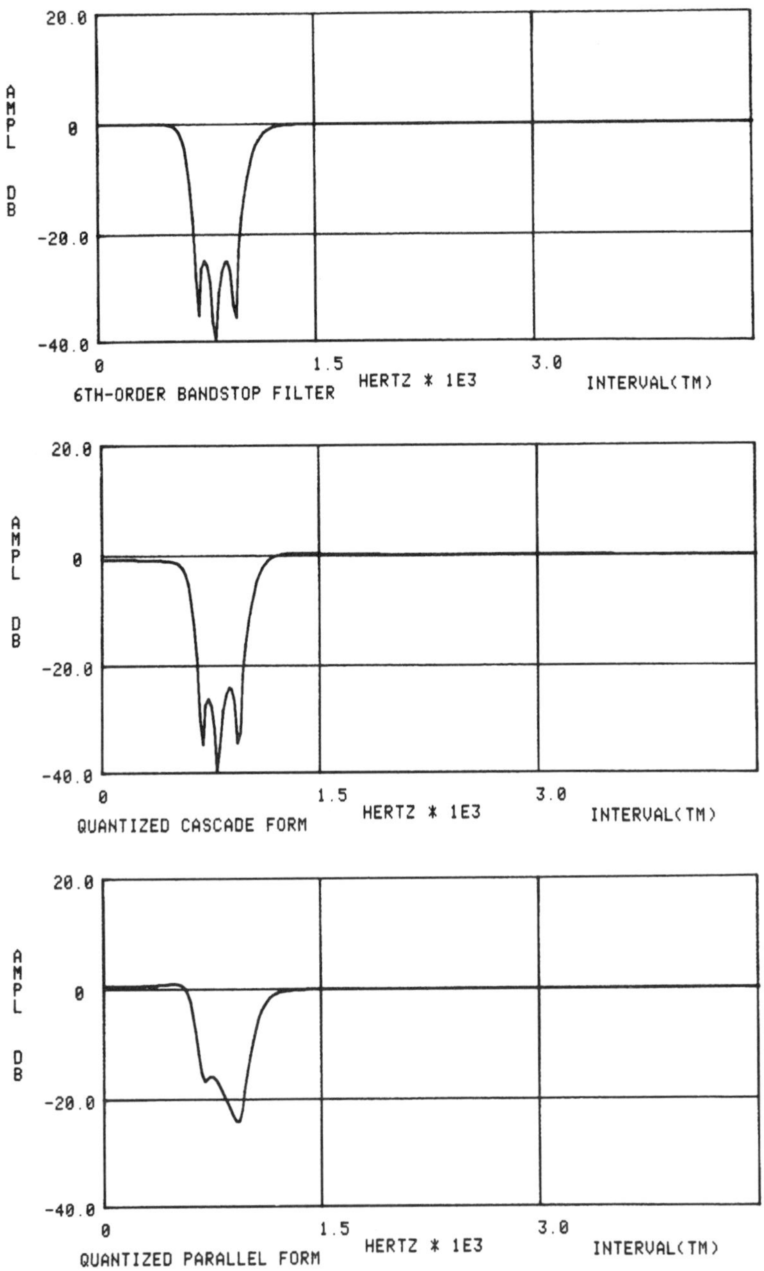

Figure 11.3. Sixth-order BSF with (a) no quantization, and quantization of coefficients in (b) cascade form and (c) parallel form to $b = 6$ bits.

same as everywhere else in the z plane, and narrowband lowpass filters are no more difficult to implement than comparable narrowband bandpass filters. The normal-form structure must, of course, have $\underline{b}$, $\underline{c}^t$, and d coefficients added to realize the zeros, as described in sections 4.4 and 11.6, but the poles are determined only by $\mathbf{A}$.

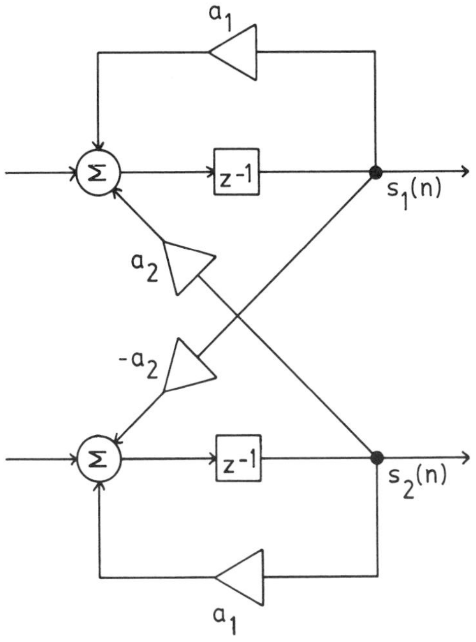

Figure 11.4. Recursive normal-form structure.

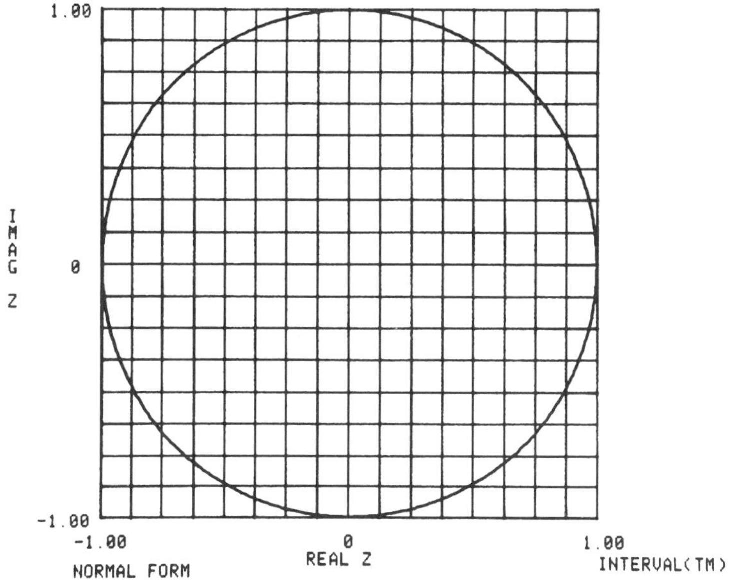

Figure 11.5. Possible pole locations for second-order normal form with $b = 3$ fractional bits.

11.2 SIGNAL QUANTIZATION

As discussed in the Introduction to this chapter, the effect of signal quantization is to add an error or noise signal $e(n)$ to the ideal output $y(n)$ from the digital filter. This noise is a composite of the errors from one or more of the following sources, as applicable:

1. The quantization error of the analog-to-digital converter at the filter input.
2. The accumulated errors resulting from the rounding or truncation of multiplication products within the filter. We will refer to this noise by the generic term *roundoff noise*, whether rounding or truncation is actually employed.
3. The quantization of the output $y(n)$ to fewer bits for input to a digital-to-analog converter or another system.

Source (3) is sometimes overlooked, but because of the accumulation of the noise in (2), we must usually allocate more bits to the internal arithmetic in the filter than are required at the output. Hence, we usually requantize the output to fewer bits. It is often reasonable for the input and output quantizations to employ the same number of bits, in which case their noise levels (powers) are the same. The input noise is, of course, filtered by the digital filter along with the signal, but for a broadband filter with unity gain, these two noise components will be comparable and will add (powerwise) to increase the noise level by about 3 dB over the level of either source alone.

Considerations such as the above for sources (1) and (3) are usually straightforward (or given in the specifications). We will therefore restrict our attenuation to the much more difficult analysis and synthesis problem of roundoff noise. This analysis requires some familiarity with the statistics of random processes. Let us first investigate the common types of product rounding and truncation and their associated errors. If we assume b fractional bits after quantization, the step size between quantization levels is $\Delta = 2^{-b}$. Actually, it is often most convenient to assume that the data is quantized to integer values (since there is no actual binary point in the data, but only an assumed one), in which case $b = 0$ and $\Delta = 1$.

The most common type of product quantization is rounding to the nearest quantization level. We then have $-\Delta/2 \leqslant e < \Delta/2$, and it is reasonable to assume that the error $e(n)$ may take on any value within this interval with equal probability. That is, the probability density $f(e)$ for the random variable e is uniform, as depicted in figure 11.6(a). The mean or average error (dc value) is thus zero, and the variance or average power of the error is

$$\sigma_0^2 = \Delta^2/12. \tag{11.2.1}$$

If, on the other hand, we simply truncate the product to b fractional bits, the resulting error depends on the type of binary arithmetic used. With one's- or two's-complement arithmetic, the product value is reduced to the next lowest quantization level, and the error is uniformly distributed over the interval $-\Delta < e \leqslant 0$, as depicted in figure 11.6(b). The mean or dc value is now $-\Delta/2$, but the variance is still given by (11.2.1). This bias or dc offset propagates through the filter (with gain) from all truncation operations to the output to produce a composite bias in the output that is easily calculated and often negligible. Separating the dc analysis from the remaining noise analysis, this truncation case becomes equivalent to the rounding case, and we need not distinguish them further.

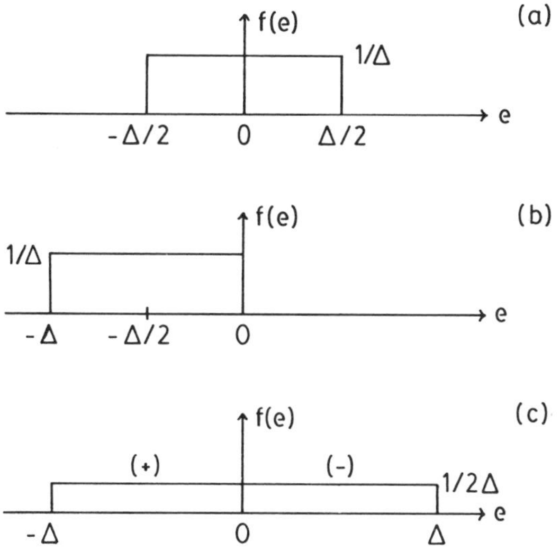

Figure 11.6. Error probability density functions for (a) rounding, (b) two's-complement truncation, and (c) magnitude truncation.

Finally, with sign-and-magnitude arithmetic, we truncate the magnitude only, and thus the error is correlated with the signal. Specifically, if the signal is positive, the error is negative, and vice versa, as indicated in figure 11.6(c). Hence, assuming that the signal is equally likely to be + or −, the overall error is uniformly distributed over $-\Delta < e < \Delta$ with zero mean, but the variance is now $4\sigma_0^2$. The increased error variance (average power) and correlation with the signal make this latter case the least attractive of the arithmetic alternatives although, as we will see later, it can eliminate the possibility of limit cycles.

To proceed with the roundoff noise analysis, we must define some transfer functions within the filter. Note that the coefficient multipliers always feed into summation nodes (possibly through delays z^{-1}), and hence we can think of the roundoff noise sources associated with the multipliers as also being inputs to the summation nodes. Let the transfer function from the error input $e_j(n)$ at the jth summation node in the filter to the output be $G_j(z)$, as depicted in figure 11.7. We will, in general, have to scale the signals within the filter to prevent overflow of the available dynamic range, and this, as we will see, changes the output noise level. After such scaling, we will denote the scaled transfer functions by $G_j^*(z)$, as in the figure.

Key simplifying assumptions are now made. First, if the signals to be rounded have sufficient amplitude and spectral content, it is reasonable to assume that the rounding error made at one sample time is uncorrelated with that made at any other time, and hence that the roundoff noise injected at that node is white. Similarly, we assume that the signals at different nodes in the filter are sufficiently different that the roundoff noise introduced at one node is uncorrelated with that injected at all other nodes within the filter. Hence, superposition holds for these noise components, and noise power spectra can be simply added. These assumptions are supported by experimental evidence.

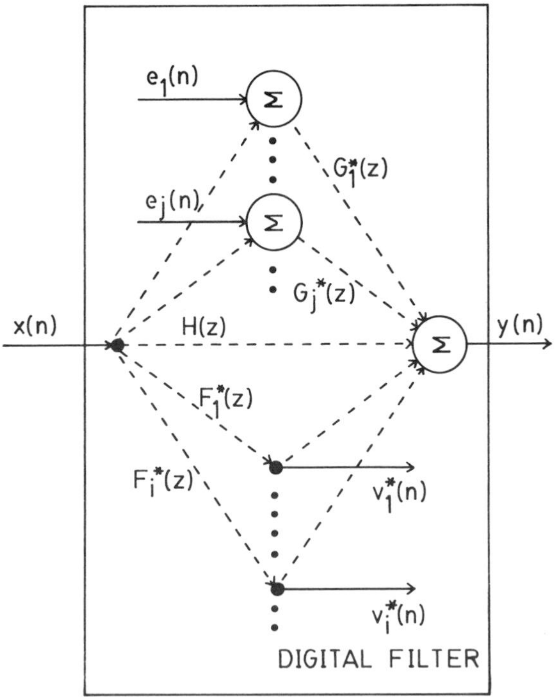

Figure 11.7. Illustration of summation and branch nodes with associated transfer functions.

To be specific, let the number of rounding operations (multipliers) directly inputting to the jth summation node be k_j. (If rounding is performed after the summation, not before, then $k_j = 1$.) The variance of the noise from each rounding operation is σ_0^2, and under the white-noise assumption, the power spectrum from each is also $N(\omega) = \sigma_0^2$. The superposition of the k_j sources inputting to the jth node thus produces a power spectrum for $e_j(n)$ of simply $N_j(\omega) = k_j\sigma_0^2$. The contribution of each source $e_j(n)$ to the output noise spectrum after scaling is then $k_j\sigma_0^2|G_j^{*\prime}(\omega)|^2$, and the spectrum of the total output roundoff noise $e(n)$ is simply the superposition of these noise contributions, i.e.,

$$N_e(\omega) = \sigma_0^2 \sum_j k_j|G_j^{*\prime}(\omega)|^2. \tag{11.2.2}$$

The variance or total average power of $e(n)$ is then

$$\sigma_e^2 = \frac{1}{\omega_s}\int_0^{\omega_s} N_e(\omega)\, d\omega. \tag{11.2.3}$$

Parseval's Theorem (see section 3.4) can be used to produce an alternate expression for σ_e^2 in the time domain. Let the impulse response (inverse transform) corresponding to $G_j^*(z)$ be $g_j^*(n)$. Substituting (11.2.2) into (11.2.3) and invoking Parseval's Theorem, we obtain

$$\sigma_e^2 = \sigma_0^2 \sum_j k_j \sum_{n=0}^{\infty} |g_j^*(n)|^2. \tag{11.2.4}$$

By setting $e_j(n) = \delta(n)$ and $x(n) = e_i(n) = 0, i \neq j$, in an implementation or simulation of the filter, we can produce $g_j(n)$ for each summation node (j) and from these estimate σ_e^2 by a finite sum of the form of (11.2.4).

11.3 DYNAMIC RANGE AND SCALING

Given the complete filter design, including the structure and scaling, the estimation of the power spectrum and variance of the output roundoff noise is relatively straightforward, as we have shown. However, we are not usually given the scaling, and often the filter structure is also not specified. Our goal, in all likelihood, is to implement the filter with maximum signal-to-noise ratio (SNR) at its output, and we must select and scale the structure to accomplish this goal. In this and following sections, we will develop appropriate scaling strategies and then compare the roundoff noise outputs from various scaled structures [39, 40]. We will also show that roundoff noise and coefficient sensitivity are, in fact, related in the sense that low noise structures have low sensitivity and, if synthesized properly, vice versa [41].

Clearly, to increase the SNR throughout the filter, the simplest thing to do is to increase the signal levels at all nodes since the roundoff noise level is fixed for a given structure and step size Δ. However, if we increase the signal levels too much, the available dynamic range of the fixed-point arithmetic will be exceeded, and the results of some calculations will *overflow*. Overflow represents a severe nonlinearity in the system, and we must either preclude its possibility or else make its probability acceptably small and minimize the resulting distortion. The latter is achieved by effecting saturation or clipping at $\pm$ full scale, as discussed in section 11.5.

It would seem, at first, that we must ensure that the signal does not overflow at any node within the network and thus that we must analyze and scale the signal level at every single node. Fortunately, however, this is not true for two reasons:

1. Some nodes are separated only by delays, and thus if the signal at the first node does not overflow, it can't at any of the subsequent nodes.
2. In the summation of more than two numbers, if the magnitude of the correct total sum is small enough not to overflow, then in arithmetic systems such as one's and two's complement, the correct total sum will be obtained regardless of the order in which the numbers are added [3]. This holds no matter how many overflows may occur in the partial sums or even if some of the inputs to the summation node have themselves overflowed as a result of being multiplied by coefficients of magnitude greater than unity. This is an extremely important property, and it may be desirable to defeat the normal overflow detection circuitry in the hardware to take advantage of it.

We must, therefore, specify certain nodes in the network where dynamic range constraints are required. These nodes are *branch nodes* having one or more outputs that feed into multipliers (since, if a node only inputs to a summation node, it represents only a partial sum). Let the signals at these branch nodes be $v_i(n)$ and the transfer functions from the input to these nodes be $F_i(z)$, or $F_i^*(z)$ after scaling, as illustrated in figure 11.7. Assume that the available dynamic range (full-scale value) for the input $x(n)$ and all $v_i(n)$ is M, i.e.,

$$|x(n)| \leqslant M. \tag{11.3.1}$$

Assuming zero initial conditions in the filter and neglecting the small error inputs $e_j(n)$, the branch node signals are given by the simple convolutions

$$v_i(n) = \sum_{k=0}^{\infty} f_i(k)x(n-k), \qquad \text{all } n. \tag{11.3.2}$$

Therefore, the $v_i(n)$ may be bounded in magnitude from (11.3.1) and (11.3.2), as follows:

$$\begin{aligned} |v_i(n)| &\leqslant \sum_{k=0}^{\infty} |f_i(k)| \cdot |x(n-k)|, \qquad \text{all } n \\ &\leqslant M \sum_{k=0}^{\infty} |f_i(k)|, \qquad \text{all } n. \end{aligned} \tag{11.3.3}$$

Thus, if the scaled $v_i^*(n)$ are also to satisfy the dynamic range constraints

$$|v_i^*(n)| \leqslant M, \tag{11.3.4}$$

we must scale the network to satisfy

$$\sum_{k=0}^{\infty} |f_i^*(k)| \leqslant 1. \tag{11.3.5}$$

This condition is not only sufficient, but also necessary, to preclude overflow (see problem 11.1).

The constraint on $f_i^*(n)$ in (11.3.5) is overly pessimistic for most signals of practical interest. Hence, using this constraint, we will be overly conservative in scaling the network and will sacrifice available dynamic range and SNR. More realistic constraints can be derived in the frequency domain, and although they do not absolutely preclude overflow as does (11.3.5), they can make it acceptably improbable while using the available dynamic range more efficiently.

To derive these constraints, we utilize the L_p norms of Fourier transforms, defined for an arbitrary $A'(\omega)$ by

$$\|A\|_p = \left[\frac{1}{\omega_s}\int_0^{\omega_s} |A'(\omega)|^p \, d\omega\right]^{1/p} \tag{11.3.6}$$

for each real $p \geqslant 1$ such that the integral converges. Note that the L_1 norm is the mean absolute value of $A'(\omega)$ over ω; while the L_2 norm is the root-mean-squared (rms) value.

For continuous transforms $A'(\omega)$, the limit of $\|A\|_p$ exists as $p \to \infty$ and is simply the peak absolute value of $A'(\omega)$, i.e.,

$$\|A\|_\infty = \max_{0 \leqslant \omega \leqslant \omega_s} |A'(\omega)|. \tag{11.3.7}$$

It can also be shown that

$$|a(n)| \leqslant \|A\|_1 \leqslant \|A\|_2 \leqslant \ldots \leqslant \|A\|_\infty. \tag{11.3.8}$$

Assume for now that the input $x(n)$ is deterministic and that its transform $X'(\omega)$ satisfies the (mild) L_1 condition

$$\|X\|_1 \leqslant M, \tag{11.3.9}$$

which from (11.3.8) ensures that $|x(n)| \leqslant M$. Since $V_i(z) = F_i(z)X(z)$, a frequency-domain expression (inverse transform) for $v_i(n)$ is

$$v_i(n) = \frac{1}{\omega_s} \int_0^{\omega_s} F_i'(\omega)\, X'(\omega)\, e^{jn\omega T}\, d\omega. \tag{11.3.10}$$

Therefore,

$$\begin{aligned} |v_i(n)| &\leqslant \frac{1}{\omega_s} \int_0^{\omega_s} |F_i'(\omega)| \cdot |X'(\omega)|\, d\omega \\ &\leqslant \|F_i\|_\infty \frac{1}{\omega_s} \int_0^{\omega_s} |X'(\omega)|\, d\omega \end{aligned} \tag{11.3.11}$$

or

$$|v_i(n)| \leqslant \|F_i\|_\infty \cdot \|X\|_1. \tag{13.3.12}$$

The desired dynamic range constraint (11.3.4) after scaling is thus provided by (11.3.9) and (11.3.12) if

$$\|F_i^*\|_\infty \leqslant 1. \tag{11.3.13}$$

In words, for inputs satisfying (11.3.9), overflows cannot occur if the peak gains to all relevant branch nodes are scaled so as not to exceed unity.

Clearly, this constraint is appropriate for sinusoidal inputs since the amplitude of the resulting sinusoid $v_i(n)$ at each branch node is determined by the gain $F_i'(\omega)$ at the input frequency ω, and the gain must be everywhere less than or equal to one if ω is not specified. But (11.3.9) is, in fact, the only condition of the form $\|X\|_q \leqslant M$ that a sinusoidal $x(n)$ can satisfy because $X'(\omega)$ then consists of two impulses in frequency and $\|X\|_q$ does not converge for $q > 1$ (see problem 11.2). Although (11.3.13) does not absolutely preclude overflows as did (11.3.5), it is more realistic and less conservative. It is, however, the most conservative of the frequency-domain constraints because the input power is assumed to be concentrated at a single frequency. Hence, overflows are very unlikely indeed if the network is scaled to satisfy (11.3.13).

Even this constraint is overly conservative for many broadband input signals. It can be shown [39] from Holder's inequality that

$$|v_i(n)| \leqslant \|F_i\|_p \cdot \|X\|_q \tag{11.3.14}$$

for all $p, q > 1$ satisfying $1/p + 1/q = 1$. Note that (11.3.12) is indeed of this form for $p = \infty$ and $q = 1$. However, of the many possible constraints provided by (11.3.14), the only other one besides (11.3.12) of significant practical interest is the case where $p = q = 2$. In that case, the deterministic input $x(n)$ is assumed to have finite energy $E = \|X\|_2^2$ (see Parseval's Theorem), and hence with $\|X\|_2 = \sqrt{E} \leqslant M$, overflow is avoided if

$$\|F_i^*\|_2 \leqslant 1. \tag{11.3.15}$$

In words, the rms value of $F^{*\prime}(\omega)$ over ω must be bounded by unity. This constraint can also be derived from the Schwarz inequality (see problem 11.3) or by considering $x(n)$ to be a random process [39]. In the latter case, the variance or average power of $v_i(n)$ will not exceed that of the input $x(n)$ if (11.3.15) is satisfied. If $x(n)$ is Gaussian, this means that the probability of overflow for $v_i(n)$ is no greater than that for $x(n)$, which is assumed to be acceptably small. Hence, (11.3.15) is another widely applicable scaling condition.

There remains only the question of how to satisfy either (11.3.5), (11.3.13), or (11.3.15). Let the cumulative scaling applied to $v_i(n)$ be c_i, i.e.,

$$F_i^*(z) = c_i F_i(z). \tag{11.3.16}$$

Then, since the frequency-domain constraints are both of the form

$$\|F_i^*\|_p = c_i \|F_i\|_p \leqslant 1 \tag{11.3.17}$$

we must introduce into the network scaling that satisfies

$$c_i \leqslant \frac{1}{\|F_i\|_p} \tag{11.3.18}$$

for $p = 2$ or ∞, as appropriate. On the other hand, rewriting (11.3.5) as $\|f_i\| \leqslant 1$, we have

$$c_i \leqslant \frac{1}{\|f_i\|} \tag{11.3.19}$$

if that absolute bound is to be employed. If complete b-bit multipliers are introduced into the network to implement scaling, we will choose to satisfy (11.3.18) or (11.3.19) with equality to maximize the SNR at each node. If, however, we only use shift operators for scaling, then $c_i = 2^{-r}$ for integer r, and we can only ensure that

$$\tfrac{1}{2} < \|F_i^*\|_p \leqslant 1 \tag{11.3.20}$$

or the equivalent for (11.3.19).

11.4 PARALLEL AND CASCADE FORMS

Important examples of the two previous sections are provided by the parallel and cascade forms. Since the parallel form is the simplest to scale, we analyze that first. Consider parallel form II shown in figure 11.8. The branch node signals $v_i^*(n)$ that must be constrained are indicated in the figure, along with the corresponding $F_i^*(z)$. Note that each of these signals feeds one multiplier directly and three others indirectly through delays. Hence, we need only one dynamic-range constraint per section. It is assumed that $H(z)$ is designed such that the output will not overflow, and thus we do not additionally constrain $y(n)$. A scaling multiplier c_i has been introduced at the input to each section; and the output multipliers γ_{0i} and γ_{1i} must, therefore, be oppositely scaled to maintain the original gain through each section, i.e.,

$$\gamma_{0i}^* = \gamma_{0i}/c_i \qquad \text{and} \qquad \gamma_{1i}^* = \gamma_{1i}/c_i. \tag{11.4.1}$$

Note that $F_i(z)$ is simply the second-order all-pole function

$$F_i(z) = \frac{1}{1 + \alpha_{1i}z^{-1} + \alpha_{2i}z^{-2}} \tag{11.4.2}$$

and thus the gain $\|F_i\|_p$ before scaling is potentially large.

The roundoff noise inputs $e_i(n)$ are also indicated in figure 11.8, and we note that there is only one such input per section, plus one $e_{L+1}(n)$ at the output summation node. If rounding is performed immediately after multiplication, and hence before summation, $k_i = 3$, $i = 1, \ldots, L$, and $k_{L+1} = 2L + 1$; while rounding after summation implies that $k_i = 1$ for all i. The $G_i^*(z)$, $i = 1, \ldots, L$, are given by

$$G_i^*(z) = \frac{\gamma_{0i}^* + \gamma_{1i}^* z^{-1}}{1 + \alpha_{1i}z^{-1} + \alpha_{2i}z^{-2}} \tag{11.4.3}$$

and, of course, $G_{L+1}^*(z) = 1$. If (11.3.18) is satisfied with equality, (11.4.1) implies that

$$G_i^*(z) = \frac{1}{c_i} G_i(z) = \|F_i\|_p G_i(z). \tag{11.4.4}$$

From (11.2.2) the roundoff-noise spectrum is thus given by

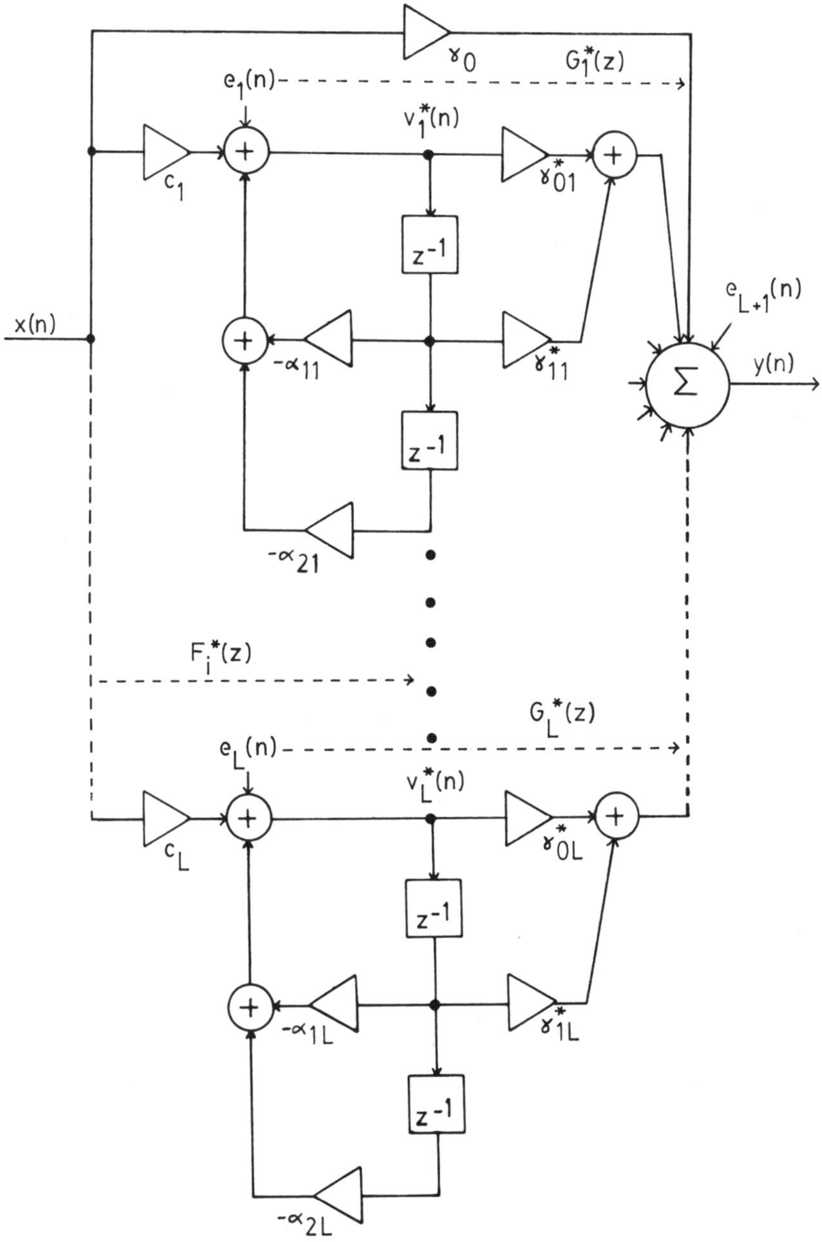

Figure 11.8. Parallel form II with scaling coefficients and noise sources.

$$N_e(\omega) = \sigma_0^2 \left[k_{L+1} + \sum_{i=1}^{L} k_i \| F_i \|_p^2 \cdot |G_i'(\omega)|^2 \right]. \tag{11.4.5}$$

Note how the scaling has directly affected (usually increased) the roundoff noise through the factors $\| F_i \|_p$, which for poles near the unit circle will be quite large. This explains why we want to avoid the pessimistic time-domain norm in (11.3.5), if possible, and to use $p = 2$ versus $p = \infty$ for $\| F_i \|_p$, if appropriate.

A very compact expression providing additional insight is readily derived for the total output noise power σ_e^2. Note first that

$$\| G_i \|_2^2 = \frac{1}{\omega_s} \int_0^{\omega_s} |G_i'(\omega)|^2 \, d\omega$$

and thus from (11.2.3), we have

$$\sigma_e^2 = \sigma_0^2 \left[k_{L+1} + \sum_{i=1}^{L} k_i \| F_i \|_p^2 \cdot \| G_i \|_2^2 \right]. \tag{11.4.6}$$

We see from this expression the contributions $F_i(z)$ and $G_i(z)$ each make to the output roundoff noise. In particular, if $p = 2$, their contributions to σ_e^2 are exactly the same!

The scaled version of parallel form I is shown in figure 11.9. In this case, the scaling gains c_i are incorporated directly into $\gamma_{0i}^* = c_i \gamma_{0i}$ and $\gamma_{1i}^* = c_i \gamma_{1i}$, and additional multipliers $1/c_i$ are required at the section outputs to compensate for these gains. The previous analysis for form II still holds for form I except that $F_i^*(z)$, not $G_i^*(z)$, is now given by (11.4.3) and $G_i(z)$, not $F_i(z)$, is given by (11.4.2). This is a general property of transpose networks; that is, the $F_i(z)$ and $G_i(z)$ are interchanged by transposition [39]. In particular, expressions (11.4.5) and (11.4.6) still hold for $N_e(\omega)$ and σ_e^2, respectively, the only difference being the form of the functions $F_i(z)$ and $G_i(z)$ and the values for k_i. Note that for parallel form I, $k_i = 4$, $i = 1, \ldots, L$, and $k_{L+1} = L + 1$ for rounding before summation. Clearly, for $p = 2$, this is the only difference in σ_e^2 between forms I and II since we then have L_2 norms for both $F_i(z)$ and $G_i(z)$ in (11.4.6); and hence the noise level for form I will be about $\frac{4}{3}$ that for form II, or 1.25 dB greater. It can be shown [40] that this is about the difference between these forms for $p = \infty$, as well. Hence, there is not really much difference between them, but all else being equal, we will choose form II.

We turn now to the cascade forms. Here, there is great design flexibility which can produce substantial variations in the roundoff noise level. Specifically, in addition to choosing between forms I and II, we must now pair the numerator and denominator factors into second-order sections and then sequentially order these sections. Rules of thumb will be given for pairing and ordering to minimize approximately the roundoff noise.

Cascade form II is shown in figure 11.10. New multipliers β_{0i}^* have been introduced so that scaling can be accomplished in the feedforward multipliers of each section, plus the input multiplier b_0^*. For zeros on the unit circle, we recall that $\beta_{2i} = \pm 1$, and hence we may choose to scale such that $\beta_{0i}^* = \pm \beta_{2i}^* = 2^{-r}$ to avoid two additional full multipliers per section. To simplify our analysis, however, we will assume that these are full multipliers.

Note that $F_i^*(z)$ and $G_i^*(z)$ for the ith section contain the cascaded transfer functions preceding and following that section, respectively. That is,

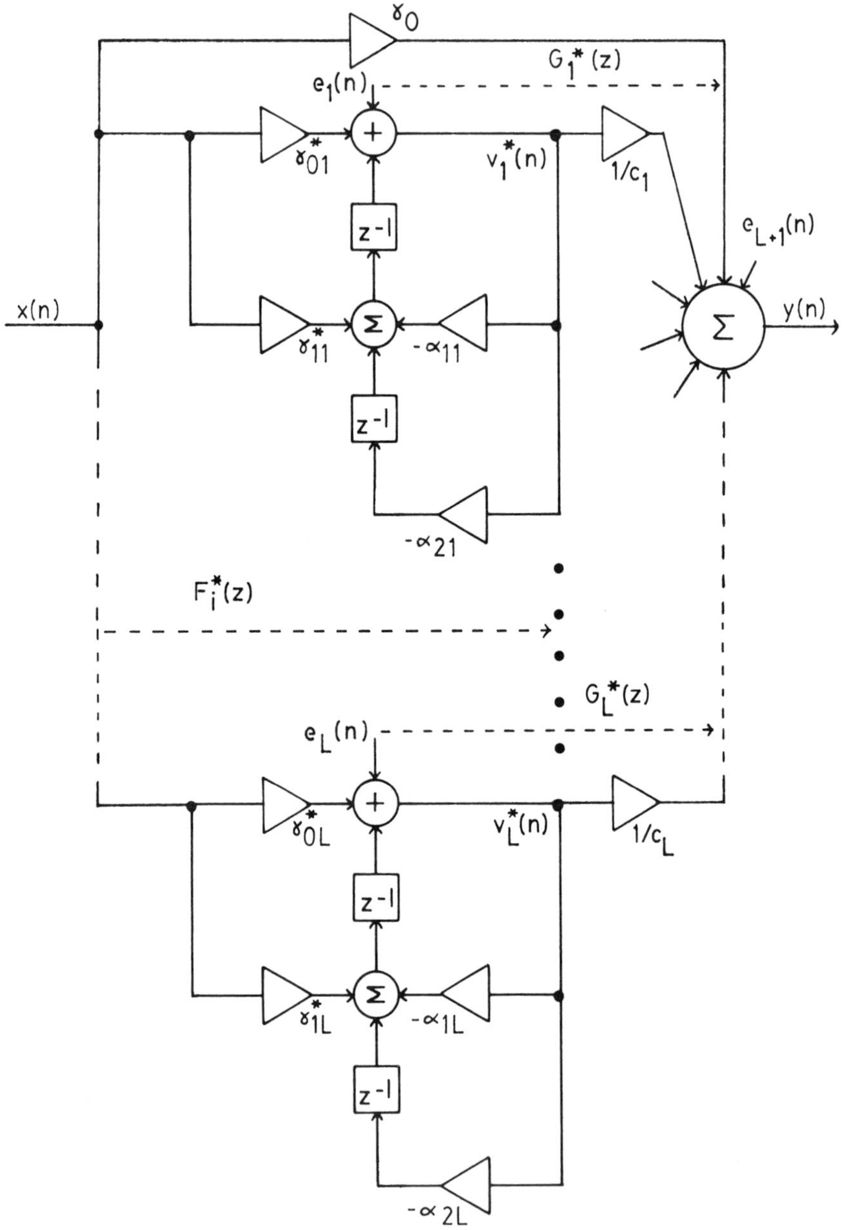

Figure 11.9. Parallel form I with scaling coefficients and noise sources.

$$F_i^*(z) \;=\; \frac{b_0^*}{1 + \alpha_{1i}z^{-1} + \alpha_{2i}z^{-2}} \prod_{k=1}^{i-1} H_k^*(z)$$

and (11.4.7)

$$G_i^*(z) \;=\; \prod\nolimits_{k=i}^{L} H_k^*(z), \qquad i = 1,\ldots, L,$$

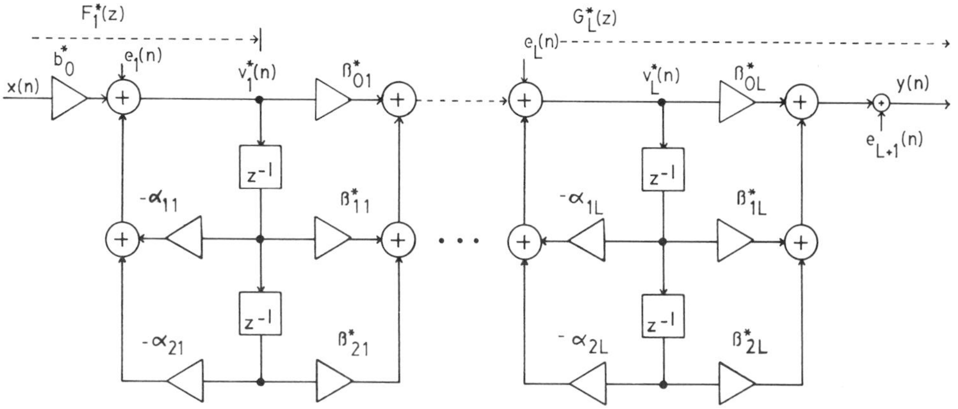

Figure 11.10. Cascade form II with scaling coefficients and noise sources.

where $H_k^*(z)$ is the transfer function of the kth section. In addition, there is a noise input at the output node with $G_{L+1}^*(z) = 1$. Scaling such that $c_i = 1/\|F_i\|_p$, we modify the feedforward multipliers to satisfy

$$
\begin{aligned}
b_0^* &= c_i \\
\beta_{ki}^* &= \frac{c_{i+1}}{c_i}\beta_{ki}, \qquad i = 1, \ldots, L
\end{aligned}
\tag{11.4.8}
$$

with $c_{L+1} = b_0$ to achieve the desired overall gain. The unscaled $F_i(z)$ are simply

$$
F_i(z) = \frac{1}{1 + \alpha_{1i}z^{-1} + \alpha_{2i}z^{-2}} \prod_{k=1}^{i-1} H_k(z). \tag{11.4.9}
$$

It is straightforward then to show that $N_e(\omega)$ and σ_e^2 satisfy (11.4.5) and (11.4.6). (See problem 11.4.) If rounding is performed before summation,

$$
\begin{aligned}
k_1 &= 3 \\
k_i &= 5, \qquad i = 2, \ldots, L \\
k_{L+1} &= 3.
\end{aligned}
\tag{11.4.10}
$$

Note, in particular, that overflow is permissible at the output of any section except the last one because these outputs are only partial sums.

Cascade form I is shown in figure 11.11. Since this is the transpose of cascade form II, the $F_i^*(z)$ and $G_i^*(z)$ in (11.4.7) are interchanged, as discussed previously; but instead of

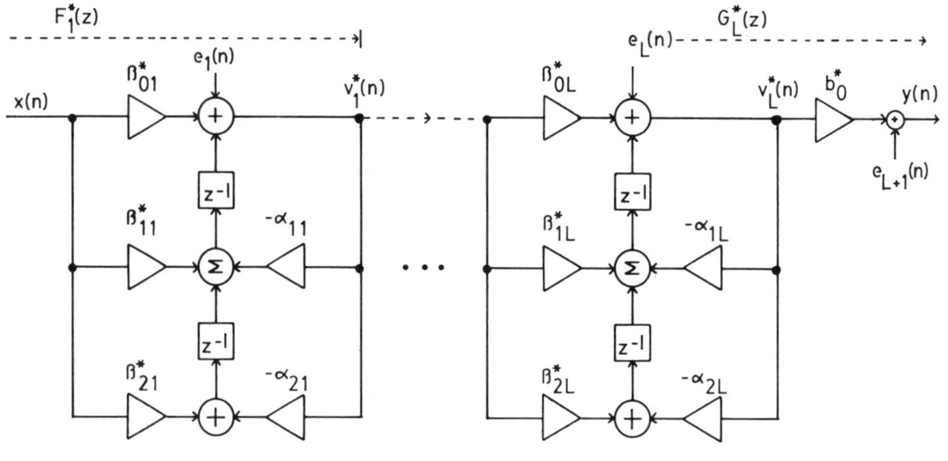

Figure 11.11. Cascade form I with scaling coefficients and noise sources.

reversing the order of the sections as this implies, we will write them in the more natural form

$$
\begin{aligned}
F_i^*(z) &= \prod_{k=1}^{i} H_k^*(z) \\
G_i^*(z) &= \frac{b_0^*}{1 + \alpha_{1i} z^{-1} + \alpha_{2i} z^{-2}} \prod_{k=i+1}^{L} H_k^*(z).
\end{aligned}
\tag{11.4.11}
$$

The feedforward coefficients must then be scaled to satisfy

$$
\begin{aligned}
\beta_{ki}^* &= \frac{c_i}{c_{i-1}} \beta_{ki}, \qquad i = 1, \ldots, L \\
b_0^* &= b_0 / c_L
\end{aligned}
\tag{11.4.12}
$$

with $c_0 = 1$, and the weights k_i are

$$
\begin{aligned}
k_i &= 5, \qquad i = 1, \ldots, L \\
k_{L+1} &= 1.
\end{aligned}
\tag{11.4.13}
$$

This form also satisfies (11.4.5) and (11.4.6).

Considering the pairs $F_i(z)$ and $G_i(z)$ (before scaling) in (11.4.7) and (11.4.11), we note that every term in the noise expressions (11.4.5) and (11.4.6) has every section transform $H_k(z)$, $k = 1, \ldots, L$, appearing in either $F_i(z)$ or $G_i(z)$. Hence, it is reasonable to assume

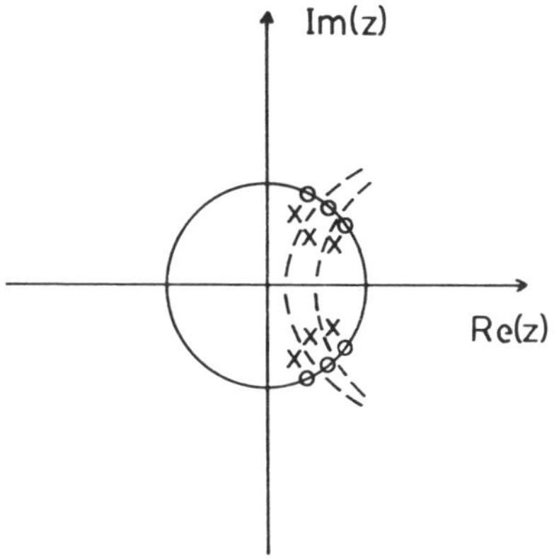

Figure 11.12. Pole/zero pairing for minimum roundoff noise in BSF.

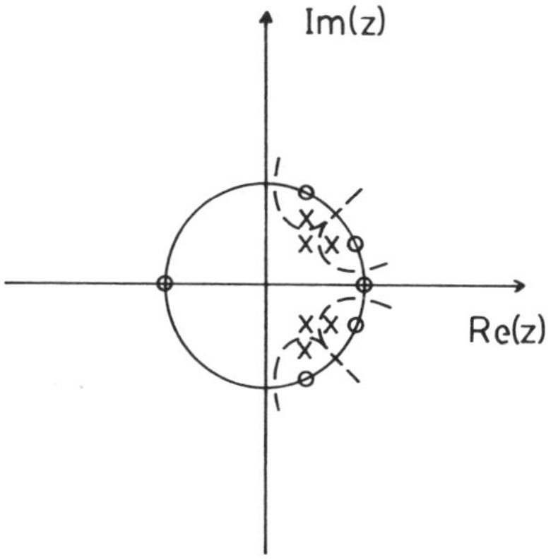

Figure 11.13. Pole/zero pairing for minimum roundoff noise in BPF.

that we need to minimize some norm of the individual $H_k(z)$. A rule of thumb has been developed for pairing the numerator and denominator factors to accomplish this [40], and it has been found to be quite effective. This rule is based on the simple observations that poles provide gain (which is bad in this context), the poles closest to the unit circle provide the highest gain, and the zeros provide attenuation which can be used to counter this gain. Hence, the rule says to pair the poles closest to the unit circle with the zeros nearest to them to minimize the worst of the gains. Then pair the next closest poles to the unit circle with the remaining zeros nearest to them, and so forth until all the poles and zeros have been paired.

While this qualitative rule may seem at first to be imprecise, it is readily quite straightforward and clear in practice. Consider the example of a sixth-order bandstop filter in figure 11.12. The poles at the band edges are the closest to the unit circle and are thus paired with the nearest zeros, leaving the central poles to be paired with the central (and nearest) zeros. Hence, the gain $H_k'(\omega)$ of each of the three sections will not peak up much above unity. The sixth-order bandpass filter in figure 11.13 is only slightly less obvious.

Again the poles at the band edges are paired with the nearest zeros, leaving the more-damped central poles to be paired with the remaining zeros at $z = \pm 1$.

Next, the paired sections must be sequentially ordered. Here, the question is basically whether we want a particular $H_k(z)$ to occur more often in $F_i(z)$ or $G_i(z)$. That is, should the most troublesome (peaked) of the $H_k(z)$ be placed toward the input or output of the filter? This clearly depends upon the norms applied to the $F_i(z)$ and $G_i(z)$. Considering (11.4.6) for σ_e^2, we note that for $p = 2$ it really does not matter much because both norms are L_2 norms. However, for $p = \infty$, we want the most peaked of the sections (typically those with poles closest to the unit circle) to fall mostly into the $G_i(z)$, not $F_i(z)$, and thus these should be placed toward the end of the cascade. That is, we should order the sections from least-peaked to most-peaked. It is left to the reader to argue that if the peak noise $\|N_e\|_\infty$ is of most concern to us and $p = 2$, the opposite is true—we should order the sections from most-peaked to least-peaked; while if $p = \infty$, it does not matter much again.

Finally, we must choose between forms I and II. As above, this is a significant question only if the scaling norm is different from the measure of the roundoff noise of most concern to us. The basic question here is whether we want the extra denominator factors (with their associated gains) in (11.4.7) and (11.4.11) to occur in $F_i(z)$ as with form II, or in $G_i(z)$ as with form I. If $p = \infty$ and we want to minimize σ_e^2, then the extra factors will have less effect in $G_i(z)$, where an L_2 norm is applicable, and form I is thus best. On the other hand, if $p = 2$ and we want to minimize $\|N_e\|_\infty$, then the extra factors belong in $F_i(z)$, and form II is best.

By this time, one of the virtues of the parallel form has become obvious—it can be a lot easier to design because there is no pairing or ordering to consider and the scaling gains c_i are much easier to calculate. Moreover, experimental results indicate that the noise level of the parallel form is comparable typically to the best of the cascade-form designs. Therefore, why not avoid this extra design effort and choose the parallel form to begin with? Because, as discussed earlier, we can often save 25 percent, and sometimes 50 percent, of the multipliers using the cascade form, and the zeros do not move off the unit circle with coefficient quantization. Hence, the cascade form is often the practical choice. The usefulness of the pairing and ordering rules given above becomes evident when one considers that there are $L!$ ways to pair the sections and another $L!$ ways to order them. Therefore, unless L is relatively small, complete enumeration and analysis of all $(L!)^2$ possibilities becomes prohibitive.

Lower bounds have been derived on the roundoff noise for the parallel and cascade forms in terms of the sensitivity of the system function $H(z)$ to variations in the feedback coefficients α_{ki} [41]. These bounds are especially useful for the cascade form because they are independent (or nearly so) of pairing, ordering, and transposition and can be computed before the network is scaled. Therefore, since the bounds have proven to be reasonably tight (within 6 dB), they can be used to obtain a reasonable estimate of the achievable noise level before much effort is expended in detailed synthesis and scaling. The coefficient sensitivities themselves, of course, are also useful in anticipating difficulties with coefficient quantization.

Considering any of the four parallel and cascade forms in figures 11.8 through 11.11, we note that each feedback coefficient α_{ki} has as its input $v_i(n)$ (or a delayed version thereof), and its roundoff noise output is a component of $e_i(n)$. Hence, the transfer functions to and from the α_{ki} multiplier are $z^{-q}F_i(z)$ and $z^{-r}G_i(z)$, respectively, where $q + r = k$. Assume now a small perturbation $\Delta\alpha_{ki}$ in this coefficient. Neglecting the small change that this perturbation makes in $F_i(z)$ and $G_i(z)$, the change in the overall system function $H(z)$ is approximately

$$\Delta H(z) \approx z^{-k}F_i(z)G_i(z)\Delta\alpha_{ki}. \tag{11.4.14}$$

Dividing through by $\Delta\alpha_{ki}$ and taking the limit as $\Delta\alpha_{ki} \to 0$, this expression becomes exact, and we find that the sensitivity of $H(z)$ to a variation of α_{ki} is simply

$$\frac{\partial H(z)}{\partial \alpha_{ki}} = z^{-k}F_i(z)G_i(z). \tag{11.4.15}$$

Since the delay factor z^{-k} does not affect any L_p norm of this sensitivity, we define a corresponding delay-free sensitivity function by

$$S_i(z) = F_i(z)G_i(z). \tag{11.4.16}$$

From Holder's or Schwarz's inequalities, we then have

$$\|S_i\|_1 = \|F_iG_i\|_1 \leqslant \|F_i\|_2 \cdot \|G_i\|_2 \tag{11.4.17}$$

and thus from (11.4.6)

$$\sigma_e^2 \geqslant \sigma_0^2\left[k_{L+1} + \sum_{i=1}^{L} k_i\|S_i\|_1^2\right], \qquad (\text{for } p = 2). \tag{11.4.18}$$

That is, for L_2 scaling, this weighted sum of L_1 sensitivity norms provides a lower bound on the achievable roundoff noise power after scaling. On the other hand, we note that

$$\|S_i\|_2 = \|F_iG_i\|_2 \leqslant \|F_i\|_\infty \cdot \|G_i\|_2 \tag{11.4.19}$$

and thus for L_∞ scaling, we have the similar bound

$$\sigma_e^2 \geqslant \sigma_0^2\left[k_{L+1} + \sum_{i=1}^{L} k_i\|S_i\|_2^2\right], \qquad (\text{for } p = \infty). \tag{11.4.20}$$

Also,

$$|S_i'(\omega)| = |F_i'(\omega)G_i'(\omega)| \leqslant \|F_i\|_\infty \cdot |G_i'(\omega)| \tag{11.4.21}$$

and thus the power density spectrum of the roundoff noise for L_∞ scaling is bounded by

$$N_e(\omega) \geqslant \sigma_0^2 \left[k_{L+1} + \sum_{i=1}^{L} k_i |S_i'(\omega)|^2 \right], \qquad (\text{for } p = \infty). \tag{11.4.22}$$

Since the sensitivity functions are independent of pairing, ordering, and transposition, the only effects of these operations on the bounds in (11.4.18), (11.4.20) and (11.4.22) result from the small differences in the k_i (for rounding before summation). To simplify the computation of these bounds for the cascade form, we note from (11.4.7) and (11.4.11) that

$$S_i(z) = \frac{H(z)}{1 + \alpha_{1i} z^{-1} + \alpha_{2i} z^{-2}}. \tag{11.4.23}$$

11.5 LIMIT-CYCLE OSCILLATIONS

The statistical analysis of roundoff noise in the preceding sections was based on the assumption of uncorrelated rounding errors. For most data signals—those with sufficient amplitude and spectral content—this assumption is valid, and experimental noise spectra closely match the predicted $N_e(\omega)$. However, for low-level inputs or periodic input signals that are synchronized to the sampling rate, the rounding errors are clearly correlated, and the roundoff noise cannot be assumed to be white.

An extreme and important example is the case of a zero or constant input signal. Ideally, the output of a stable discrete-time filter would asymptotically approach zero or a constant for such an input, but with quantization, we often find instead that relatively small *limit-cycle oscillations* occur. Although they are small, these limit cycles can be troublesome. For example, the dynamic range of the human ear is about 60 dB, and thus in a digitally implemented telephone system, the "idle channel" noise when no one is talking could be small, but audible, tones due to limit cycles from the digital filters. Hence, we must either allocate enough bits to the filters so that the limit cycles cannot affect the D/A output, or we must find structures that suppress these limit cycles. The other type of oscillation due to the nonlinearity of digital filters is *overflow oscillation*, described later in this section.

As a simple example of limit cycles, consider the first-order filter satisfying the difference equation $y(n) = x(n) + 0.9y(n-1)$. The ideal zero-input response (without quantization) to an initial condition of $y(0) = 10$ would be the sequence 10, 9, 8.1, 7.29, . . . , decaying asymptotically to zero. However, with rounding to the nearest integer, the output sequence is instead 10, 9, 8, 7, 6, 5, 5, 5, 5, . . . , and thus does not decay to zero. This has been called the *deadband effect* for the deadband of values $[-5, 5]$, which are self-sustaining. For a general first-order filter $y(n) = x(n) + ay(n-1)$, the deadband $[-K, K]$ is given by

$$K = \left[\frac{0.5}{1 - |a|} \right]_{\text{int}} \tag{11.5.1}$$

where int denotes "the integer part of". Note that for $a > 0$ these limit cycles are constant (dc); while for $a < 0$ the signs alternate, and the frequency of oscillation is thus $\omega_s/2$.

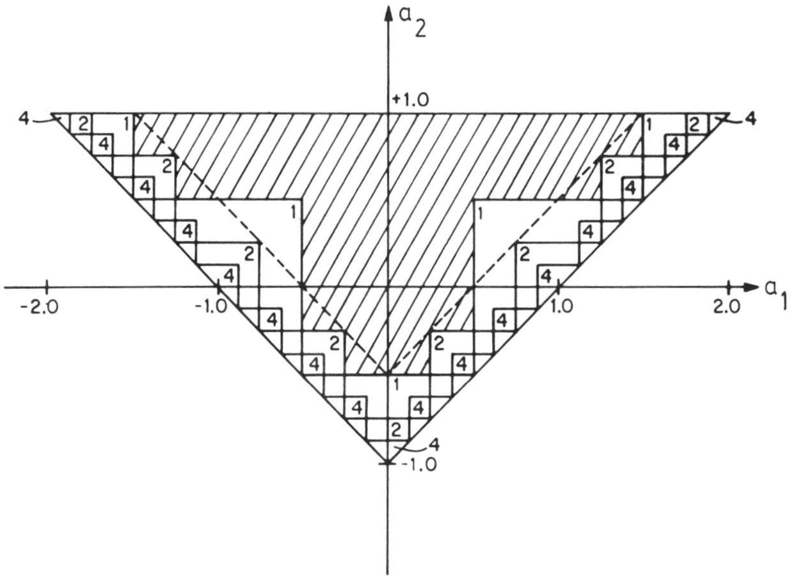

Figure 11.14. Regions of (a_1, a_2) for first-order-mode limit cycles in second-order digital filters.

The second-order case is somewhat more complicated, and there are two distinct limit-cycle modes. The first mode is analogous to the first-order case and produces either dc or $\omega_s/2$ limit cycles. For the quantized second-order difference equation

$$y(n) \;=\; x(n) \;-\; [a_1 y(n\;-\;1)]_{\text{int}} \;-\; [a_2 y(n\;-\;2)]_{\text{int}}, \tag{11.5.2}$$

such a limit cycle of amplitude y_0 can be sustained with zero input if

$$y_0 \;=\; \pm\,[a_1 y_0]_{\text{int}} \;-\; [a_2 y_0]_{\text{int}}. \tag{11.5.3}$$

Regions of (a_1, a_2) values satisfying this nonlinear equation and the corresponding values y_0 are indicated within the stability triangle in figure 11.14. The dotted line corresponds to the necessary, but not sufficient, condition for the existence of such first-order limit cycles (for $y_0 = 2$) given by

$$|a_1| \;\geqslant\; \frac{y_0 - 1}{y_0} \;+\; a_2. \tag{11.5.4}$$

Note that the closer (a_1, a_2) is to either side of the stability triangle, the larger the limit cycles can be.

The second mode corresponds to truly second-order oscillations; that is, the network behaves as if there were two complex-conjugate poles on the unit circle. An *effective-value* linear model [42] has been proposed to estimate the amplitude and frequency of these second-order limit cycles, with good results in most cases. This model is based upon the simplifying approximation that, for "effective poles" on the unit circle, the effective value

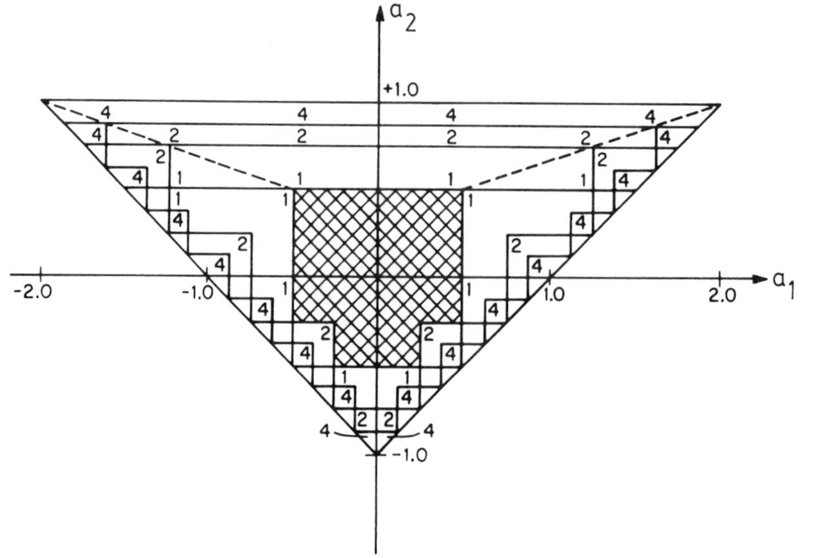

Figure 11.15. Regions of (a_1, a_2) for first- and second-order-mode limit cycles in second-order digital filters.

of a_2 must be unity, i.e., $[a_2 y(n - 2)]_{\text{int}} = y(n - 2)$. But this can happen only for values of $y(n - 2)$ within the deadband $[-K, K]$ where

$$K = \left[\frac{0.5}{1 - a_2}\right]_{\text{int}} \tag{11.5.5}$$

Hence, we estimate that the limit cycles will not exceed the amplitude K. The value of a_1, on the other hand, determines the limit-cycle frequency ω_0 by the approximation

$$a_1 \approx -2 \cos \omega_0 T \tag{11.5.6}$$

where ω_0 is a rational fraction of ω_s. Figure 11.15 shows the addition of these second-order deadbands to those of figure 11.14.

The above estimates of the limit-cycle amplitudes are quite accurate in most cases. Occasionally, for values $a_2 > 0.9$, the peak amplitude K in (11.5.5) is exceeded by one. In even rarer cases, a different limit-cycle mode is encountered with amplitudes up to $3K$. These larger limit cycles are called *rolling pin* limit cycles [43], and their possibility is best checked by simulation. Numerous bounds have also been derived on limit-cycle amplitudes [44, 45], and they are generally somewhat loose to accomodate the possibility of rolling-pin limit cycles. Magnitude truncation (instead of rounding) can eliminate many, but not all, limit cycles [46–48]; but as discussed in section 11.2, this produces correlated noise and increased variance in the general roundoff-noise case.

The rounding limit cycles described above can always be reduced to acceptable levels by the addition of a sufficient number of bits to the signal representation. Another form of autonomous oscillation is produced by the overflow characteristic of one's- or two's-

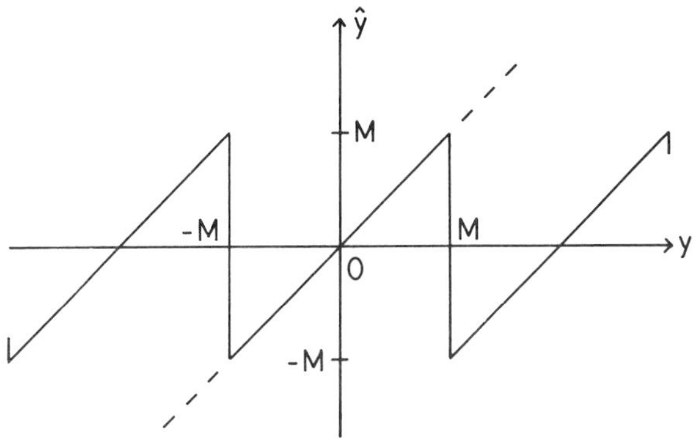

Figure 11.16. One's- or two's-complement overflow characteristic.

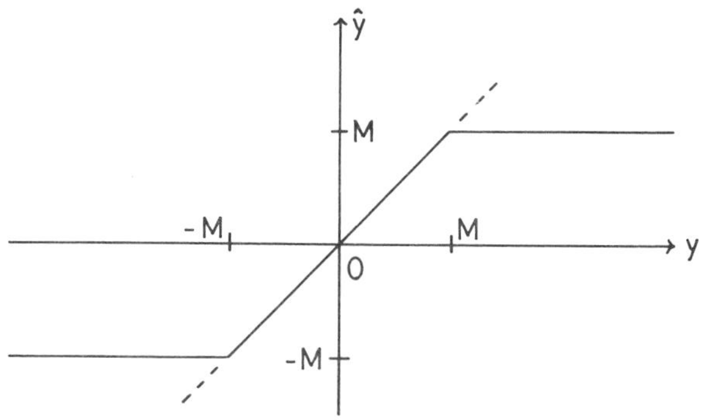

Figure 11.17. Overflow characteristic of saturation arithmetic.

complement arithmetic. These overflow oscillations consume virtually the entire dynamic range of the filter and, therefore, must be precluded altogether. The overflow characteristic of a one's- or two's-complement adder is depicted in figure 11.16, with y being the correct output sum, $\hat{y}$ the result of overflow, and M the full-scale value. For example, if the value $1.1(2M/3)$ is added to $0.9(2M/3)$, the overflowed result will be $-2M/3$ instead of the correct $4M/3$. Hence, the zero-input difference equation $y(n) = 1.1y(n-1) - 0.9y(n-2)$ can support an overflow oscillation of amplitude $2M/3$ and frequency $\omega_s/2$. Large oscillations at many other frequencies are also possible.

One might argue that with zero initial conditions and appropriate signal scaling to prevent overflow, such oscillations could not occur. However, we usually cannot control the initial conditions when the system power is turned on or whether other transients occur, and we usually scale only to minimize the probability of overflow, not to preclude it entirely. Therefore, we must take other steps to ensure that overflow oscillations cannot occur.

The most common solution to this problem is to modify the adder characteristic to implement clipping or *saturation arithmetic*. That is, one's- or two's-complement addition is performed normally unless an overflow is detected in the resulting total (not partial) sum. When such an overflow is detected, the full-scale amplitude $\pm M$ is substituted for the overflowed result with the appropriate sign. Hence, the adder characteristic becomes that depicted in figure 11.17, which has been shown to preclude overflow oscillations in second-order direct-form filters [49], and thus in cascade- or parallel-form filters. This does not, however, hold for direct-form filters of greater than second order [50].

11.6 STATE-SPACE STRUCTURES

We have seen in sections 11.1 through 11.4 that the structure of a digital filter has a profound effect on its sensitivity to coefficient quantization and on its roundoff noise output. The question naturally arises as to whether there are systematic ways to vary the structure analytically to minimize these effects. A completely general technique for this does not yet exist; but state-variable techniques have been used to accomplish much of this goal. In particular, given any design with state matrices $(\mathbf{A}, \underline{b}, \underline{c}^t, d)$, a state transformation can be made to produce a new *state-space* structure that has minimum roundoff noise and coefficient sensitivity. Note that although any structure has a state-variable description, the term "state-space structure" is reserved for a network that implements the state equations directly, i.e., each coefficient multiplier corresponds to a single element of $(\mathbf{A}, \underline{b}, \underline{c}^t, d)$. In general, however, the number of multipliers is increased dramatically by transformation, so this approach is usually restricted to second-order structures where the increase is moderate.

We found in section 4.4 that the state transformation $\hat{\underline{s}}(n) = \mathbf{T}\underline{s}(n)$ produces the transformed state matrices $(\mathbf{TAT}^{-1}, \mathbf{T}\underline{b}, \underline{c}^t\mathbf{T}^{-1}, d)$. Let the transfer function to state $s_i(n)$ be $F_i(z)$ and the transfer function from the error input $e_i(n)$ associated with rounding of $s_i(n+1)$ or its components be $G_i(z)$. In vector form, these transfer functions are readily shown to be given by

$$\underline{F}(z) = \begin{bmatrix} F_1(z) \\ F_2(z) \\ \cdot \\ \cdot \\ \cdot \end{bmatrix} = (z\mathbf{I} - \mathbf{A})^{-1}\underline{b}$$

and (11.6.1)

$$\underline{G}(z) = \begin{bmatrix} G_1(z) \\ G_2(z) \\ \cdot \\ \cdot \\ \cdot \end{bmatrix} = (z\mathbf{I} - \mathbf{A}^t)^{-1}\underline{c}.$$

(See problem 11.10.) Therefore, the effect of the transformation $\mathbf{T}$ on these vector transfer functions is

$$\hat{\underline{F}}(z) = \mathbf{T}\underline{F}(z) \quad \text{and} \quad \hat{\underline{G}}(z) = \mathbf{T}^{-t}\underline{G}(z). \tag{11.6.2}$$

For example, the network may be scaled to satisfy the L_p overflow constraint $\| \hat{F}_i \|_p = 1$ via the diagonal transformation

$$\mathbf{T} = \begin{bmatrix} \|F_1\|_p^{-1} & 0 & \cdots \\ 0 & \|F_2\|_p^{-1} & \cdots \\ \cdot & \cdot & \cdot \\ \cdot & \cdot & \cdot \\ \cdot & \cdot & \cdot \end{bmatrix}, \tag{11.6.3}$$

in which case

$$\underline{\hat{G}}(z) = \begin{bmatrix} \|F_1\|_p G_1(z) \\ \|F_2\|_p G_2(z) \\ \cdot \\ \cdot \\ \cdot \end{bmatrix}. \tag{11.6.4}$$

Since the network is scaled such that the dynamic ranges of all states are the same, it is reasonable to assume that it is desirable to balance the noise contributions from all states as well, i.e.,

$$\|\hat{G}_i\|_2 = \|\hat{G}_j\|_2 \qquad \text{for all } i, j. \tag{11.6.5}$$

It has indeed been shown that the minimum roundoff-noise variance σ_e^2 for L_2 scaling is produced by networks satisfying (11.6.5) [51, 52]. In the second-order case, one such network is the state-space structure satisfying [53]

$$\mathbf{A}^t = \mathbf{JAJ} = \begin{pmatrix} a_{22} & a_{21} \\ a_{12} & a_{11} \end{pmatrix} \quad \text{and} \quad \underline{c} = \varrho \mathbf{J}\underline{b} = \varrho \begin{pmatrix} b_2 \\ b_1 \end{pmatrix}, \tag{11.6.6}$$

where ϱ is an arbitrary scalar and $\mathbf{J}$ is the permutation matrix

$$\mathbf{J} = \begin{pmatrix} 0 & 1 \\ 1 & 0 \end{pmatrix}.$$

Simply stated, (11.6.6) requires that

$$a_{11} = a_{22} \tag{11.6.7a}$$

and

$$\frac{c_1}{c_2} = \frac{b_2}{b_1}. \tag{11.6.7b}$$

Then,

$$\underline{G}(z) = \varrho \mathbf{J}\underline{F}(z) = \varrho \begin{pmatrix} F_2(z) \\ F_1(z) \end{pmatrix} \tag{11.6.8}$$

and when $\underline{F}(z)$ is L_2 scaled via (11.6.3), we see from (11.6.4) that (11.6.5) is indeed satisfied. For L_∞ scaling, the resulting network is not optimum in the sense of minimum σ_e^2, but then $\|\hat{G}_1\|_\infty = \|\hat{G}_2\|_\infty$ and the peak roundoff noise density $\|N_e\|_\infty$ is approximately minimized.

For a second-order filter given by

$$H(z) = d + \frac{\gamma_2 z^{-2} + \gamma_1 z^{-1}}{\alpha_2 z^{-2} + \alpha_1 z^{-1} + 1} \tag{11.6.9}$$

having complex-conjugate poles, a structure satisfying (11.6.6) is provided by [53]

$$\begin{aligned}
a_{11} &= a_{22} = -\alpha_1/2 \\
b_1 &= \tfrac{1}{2}(1 + \gamma_2) \qquad b_2 = \tfrac{1}{2}\gamma_1 \\
c_1 &= \frac{\gamma_1}{1 + \gamma_2} \qquad c_2 = 1 \\
a_{12} &= (1 + \gamma_2)(K_1 \pm K_2)/\gamma_1^2 \\
a_{21} &= (K_1 \mp K_2)/(1 + \gamma_2),
\end{aligned} \tag{11.6.10}$$

where

$$\begin{aligned}
K_1 &= \gamma_2 - \tfrac{1}{2}\alpha_1\gamma_1 \\
K_2 &= (\gamma_2^2 - \gamma_1\gamma_2\alpha_1 + \gamma_1^2\alpha_2)^{1/2}.
\end{aligned}$$

A second-order state-space structure is shown in figure 11.18.

We are interested primarily in cascade or parallel forms comprised of these second-order sections because an optimum Nth-order state-space structure would require, in general, $N^2 + 2N + 1$ multipliers to implement $(\mathbf{A}, \underline{b}, \underline{c}^t, d)$, as opposed to $4N + 1$ for a parallel state-space form and $9N/2$ for a cascade form. Remember that the canonical forms require only $2N + 1$ multipliers. Therefore, the cost for the improved performance of the cascade or parallel state-space forms is approximately a factor of two in multipliers required, which, although expensive, may be worth it if the word length is inadequate and not easily increased (such as in microprocessor systems).

We know from section 11.4 that low roundoff noise implies low coefficient sensitivities. Hence, by minimizing the roundoff noise variance, we must have minimized the sensitivities in some sense. To show that this is true, we note that

$$\frac{\partial H(z)}{\partial b_i} = G_i(z) \qquad \text{and} \qquad \frac{\partial H(z)}{\partial c_i} = F_i(z). \tag{11.6.11}$$

But after optimization and scaling, we have normalized the $\|\hat{F}_i\|_2$ to a moderate value of unity and minimized σ_e^2 which is proportional to the sum of the squared norms $\|\hat{G}_i\|_2^2$.

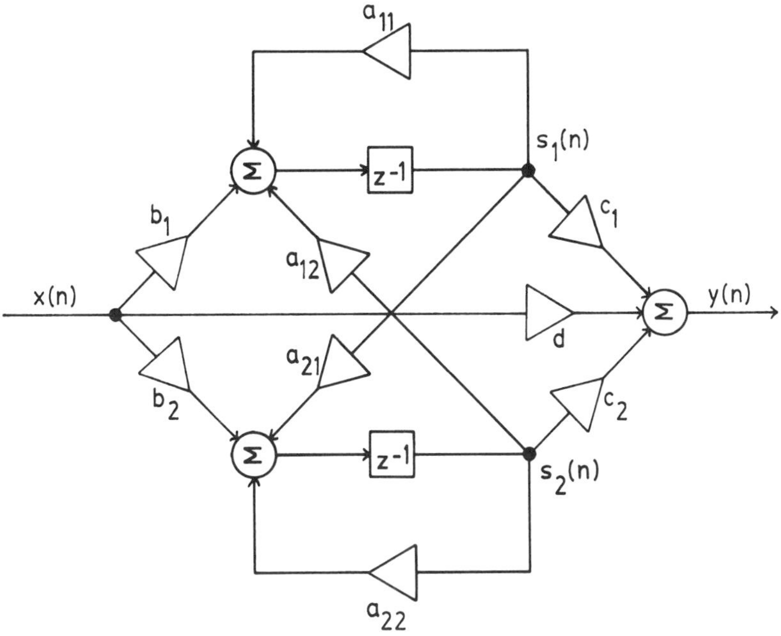

Figure 11.18. Second-order state-space structure.

Hence, via (11.6.5) we have minimized the maximum of the sensitivity norms $\|\hat{G}_i\|_2$, subject to L_2 scaling. Furthermore,

$$\frac{\partial H(z)}{\partial a_{ij}} = G_i(z)F_j(z) = S_{ij}(z) \tag{11.6.12}$$

and by the Schwarz inequality

$$\|S_{ij}\|_1 \leqslant \|G_i\|_2 \cdot \|F_j\|_2$$

or, after scaling,

$$\|\hat{S}_{ij}\|_1 \leqslant \|\hat{G}_i\|_2. \tag{11.6.13}$$

Therefore, as argued above, we have minimized the maximum of these L_1 sensitivity bounds, subject to L_2 scaling.

In section 11.1, the normal form having

$$\mathbf{A} = \begin{pmatrix} a_1 & a_2 \\ -a_2 & a_1 \end{pmatrix}$$

and complex-conjugate poles $a_1 \pm ja_2$ was also found to be a low-sensitivity recursive structure. Note that this structure satisfies (11.6.7a), and thus if we synthesize $\underline{b}$ and $\underline{c}$ to satisfy (11.6.7b) also, as described later, we will produce the optimum state-space structure after scaling via (11.6.3). However, this will destroy the normal structure of **A** and thus the uniform grid of possible pole locations depicted in figure 11.5. If we wish to scale this network without changing the normal structure of **A**, the appropriate scaling transformation is the orthogonal rotation

$$\mathbf{T} = \mu \begin{pmatrix} \cos\theta & \sin\theta \\ -\sin\theta & \cos\theta \end{pmatrix} \tag{11.6.14}$$

where the rotation angle θ is given by [53]

$$\tan 2\theta = (\|F_1\|_2^2 - \|F_2\|_2^2)/2(F_1, F_2)$$

and (F_1, F_2) is the cross-spectral norm

$$(F_1, F_2) = \frac{1}{\omega_s}\int_0^{\omega_s} F_1'(\omega)F_2'(-\omega)\,d\omega.$$

This produces $\|\hat{F}_1\|_2 = \|\hat{F}_2\|_2$, the constant μ is then set to $1/\|\hat{F}_i\|_2$ to normalize these to unity. This scaled normal structure is not optimum, but it is very close to the optimum structure in performance, being typically only 1 to 3 dB worse.

Additional insight into the normal form is provided by considering the implementation of the network in the partial-fraction-expansion form

$$H(z) = \gamma_0 + \sum_{k=1}^{N} \frac{\gamma_k z^{-1}}{1 - p_k z^{-1}} \tag{11.6.15}$$

which corresponds to a parallel combination of first-order sections with complex coefficients p_k and γ_k. For a section with complex pole p_k, the corresponding conjugate section having coefficients p_k^* and γ_k^* need not actually be implemented because its output sequence would just be the conjugate of that for this section. Hence, the sum of their output sequences can be computed as simply twice the real part of the output sequence from either one. Considering a single representative section with pole $p = p_r + jp_i$ and residue $\gamma = \gamma_r + j\gamma_i$, we note that the real and imaginary parts of the complex state $s(n)$ can be computed by the state equations

$$\begin{aligned} s_r(n+1) &= p_r s_r(n) - p_i s_i(n) + \gamma_r x(n) \\ s_i(n+1) &= p_i s_r(n) + p_r s_i(n) + \gamma_i x(n) \end{aligned} \tag{11.6.16}$$

and the real-valued output (sum of both conjugate sections) is then given by

$$y_r(n) = 2s_r(n).$$

But this is just the normal form with state matrices

$$\mathbf{A} = \begin{pmatrix} p_r & -p_i \\ p_i & p_r \end{pmatrix} \qquad \underline{b} = \begin{pmatrix} \gamma_r \\ \gamma_i \end{pmatrix} \qquad \underline{c} = (2 \quad 0), \tag{11.6.17}$$

which can be scaled by (11.6.14) to leave **A** unchanged.

If instead we wish to synthesize the optimum state-space structure using the normal form as an intermediate step, $\underline{b}$ and $\underline{c}$ must satisfy (11.6.7b) before scaling via (11.6.3). Let the residue be expressed as a product $\gamma = u \cdot v$ and

$$\begin{aligned} s_r(n+1) &= p_r s_r(n) - p_i s_i(n) + u_r x(n) \\ s_i(n+1) &= p_i s_r(n) + p_r s_i(n) + u_i x(n). \end{aligned} \tag{11.6.18}$$

Then the real-valued output is computed by

$$y_r(n) = 2v_r s_r(n) - 2v_i s_i(n)$$

and thus

$$\underline{b} = \begin{pmatrix} u_r \\ u_i \end{pmatrix} \qquad \underline{c} = \begin{pmatrix} 2v_r \\ -2v_i \end{pmatrix}. \tag{11.6.19}$$

To satisfy (11.6.7b), let $v = ju$ so that

$$v_i = u_r \qquad \text{and} \qquad v_r = -u_i$$

and hence

$$\frac{c_1}{c_2} = \frac{-v_r}{v_i} = \frac{u_i}{u_r} = \frac{b_2}{b_1}$$

as required. Then, $\gamma = ju^2$, or $u = \sqrt{-j\gamma}$ and $v = -\sqrt{j\gamma}$.

An important property of the normal and optimal second-order forms is that overflow oscillations cannot occur, even with the 1's or 2's complement overflow characteristic of figure 11.16 [54, 55]. To show this for the normal form, we use the L_2 norm (Euclidean length) of the state vector, defined by

$$\|\underline{s}(n)\| = (|s_1(n)|^2 + |s_2(n)|^2)^{1/2}.$$

With zero input, $\underline{s}(n+1) = \mathbf{A}\underline{s}(n)$ and thus

$$\|\underline{s}(n+1)\| = \|\mathbf{A}\underline{s}(n)\|. \tag{11.6.20}$$

We now ask if $\|\mathbf{A}\underline{s}(n)\| < \|\underline{s}(n)\|$? In general, this is not true, but for the normal form, we will show that it is true. Writing the poles as $p, p^* = r(\cos\theta \pm j\sin\theta)$, $\mathbf{A}$ can be written as

$$\mathbf{A} = r\begin{pmatrix} \cos\theta & -\sin\theta \\ \sin\theta & \cos\theta \end{pmatrix}$$

and, therefore, the transformation $\mathbf{A}\underline{s}(n)$ corresponds to a rotation of $\underline{s}(n)$ by θ and a scaling by r. But, since rotation of a vector does not change its length and for a stable filter $r < 1$, (11.6.20) implies that

$$\|\underline{s}(n+1)\| < \|\underline{s}(n)\|. \tag{11.6.21}$$

What happens now if one (or both) of the states $s_i(n+1)$ overflows? Note from figure 11.16 that the sum $\hat{s}_i(n+1)$ satisfies $|\hat{s}_i(n+1)| \leqslant |s_i(n+1)|$, with or without overflow, which from (11.6.21) implies that

$$\|\hat{\underline{s}}(n+1)\| < \|\underline{s}(n)\|. \tag{11.6.22}$$

Therefore, the norm of the state vector for a normal form with zero input decreases monotonically with n, even if overflows occur, and autonomous oscillation cannot be sustained. The proof for the optimal form is given by Mills, Mullis, and Roberts [55].

The same proofs show that for magnitude truncation, limit cycles are also impossible in the normal and optimal forms if the truncation is performed after summation. These proofs do not apply to truncation before summation, although no limit cycles have been observed for the normal form in that case either. However, rounding is usually preferred over truncation, as argued earlier, and limit cycles of comparable magnitude to those of the direct form do occur in the normal form in the case of rounding [56].

PROBLEMS

11.1 Show that (11.3.5) is not only sufficient, but also necessary to preclude all possibility of overflow. (*Hint*: Assume that $x(n) = \pm M$ with an appropriate sign sequence.)

11.2 Show that condition (11.3.9) is satisfied by sinusoidal input $x(n) = M\sin(\omega_0 nT + \theta)$.

11.3 Show by the Schwarz inequality that

$$|v_i(n)| \leqslant \|V_i\|_1 < \|F_i\|_2 \cdot \|X\|_2$$

which is a special case of (11.3.14).

11.4 Show that cascade form II in (11.4.7–9) satisfies equations (11.4.5) and (11.4.6) for $N_e(\omega)$ and σ_e^2.

11.5 Draw direct-form I and II networks for the highpass filter

$$H(z) = \frac{1 - z^{-1}}{1 - 0.9z^{-1}}.$$

a. Show that form I already satisfies $\| F \|_p \approx 1$, for all p, and thus requires no further scaling.

b. Calculate the scaling constants c_i for form II for both L_2 and L_∞ scaling. (*Hint*: In the L_2 case, use Parseval's relation.)

c. Sketch the noise power spectrum $N_e(\omega)$ for forms I and II and both L_2 and L_∞ scaling. Compare the peak densities.

d. Calculate the output noise variance σ_e^2 in all the cases of part (c) and compare them. (*Hint*: Use Parseval's relation.)

11.6 The allpass filter

$$H(z) = \frac{(0.8 - z^{-1})(0.8 + z^{-1})}{(1 - 0.8z^{-1})(1 + 0.8z^{-1})}$$

is built as a cascade of two first-order sections.

a. Show that form I already satisfies $\| F_i \|_p = 1$ and thus requires no further scaling.

b. Calculate the scaling constants c_i for form II for L_2 and L_∞ scaling and for worst-case scaling (absolutely no overflow). (*Hint*: In the L_2 case, use Parseval's relation.)

c. Sketch the noise power spectrum $N_e(\omega)$ for forms I and II and both L_2 and L_∞ scaling. Compare the peak densities.

d. Calculate the output noise variance σ_e^2 in all the cases of part (c) and compare them. (*Hint*: Use Parseval's relation.)

11.7 Using figure P11.7, find

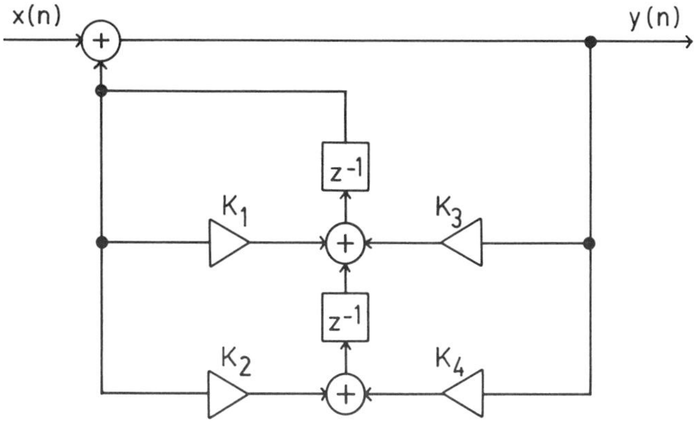

a. the $F_i(z)$ required to scale the network in the figure.

b. the $G_i(z)$ and k_i needed to determine $N_e(\omega)$.

c. $\partial H(z)/\partial K_i$, $i = 1, 3$, (without taking any partials).

d. the state matrices $\mathbf{A}$, $\underline{b}$, $\underline{c}^t$, d.

11.8 The following filter is to be realized with L_∞ scaling and approximately minimum roundoff noise power σ_e^2:

$$H(z) = \frac{N_1(z)N_2(z)N_3(z)}{D_1(z)D_2(z)D_3(z)},$$

where

$$N_1(z) = 1 + z^{-1} \qquad D_1(z) = 1 - 1.2z^{-1} + 0.7z^{-2}$$

$$N_2(a) = 1 + z^{-2} \qquad D_2(z) = 1 - z^{-1} + 0.9z^{-2}$$

$$N_3(z) = 1 - 0.7z^{-1} + z^{-2} \qquad D_3(z) = 1 - 0.7z^{-1}$$

a. Sketch the pole/zero diagram.
b. Pair the $N_i(z)$ and $D_j(z)$ to form three corresponding sections $H_k(z)$ for minimum roundoff noise.
c. Assuming that the relative magnitudes of the peak factors $\|H_k\|_\infty / \|H_k\|_2$ are determined primarily by the $D_k(z)$, give a good ordering of the $H_k(z)$.

11.9 For the ARMA lattice in figure 10.12, find the sensitivities $\partial H(z)/\partial c_p$ for the tap coefficients c_p. Which coefficient is expected to be the most sensitive to quantization? The least sensitive?

11.10 Derive $\underline{F}(z)$ and $\underline{G}(z)$ in (11.6.1), and $\hat{\underline{F}}(z)$ and $\hat{\underline{G}}(z)$ in (11.6.2).

11.11 For the second-order filter

$$H(z) = \frac{1}{1 - 0.94z^{-1} + 0.91z^{-2}}$$

a. Predict the maximum zero-input limit-cycle amplitude if each multiplication product is rounded to the nearest integer value.
b. Predict the frequency of the zero-input limit cycles.
c. Calculate the output $y(n)$ for $n = 0, 1, \ldots, 15$, given that $y(-2) = y(-1) = 8$ and $x(n) = 0$, and check your result with (a) and (b).

11.12 For $\mathbf{A}$, $\underline{b}$, $\underline{c}^t$ satisfying the second-order optimality conditions in (11.6.6), show that (11.6.8) holds, and thus by (11.6.4) that the general optimality condition in (11.6.5) is indeed satisfied after scaling.

12. DIGITAL-FILTER IMPLEMENTATION

12.0 INTRODUCTION

Digital filters and signal-processing systems are implemented in two basic ways: software in general-purpose computers, and special-purpose hardware. The software approach is relatively straightforward, given the computational structures derived in previous chapters. Questions such as the numbers of multiplications and adds, coefficient sensitivity, roundoff noise, in-place FFT computation, etc., must be addressed; but the hardware considerations in that case are restricted to how fast the computer must be, how much storage is required, whether to add a high-speed floating-point unit or array processor, and so forth. In the highest-speed applications or for production quantities, however, it is often necessary to design special-purpose hardware to satisfy the system specifications at an acceptable cost. The advent of custom VLSI (very large-scale integration) has made this approach even more attractive and powerful, and promises to provide systems of ever-increasing complexity and speed. In this chapter, we describe special-purpose hardware implementations of digital filters and related signal-processing systems that take advantage of the highly parallel computational structure of these systems. This material assumes a basic knowledge of binary number systems and digital logic.

12.1 BIT-SERIAL ARITHMETIC AND VLSI

In general-purpose computers, the instructions are executed sequentially, and hence the speed of the computer is determined by how fast each instruction is executed. Since multiplications are among the slowest instructions, fast multipliers are often incorporated

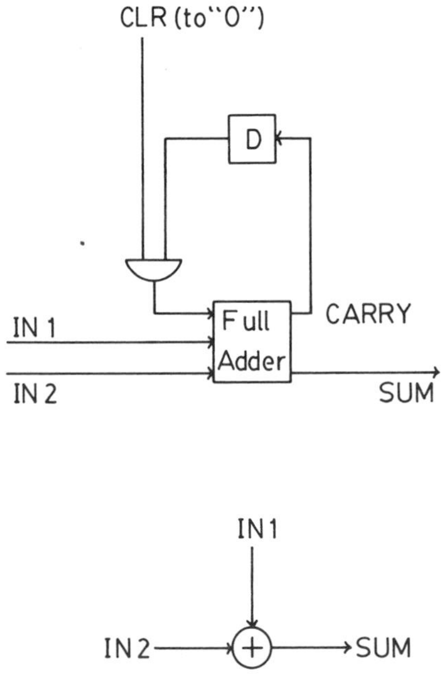

Figure 12.1. Serial two's-complement adder.

to enhance the speed of the machine. These multipliers invariably incorporate bit-parallel adders, and a variety of other schemes, to achieve the rapid accumulation of partial products. The ultimate speed is provided by the *array multiplier*, which includes a binary full adder for each bit in each partial product sum. Therefore, to multiply $(k+1)$-bit numbers by $(b+1)$-bit numbers, the array multiplier requires about $b(k+1)$ full adders.

In digital filter implementation, we are concerned about speed, too, for this determines the maximum throughput rate of the system, i.e., the real-time sampling rate (f_s) and filter order (N). However, considering any of the available filter structures, we realize that the implied operations or instructions (multiplications, additions, and stores) need not occur sequentially, but can usually be realized concurrently or *in parallel*. Hence, the throughput rate is determined not only by the speed of each operation, but also by the degree of parallelism in the implementation. Bit-serial arithmetic is then of potential interest because useful tradeoffs can be made between the complexity of each operation and the degree of parallelism. This is especially true in VLSI implementations, where considerations of modularity, local connectivity, and propagation delay make bit-serial arithmetic highly attractive [3, 57].

The key hardware component in these implementations is the bit-serial adder. Such an adder is shown in figure 12.1. The two serial inputs are assumed to be in two's-complement form and synchronized to a clock signal, with the least significant bits (LSB's) first. The full adder is a combinational logic circuit which adds the two present input bits and a carry bit to produce a sum bit and an output carry bit. The output carry is stored in a one-bit delay (D) or flip-flop to be input during the next clock cycle. After the most significant bits (MSB's) or sign bits are added to complete the addition, a new addition may begin immediately in the next clock cycle. Since a nonzero carry bit may still be stored in D from the previous addition, a *clear* (CLR) timing signal, which is zero during each LSB clock cycle, forces the input carry to zero via the AND gate.

Assuming that the data samples are $(k+1)$-bit two's-complement numbers, an addition requires $k + 1$ clock cycles. Likewise, a one-sample delay (z^{-1}) implies a $(k+1)$-bit delay and hence is realized by a shift register comprised of $k + 1$ flip-flop stages. The remaining hardware component is the multiplier, for which there are a variety of configurations. We will describe a *pipelined* serial multiplier which accepts both the data and coefficients in serial two's-complement form [58]. To be "pipelined" means that the serial data samples are introduced one after the other with no extra clock cycles in between, and the $(k+1)$-bit rounded products likewise emerge in serial pipelined form. The $(b+1)$-bit coefficients are also input in serial form, but with $(k - b)$ zero bits in between (assuming $k > b$) to synchronize the data and coefficients.

To describe the operation of this multiplier, let $a = a_k \ldots a_1 a_0$ denote a two's-complement sample having the value

$$a = -a_k 2^k + \sum_{i=0}^{k-1} a_i 2^i \tag{12.1.1}$$

and $c = c_0 . c_{-1} \ldots c_{-b}$ denote a fractional two's-complement coefficient with value

$$c = -c_0 + \sum_{i=1}^{b} c_{-i} 2^{-i}. \tag{12.1.2}$$

The product $a \cdot c$ is then produced by the following sequence of one-bit multiplications by the c_{-i} bits, shifts (delays) to multiply by 2, additions, and sign extensions:

$$\begin{array}{rl}
 & a_k \ldots a_1 a_0 \\
\times & \underline{c_0 . c_{-1} \ldots c_{-b}} \\
c_{-b} \times & a_k a_k \ldots a_1 \\
+ c_{-b+1} \times & \underline{a_k \ldots a_1 a_0} \\
 & d_s d_k \ldots d_1 \\
 & \vdots \\
 & g_s g_k \ldots g_1 \\
+ c_{-1} \times & a_k \ldots a_1 a_0 \\
 & \underline{+1} \qquad \leftarrow (\text{"1"}) \\
 & h_s h_k \ldots h_1 \\
- c_0 \times & \underline{a_k \ldots a_1 a_0} \\
 & p_k \ldots p_1 p_0
\end{array}$$

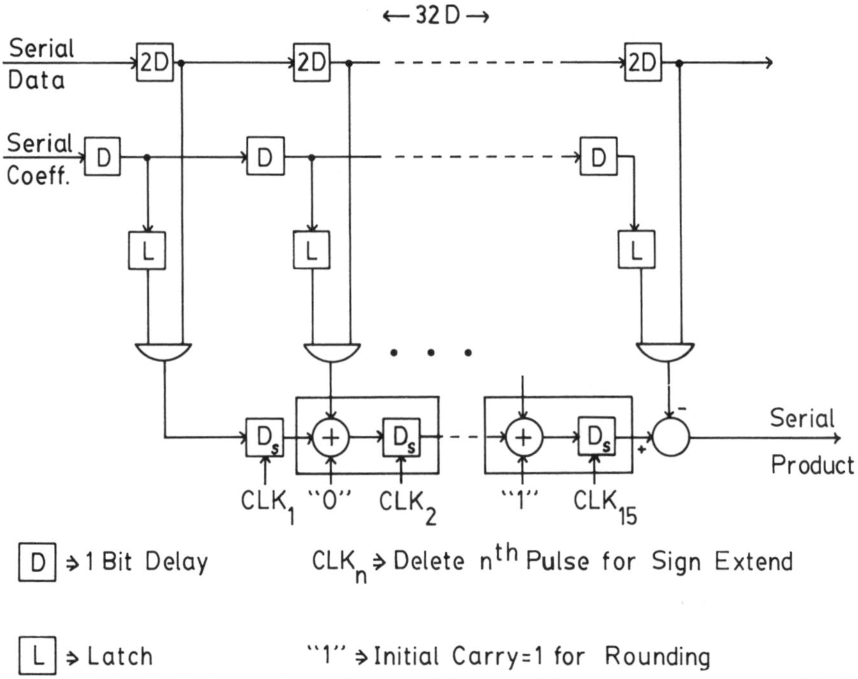

Figure 12.2. Serial pipelined two's-complement multiplier.

Note that each partial sum and the first partial product must be *sign extended* by adding a sign bit (h_s, g_s, . . . d_s, a_k) to align it with the next partial product and maintain the correct sign; while the corresponding LSB's (h_0, g_0, . . . d_0, a_0) are dropped (but with the associated carries being propagated) to maintain $(k+1)$-bit samples. The sign bit of the partial sum equals the MSB (h_k, g_k, . . . d_k) unless an overflow occurs in the partial sum, in which case it equals the complement of the MSB. An overflow is detected when the input carry to the MSB addition differs from the output carry from that addition. The LSB of each partial sum is automatically dropped by being overwritten by the corresponding sign extension from the preceding multiplication. Rounding is done by setting the input carry to "1" when generating h_0 so that an output carry bit will propagate to the h_1 addition if the product should be rounded up in value. The final operation is a subtraction if the coefficient is negative ($c_0 = 1$).

As a simple example, we multiply -30 (100010) by -0.25 (1.110) in the form of the above diagram, as follows:

```
      100 010
  ×     1.110
     000000
     100010
    110001
    100010
        +1
    101010
 -  100010
    001000
```

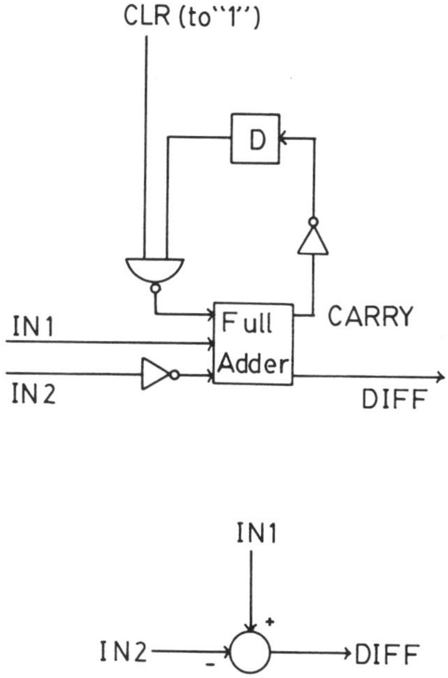

Figure 12.3. Serial two's-complement subtractor.

The rounded answer (+8) is correct). Note that the sign extensions follow the rules given above.

The pipelined multiplier is shown in figure 12.2 for $(b + 1) = 16$. The serial data sample $(a_k \ldots a_1 a_0)$ is delayed repeatedly to produce the required multiplications by powers of 2. These are 2-bit delays (2D), not 1-bit as expected, because extra delays are inserted elsewhere between adders to minimize propagation delay, and we must synchronize the data streams. The serial coefficient $(c_0 . c_{-1} \ldots c_{-b})$ is converted to parallel form and stored in the *latches* (L), starting with c_{-b} in the left-most latch. The latches are loaded from left to right every two clock cycles to synchronize each bit c_{-i} with its respective delayed data sample. Hence, the latches are *strobed* (loaded) every $(k + 1)$ cycles, but in staggered form. The one-bit multiplications $c_{-i}(a_k \ldots a_1 a_0)$ are performed by the AND gates, and these partial products are accumulated by the serial adders, with extra delays (D_s) in between to reduce the maximum propagation delay to that for one adder. The D_s notation indicates that sign extensions are performed in conjunction with these delays. Note the final subtractor to accumulate $-c_0(a_k \ldots a_1 a_0)$ and the adder with an initial carry input of "1" for rounding.

Delays may be cascaded with the multiplier to increase the coefficient magnitude by powers of two (as for the α_{1i} and β_{1i} coefficients of the cascade form, which can have magnitudes of up to two); but then the possibility of product overflow exists and must be included in the overflow check for the summation following the multiplier. Remember that overflows are acceptable in partial sums (and hence products) in the network as long as they cancel out in the final sums.

A serial subtractor is shown in figure 12.3. Subtraction is simply the negation of the

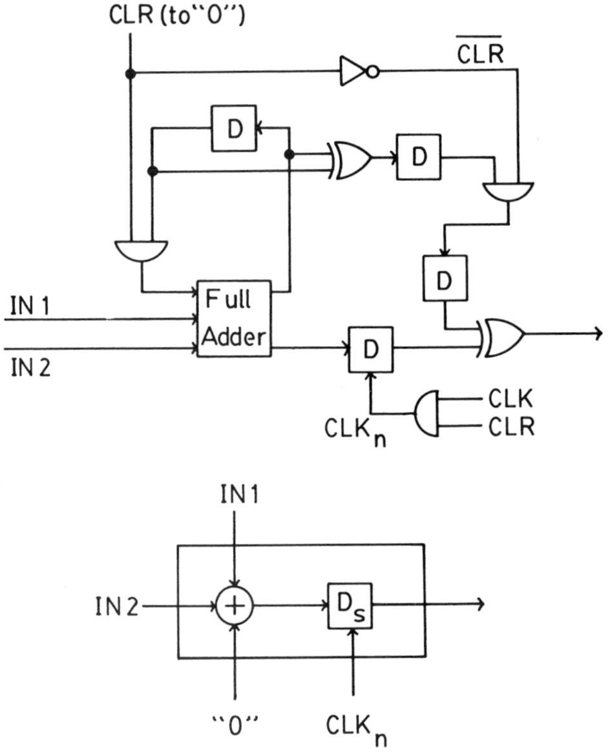

Figure 12.4. Adder/delay combination with sign extension for pipelined multiplier.

subtrahend (IN2), followed by addition. Negation of a two's-complement number is implemented by complementing all the bits of the number and adding "1" in the LSB. Hence in the subtractor, an inverter complements all bits of IN2, and the carry loop is modified to cause an initial carry input of "1" during the LSB cycle.

An implementation of the adder/delay combination with sign extension is shown in figure 12.4. The MSB is extended by deleting the clock pulse to the delay where it is stored during the following LSB (CLR) cycle. Thus the LSB is dropped, but not the carry from the LSB cycle. An exclusive-OR gate compares the input and output carries of the adder. If these carries differ during the MSB cycle (LSB cycle after one delay), a "1" is generated by the AND gate on the right to complement the extended sign bit. For the final adder/delay stage in the multiplier, which needs an input carry of "1" to implement rounding, the carry loop is modified as in the subtractor of figure 12.3.

Let us now consider how to combine bit-serial adders, delays, and multipliers into a digital filter network. The pipelined multiplier has a delay of $2(b + 1)$ bits or clock cycles from the LSB of the serial data input to the LSB of the serial product output. This is seen most easily in figure 12.2 from the upper path of $2(b + 1)$ delays to the $(-)$ input of the subtractor. Hence, if $b = k$, this represents a two-sample delay (z^{-2}) in the network (without multiplexing). If, however, $(b + 1) = (k + 1)/2$, which is not unreasonable in many applications, the multiplier represents only a one-sample delay (z^{-1}). For

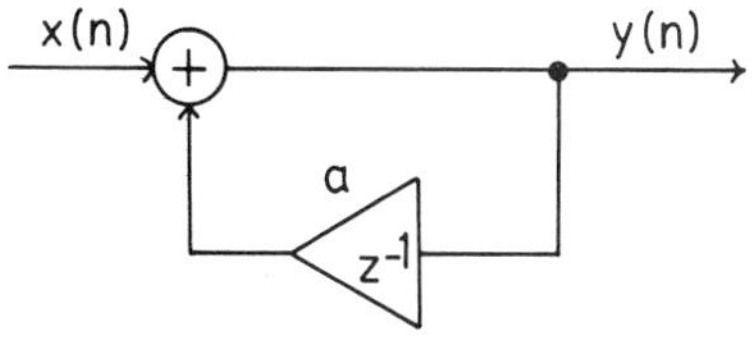

Figure 12.5. First-order filter showing delay inherent in serial multiplier.

nonrecursive (FIR) filters, this delay is included in all feedforward paths, and thus in the overall system function, to produce $z^{-1}H(z)$ or $z^{-2}H(z)$. In recursive (IIR) filters, however, the multiplier delays must be absorbed into the feedback loop delays so that each loop delay is unchanged. If the multiplier contains a one-sample delay or less, this can be done, but not if it has more than a one-sample delay (without multiplexing).

A simple first-order recursive filter where the multiplier incorporates a one-sample delay is shown in figure 12.5. Note that this circuit contains $b + 1$ serial adders (b in the multiplier), each of which is usefully working 100 percent of the time because of pipelining and functional parallelism. This high degree of parallelism is the key to the speed of bit-serial implementations of digital filters and signal processors, even though each addition requires $k + 1$ clock cycles. Thus, for example, a filter with $k + 1 = 16$ and a 16-MHz clock can process data at a 1-MHz sampling rate, which is quite respectable.

A further complication arises in the second-order recursive case, and that is the possibility of overflow oscillation. If optimal or normal state-space structures are employed, these oscillations are precluded, but at the cost of approximately twice the number of multipliers. In compensation, however, the required number of coefficient bits ($b + 1$) and data bits ($k + 1$) are both reduced (see chapter 11). If, on the other hand, we choose the canonical cascade or parallel forms, overflow oscillations must be explicitly precluded by, for example, implementing saturation arithmetic. We will not go into the circuit details here except to note that one cannot set the overflowed value to $\pm$ full scale until an overflow is, in fact, detected. This implies waiting for all $k + 1$ bits of the sum to emerge from the summation and thus represents another $(k+1)$-bit delay (z^{-1}). Hence, saturation arithmetic cannot be implemented in second-order structures unless zero samples (or other multiplexed samples) are inserted between the input data samples to the network.

Bit-serial implementations are readily multiplexed to process either multiple input channels or multiple sections of a single filter, or both [3, 59]. This usually alleviates all problems of insufficient loop delay and, more importantly, provides further efficiency of circuit utilization. For example, if M parallel channels of the same filter are needed and the clock rate can be increased by the factor M, the samples of the M inputs can be interleaved (sample by sample) into a single serial-data input to the filter, and the outputs will emerge in a similarly interleaved data stream. The only changes required in the implementation are the increase in the clock rate to $M(k + 1)f_s$ and a lengthening of all z^{-1} delays to $M(k + 1)$ bits ($k + 1$ for each multiplexed sample). If different filters are required for the different channels, the coefficients of the filter are changed (cycled) accordingly. Clearly, this also applies to the multiplexing of the L sections of the parallel form, where the inputs to all sections are the same, and the L multiplexed outputs are accumulated to form a single output.

For the cascade form, the output of each section is the input to next section, and hence a feedback loop is required to feed the multiplexed outputs back to the input, except for the final output. Again, for L sections, the clock rate is increased to $L(k + 1)f_s$, and the z^{-1} delays are $L(k + 1)$ bits long. Combinations of input multiplexing and section multiplexing are straightforward and result in a clock rate of $ML(k + 1)f_s$ and delays of $ML(k + 1)$ bits. For example, for a clock rate of 16 MHz and $(k + 1) = 16$ bits, a second-order section in a speech-processing system with a sampling rate of $f_s = 8$ kHz can be multiplexed by a factor of 125 to process 125 channels, or to implement tenth-order filters (5 sections) for 25 channels, and so forth.

As mentioned earlier, the bit-serial approach described in this section is well suited to VLSI implementation [57]. We will briefly discuss the features of this approach using terminology developed to characterize VLSI circuits [60, 61]. The implementation is *flow-simple* (F) because each circuit element is used only once per "problem instance," i.e., clock cycle. It is *cell-simple* (C) because each circuit element is, in fact, usefully employed during each clock cycle. These two properties together imply that the circuit is *completely pipelined* (P), i.e., each circuit element is used once and only once per clock cycle. The circuit will be *locally connected* (L) if the circuit elements are connected only to neighboring elements, except for the *broadcast* distribution of power, ground, and the basic clock. This property (L) is desirable because it eases circuit layout and minimizes wiring crossovers. If implemented, it prohibits the broadcast distribution of timing signals, such as the CLR's and latch clocks. Instead, a few basic timing signals (e.g., a single CLR) are routed through intervening delays around the circuit, and are further delayed and/or gated locally to form the needed timing signals. The circuit is *systolic* if it is flow-simple and locally connected. To summarize, therefore,

$$P = F \cap C \quad \text{and} \quad S = F \cap L. \tag{12.1.3}$$

The circuits we have described are completely pipelined (P) and are readily designed to be locally connected (L), in which case they are also systolic (S).

12.2 DISTRIBUTED ARITHMETIC

There are, of course, many other ways to implement digital filters in hardware, with each having its pros and cons. One that also takes advantage of the fixed algorithmic structure of filtering is called *distributed arithmetic*—it trades memory (often read-only memory (ROM)) for computation [62, 63].

Consider the implementation of a second-order section in the direct form of figure 12.6. The delays to the right are shown in dotted lines because, in a cascade-form network, they can be shared with the following section in all but the last section. For notational convenience, we denote the bit-serial two's-complement data by $a_1 = x(n)$, $a_2 = x(n - 1)$, $a_3 = x(n - 2)$, $a_4 = y(n - 1)$, and $a_5 = y(n - 2)$, as shown. Let the jth bit of the integer a_i be a_{ij}, that is, $a_i = (a_{ik} \ldots a_{i0})$. The corresponding coefficients are c_i,

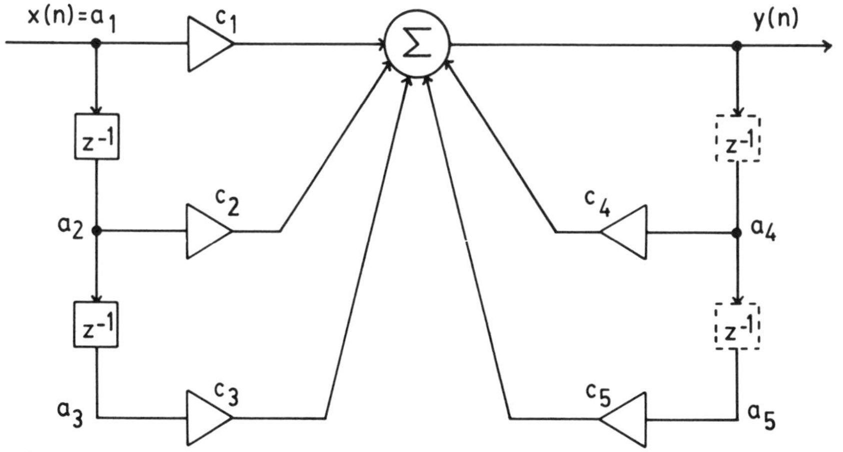

Figure 12.6. Second-order filter in direct form.

$i = 1, \ldots, 5$, and may have magnitudes $|c_i| < 2$ for c_2 and c_4, in particular. The output $y(n)$ is thus given by

$$y(n) = \sum_{i=1}^{5} c_i a_i$$

$$= \sum_{i=1}^{5} c_i \left(-a_{ik} 2^k + \sum_{j=0}^{k-1} a_{ij} 2^j \right). \tag{12.2.1}$$

Reversing the order of the summations, we have

$$y(n) = 2^k \left(-s_k + \sum_{j=0}^{k-1} 2^{j-k} s_j \right) \tag{12.2.2}$$

where

$$s_j = \sum_{i=1}^{5} c_i a_{ij}.$$

Since the coefficients c_i are fixed, there are only $2^5 = 32$ possible values for the sums s_j, depending upon the values of the five data bits a_{ij}, $i = 1, \ldots 5$. Hence, if the 32 possible values are precomputed and stored in a memory (ROM or RAM), we can look them up as required and need not compute these sums of products over and over again in the circuit. The five data bits from the five-bit address to the memory.

Actually, it is the scaled sums $2^k s_j$ that are stored in the memory in two's complement. These are then rescaled (iteratively) by $2^{j-k}, j = 0, 1, \ldots k$, and accumulated to form $y(n)$, with the final sum $2^k s_k$ being subtracted. This is accomplished by a fast $(k+1)$-bit parallel

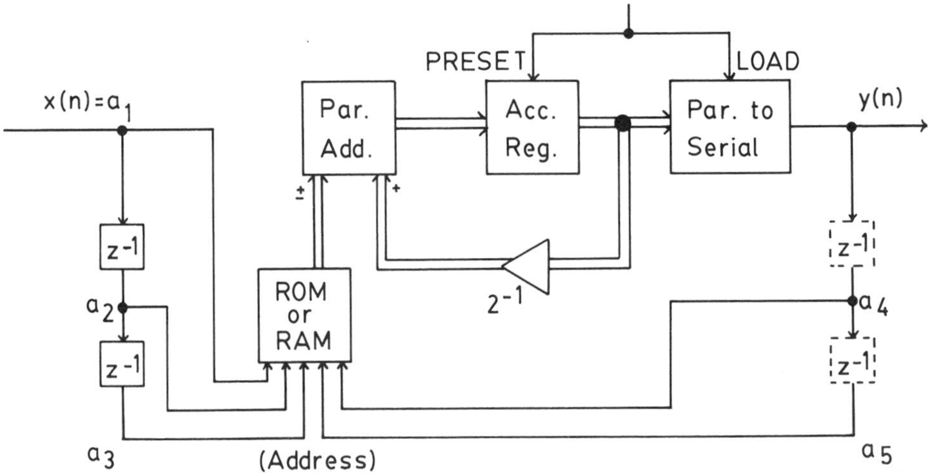

Figure 12.7. Distributed-arithmetic implementation of the second-order filter in figure 12.6.

accumulator with built-in scaling by 2^{-1}, as depicted in figure 12.7. Bit-parallel signals are indicated by double lines.

The operation of this circuit proceeds as follows: During the LSB clock cycle, the a_{i0} bits form the memory address, and the resulting memory output $2^k s_0$ is added to the preset value in the accumulator register (to implement rounding). During the next clock cycle, the a_{i1} bits cause $2^k s_1$ to be added to $2^{k-1} s_0$. Next, $2^k s_2$ is added to $2^{k-1} s_1 + 2^{k-2} s_0$. This accumulation continues until $2^k s_k$ is subtracted from the accumulated total during the $(k+1)$st (MSB) cycle to produce the output $y(n)$. At the beginning of the next LSB cycle, $y(n)$ is transferred to the parallel-to-serial register; while the accumulator register is preset to begin the new accumulation of $y(n + 1)$. Rounding is effected by presetting the accumulator register to 010. . . 0 and dropping the LSB of each $(k+1)$-bit accumulated partial sum as it is scaled by 2^{-1} (shifted one place to the right).

Note that the parallel-to-serial conversion represents a $(k+1)$-bit delay in $y(n)$, which must be deducted from the following z^{-1} delay. Comparing this implementation to that of the preceding section, we note that there are only $k + 1$ full adders here (in the parallel adder) versus $3(b + 1)$ to $5(b + 1)$ in a typical second-order section there. However, we have added some memory, and there must be additional circuitry (carry-lookahead, etc.) to speed up the parallel adder. Even then, the clock rate, and thus the throughput rate, cannot be as high here because of the carry propagation delays in the parallel adder.

We conclude this section by mentioning several other possible applications of distributed arithmetic. In the second-order state-space structure of figure 11.8, there are three 3-input summation notes which can be implemented by three memory/accumulator circuits, each requiring only 8 words of memory. On the other hand, an Mth-order direct-form FIR structure using distributed arithmetic would require a memory of 2^{M+1} words if implemented by a single memory/accumulator, which would often be prohibitively large. However, for $M + 1 = KL$, it may be practical to employ K memory/accumulator circuits having 2^L words of memory each, followed by $K - 1$ bit-serial adders.

12.3 FIR IMPLEMENTATION VIA THE DFT

Implementation of an FIR filter corresponds to a computation of the linear convolution of the data sequence $x(n)$ and the finite-duration impulse response $h(n)$ of order M. If the data sequence also has finite duration (N_1), the linear convolution can be implemented by multiplying the corresponding DFT's with zero padding, as described in Section 7.2. The potential advantage of this approach is that the FFT algorithm can be used to compute the DFT's to reduce the number of multiplications and additions. Specifically, direct computation of the convolution would require about MN_1 real computations (multiplications and additions); while the FFT of the zero-padded data involves about $N_2 \log N_2$ complex computations for $N_2 \geqslant N_1 + M$, the product of the DFT's takes N_2 complex multiplications, and the inverse FFT of the product requires another $N_2 \log N_2$ complex computations. (We assume that the zero-padded DFT of $h(n)$ is already available and need not be computed.) This approach has been called *high-speed convolution* [64].

Often, the data length N_1 is too large to permit the computation of a single DFT of the data. This is especially true in a real-time system where the input data sequence is of indefinite duration. In such cases, the data sequence can be divided into shorter segments of convenient length, and each segment convolved with $h(n)$ via DFT's. However, there will be end effects from each convolution, as described in sections 7.1 and 7.2, which must be accounted for as the segments are recombined to produce the output sequence.

There are two procedures to effect this segmentation and recombination of the data. In the *overlap-add method*, the input sequence is divided into nonoverlapping segments $x_k(n)$ of length $N_1 = N_2 - M$, where N_2 is the desired length of the FFT's, and each zero-padded segment is convolved with $h(n)$ to produce an output segment $y_k(n)$ of length N_2. Since

$$x(n) = \sum_k x_k(n) \tag{12.3.1}$$

and convolution is a linear operation, the output sequence $y(n)$ is simply

$$y(n) = \sum_k y_k(n). \tag{12.3.2}$$

However, each output segment $y_k(n)$ overlaps the following segment $y_{k+1}(n)$ by M samples and thus (12.3.2) implies actual addition of the segments, not simply concatenation as in (12.3.1). Hence, the output segments are "overlapped" and "added." This is illustrated in figure 12.8.

In the *overlap-save method*, on the other hand, it is the input segments that are overlapped; while the output segments are truncated to be nonoverlapping and then concatenated. In particular, input segments $x_k(n)$ of length N_2 are taken with an overlap of M samples between segments. Circular convolution of the $x_k(n)$ and $h(n)$ via N_2-point DFT's produce output segments $y_k(n)$ whose first M samples are not usable because they are circularly wrapped and superimposed, as discussed in section 7.1. Hence, the first M samples of each $y_k(n)$ are discarded, as depicted in figure 12.9, and the resulting N_1-point segments are concatenated to produce $y(n)$.

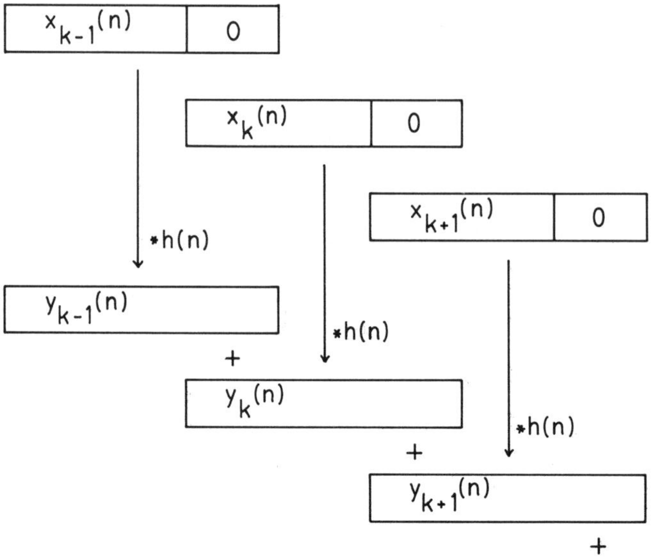

Figure 12.8. Illustration of overlap-added method.

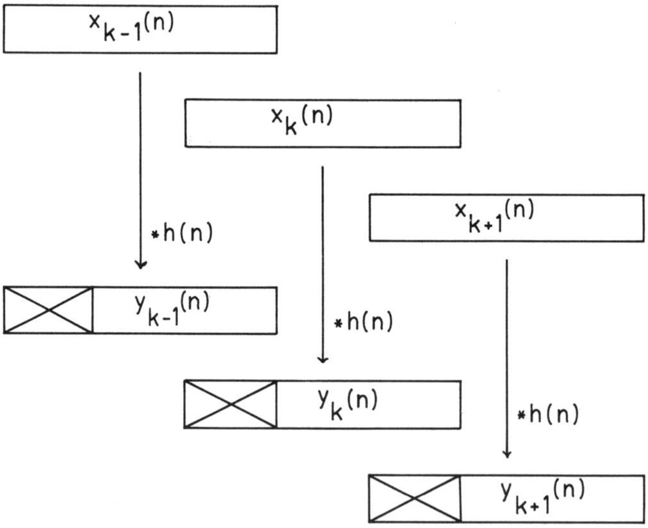

Figure 12.9. Illustration of overlap-save method.

In both methods, therefore, the generation of N_1 output samples requires about $2N_2 \log N_2$ complex computations. Clearly, to avoid excessive overlap, we will usually choose $N_1 > M$ and often $N_1 \gg M$. Stockham [64] has found that for an actual increase in the speed of computation by "high-speed" convolution, M must be at least 32. Approximate factors for the increase in speed, as estimated by Stockham, are given below:

Filter Order (M):	16	32	64	128	256	512	1024
Speed-up Factor:	0.7	1.3	2.3	4	6	13	24

For large values of M, the impulse response $h(n)$ may also be segmented to produce DFT's of the desired length.

12.4 BLOCK IIR IMPLEMENTATIONS

In the preceding section, we have seen how to implement finite convolutions or FIR filters by processing segments or "blocks" of data. It is also possible to implement IIR filters in block form, with several potential advantages. One of these, however, is usually not a reduction in computation; for although FFT's may be utilized in certain cases, the order (N) of the IIR filter must exceed 20 to 40 (depending upon the algorithm) for this to result in reduced computation [65]. Instead, one advantage is that the computations are highly parallel and, with block length L, are performed at the slower rate f_s/L. Therefore, block implementation is well suited to array processors. Another advantage is that, for the purposes of quantization analysis, the poles of the filter are effectively raised to the Lth power, and thus the quantization effects of coefficient sensitivity, roundoff noise, and limit cycles are all reduced [67, 68].

Anticipating some of the block structures to be derived, we first consider the first-order difference equation

$$a_0 y(n) + a_1 y(n-1) = b_0 x(n) + b_1 x(n-1). \tag{12.4.1}$$

The output $y(n)$ may be computed by

$$y(n) = a_0^{-1}[-a_1 y(n-1) + b_0 x(n) + b_1 x(n-1)], \tag{12.4.2}$$

which corresponds to the direct-form structure in figure 12.10(a), where D denotes a one-sample delay. One delay is eliminated by using direct form I, shown in figure 12.10(b).

Now consider an Nth-order digital filter with system function

$$\frac{Y(z)}{X(z)} = H(z) = \frac{B(z)}{A(z)} \tag{12.4.3}$$

where $B(z)$ and $A(z)$ are Nth-order polynomials in z^{-1}. In terms of convolution, we then have

$$a(n) * y(n) = b(n) * x(n) \tag{12.4.4}$$

or, in matrix form for causal sequences $x(n)$ and $y(n)$,

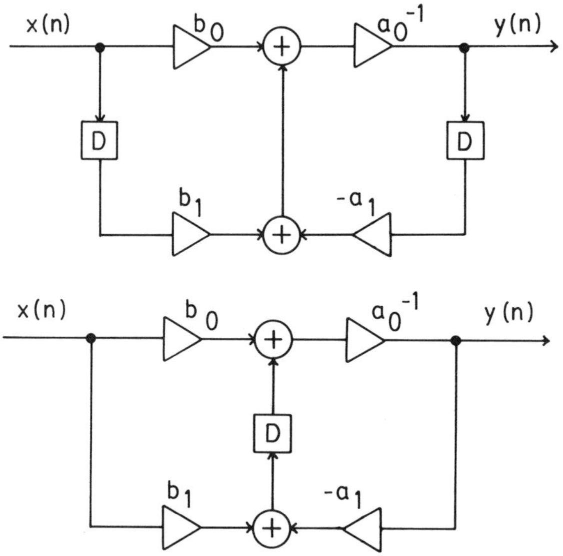

Figure 12.10. First-order digital filter structures.

$$\begin{bmatrix} 1 & 0 & 0 \cdots \\ a_1 & 1 & 0 \cdots \\ a_2 & a_1 & 1 \cdots \\ \vdots & & \ddots \\ a_N & a_{N-1} & \cdots \\ 0 & a_N & \ddots \\ \vdots & & \ddots \\ 0 & 0 & \cdots \end{bmatrix} \begin{bmatrix} y_0 \\ y_1 \\ y_2 \\ \vdots \\ y_N \\ \vdots \end{bmatrix} = \begin{bmatrix} b_0 & 0 & 0 \cdots \\ b_1 & b_0 & 0 \cdots \\ b_2 & b_1 & b_0 \cdots \\ \vdots & & \ddots \\ b_N & b_{N-1} & \cdots \\ 0 & b_N & \ddots \\ \vdots & & \ddots \\ 0 & 0 & \cdots \end{bmatrix} \begin{bmatrix} x_0 \\ x_1 \\ x_2 \\ \vdots \\ x_N \\ \vdots \end{bmatrix} \tag{12.4.5}$$

Segmenting the $x(n)$ and $y(n)$ sequences (vectors) into nonoverlapping blocks (vectors) $\underline{x}_k$ and $\underline{y}_k$ of length L, (12.4.5) becomes [65]

$$\begin{bmatrix} \mathbf{A}_0 & 0 & 0 \cdots \\ \mathbf{A}_1 & \mathbf{A}_0 & 0 \cdots \\ 0 & \mathbf{A}_1 & \mathbf{A}_0 \cdots \\ \vdots & & \ddots \end{bmatrix} \begin{bmatrix} \underline{y}_0 \\ \underline{y}_1 \\ \underline{y}_2 \\ \vdots \end{bmatrix} = \begin{bmatrix} \mathbf{B}_0 & 0 & 0 \cdots \\ \mathbf{B}_1 & \mathbf{B}_0 & 0 \cdots \\ 0 & \mathbf{B}_1 & \mathbf{B}_0 \cdots \\ \vdots & & \ddots \end{bmatrix} \begin{bmatrix} \underline{x}_0 \\ \underline{x}_1 \\ \underline{x}_2 \\ \vdots \end{bmatrix} \tag{12.4.6}$$

where $\mathbf{A}_0$, $\mathbf{A}_1$, $\mathbf{B}_0$, and $\mathbf{B}_1$ are $L \times L$ Toeplitz submatrices given by

$$\mathbf{A}_0 = \begin{bmatrix} 1 & 0 & 0 & \cdots & \cdots & 0 \\ a_1 & 1 & 0 & \cdots & \cdots & 0 \\ \vdots & \ddots & \ddots & & & \vdots \\ a_N & \cdots & a_1 & 1 & 0 \cdots & 0 \\ 0 & \ddots & & \ddots & \ddots & \vdots \\ \vdots & & \ddots & & \ddots & \vdots \\ 0 & \cdots & 0 & a_N \cdots & a_1 & 1 \end{bmatrix} \qquad \mathbf{B}_0 = \begin{bmatrix} b_0 & 0 & 0 & \cdots & \cdots & 0 \\ b_1 & b_0 & 0 & \cdots & \cdots & 0 \\ \vdots & \ddots & \ddots & & & \vdots \\ b_N & \cdots & b_1 & b_0 & 0 \cdots & 0 \\ 0 & \ddots & & \ddots & \ddots & \vdots \\ \vdots & & \ddots & & \ddots & \vdots \\ 0 & \cdots & 0 & b_N \cdots & b_1 & b_0 \end{bmatrix}$$

$$\mathbf{A}_1 = \begin{bmatrix} 0 & \cdots & 0 & a_N & \cdots & a_2 & a_1 \\ 0 & & & & \ddots & & a_2 \\ \vdots & & & & & \ddots & \vdots \\ & & & & & & a_N \\ & & & & & & 0 \\ & & & & & & \vdots \\ 0 & \cdots & & & & & 0 \end{bmatrix} \qquad \mathbf{B}_1 = \begin{bmatrix} 0 & \cdots & 0 & b_N & \cdots & b_2 & b_1 \\ 0 & & & & \ddots & & b_2 \\ \vdots & & & & & \ddots & \vdots \\ & & & & & & b_N \\ & & & & & & 0 \\ & & & & & & \vdots \\ 0 & \cdots & & & & & 0 \end{bmatrix}$$

Given the initial conditions $\underline{x}_{-1} = \underline{0}$ and $\underline{y}_{-1} = \underline{0}$, each row of (12.4.6) is of the form

$$\mathbf{A}_0 \underline{y}_k + \mathbf{A}_1 \underline{y}_{k-1} = \mathbf{B}_0 \underline{x}_k + \mathbf{B}_1 \underline{x}_{k-1} \tag{12.4.7}$$

and thus $\underline{y}_k$ may be computed as

$$\underline{y}_k = \mathbf{A}_0^{-1}[-\mathbf{A}_1 \underline{y}_{k-1} + \mathbf{B}_0 \underline{x}_k + \mathbf{B}_1 \underline{x}_{k-1}]. \tag{12.4.8}$$

Note that (12.4.7) and (12.4.8) are simply vector versions of (12.4.1) and (12.4.2) with matrix coefficients; and that (12.4.8) implies analogous block implementations, shown in figure 12.11, where Δ denotes a one-block delay [66].

All of the matrix multiplications in (12.4.8) can be performed using either direct multiplication or finite convolution via zero-padded FFT's [65]. However, for most practical IIR filters ($N < 20$), direct multiplication implies fewer multiplications. Many other structures for block implementation can also be derived [65–68], with various computational properties.

A particular class of block structures of special interest is based upon state-variable analysis [67, 68]. Given the state equations for the filter,

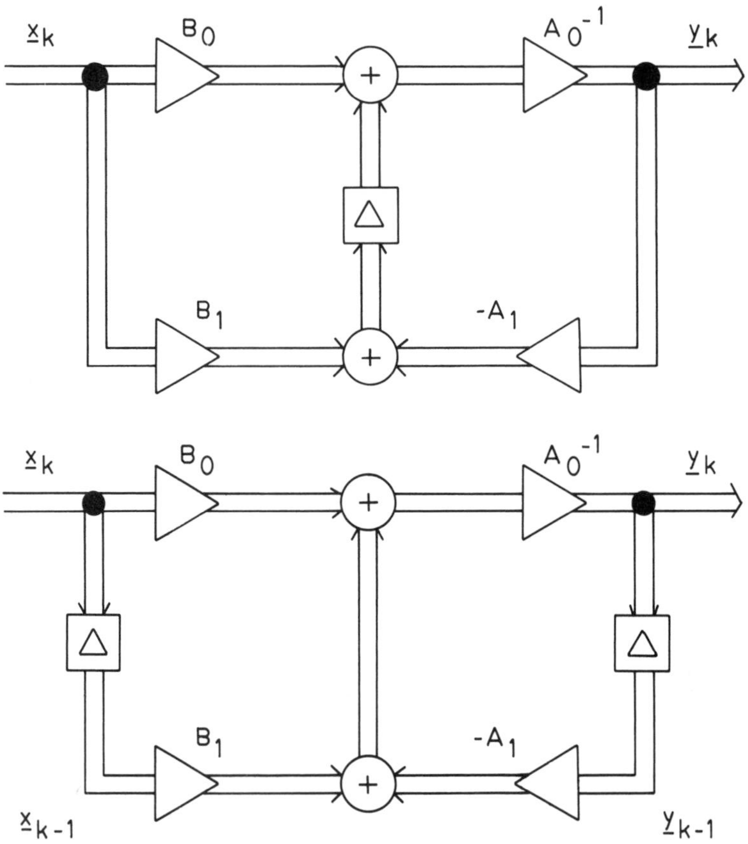

Figure 12.11. Block implementation structures.

$$\begin{aligned} \underline{s}(n+1) &= \mathbf{A}\underline{s}(n) + \underline{b}x(n) \\ y(n) &= \underline{c}^t\underline{s}(n) + dx(n), \end{aligned} \tag{12.4.9}$$

we can, if we choose, compute instead the state $\underline{s}(n+L)$ by

$$\underline{s}(n+L) = \mathbf{A}^L\underline{s}(n) + \mathbf{A}^{L-1}\underline{b}x(n) + \ldots + \underline{b}x(n+L-1) \tag{12.4.10}$$

and the outputs $y(n+i)$, $i = 0, 1, \ldots L-1$, by

$$y(n+i) = \underline{c}^t\mathbf{A}^i\underline{s}(n) + \underline{c}^t\mathbf{A}^{i-1}\underline{b}x(n) + \ldots + \underline{c}^t\underline{b}x(n+i-1) + dx(n+i). \tag{12.4.11}$$

Defining the input and output data blocks to be, as before,

$$\begin{aligned} \underline{x}_k &= [x(n), \quad x(n+1), \ldots, x(n+L-1)]^t \\ \underline{y}_k &= [y(n), \quad y(n+1), \ldots, y(n+L-1)]^t \end{aligned}$$

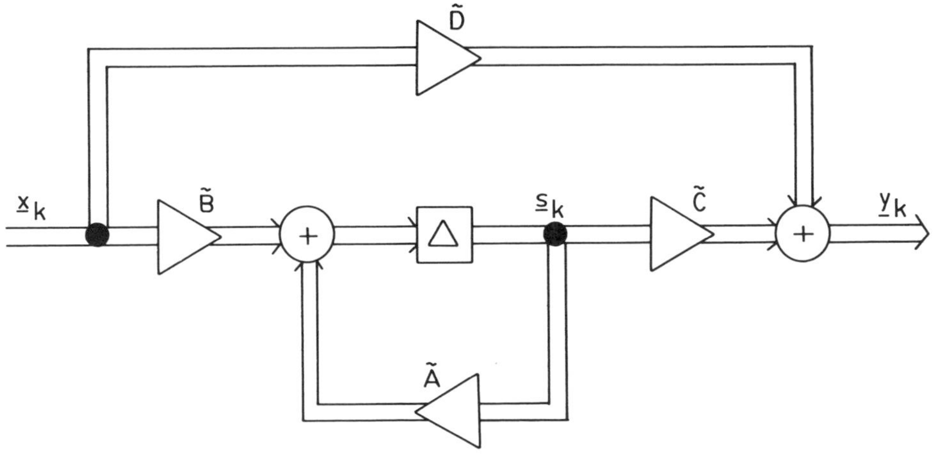

Figure 12.12. Block state-space structure.

for $n = kL$, (12.4.10) and (12.4.11) imply the block state equations

$$\underline{s}_{k+1} = \tilde{\mathbf{A}}\underline{s}_k + \tilde{\mathbf{B}}\underline{x}_k \quad \text{and} \quad \underline{y}_k = \tilde{\mathbf{C}}\underline{s}_k + \tilde{\mathbf{D}}\underline{x}_k \tag{12.4.12}$$

where $\underline{s}_k = \underline{s}(kL)$ and

$$\tilde{\mathbf{A}} = \mathbf{A}^L \qquad \tilde{\mathbf{B}} = [\mathbf{A}^{L-1}\underline{b} \mid \ldots \mid \mathbf{A}\underline{b} \mid \underline{b}]$$

$$\tilde{\mathbf{C}} = \begin{bmatrix} \underline{c}^t \\ \hline \underline{c}^t\mathbf{A} \\ \hline \cdot \\ \cdot \\ \cdot \\ \hline \underline{c}^t\mathbf{A}^{L-1} \end{bmatrix} \qquad \tilde{\mathbf{D}} = \begin{bmatrix} d & 0 & 0 \ldots\ldots 0 \\ \underline{c}^t\underline{b} & d & 0 \ldots\ldots 0 \\ \vdots & \ddots & \vdots \\ \underline{c}^t\mathbf{A}^{L-2}\underline{b} & \ldots & \underline{c}^t\underline{b} \quad d \end{bmatrix} \tag{12.4.13}$$

This implementation has also been termed *state decimation* [68] because we only compute every Lth state vector directly. Note that only the $\tilde{\mathbf{D}}$ matrix is Toeplitz, and thus only the multiplication of $\underline{x}_k$ by $\tilde{\mathbf{D}}$ could be computed as a convolution via FFT's although, as before, this will not usually save multiplications. The block state-space structure is depicted in figure 12.12.

Zeman and Lindgren [68] have shown that the number of multiplications is minimized for a general $(\tilde{\mathbf{A}}, \tilde{\mathbf{B}}, \tilde{\mathbf{C}}, \tilde{\mathbf{D}})$ by the block length

$$L_{\text{opt}} = \sqrt{2}N, \tag{12.4.14}$$

in which case the number of multiplications per output sample is approximately $3.41N + 0.5$. This includes the case of state decimation of the optimum Nth-order state-space structure having minimum roundoff noise [51, 52], which otherwise requires $N^2 + 2N + 1$ multiplications per output sample. Therefore, block implementation with $L \approx L_{opt}$ causes the multiplications to increase only linearly with N, not as $(N + 1)^2$.

Further savings in multiplications can be made by block implementation of a parallel combination of optimal or normal second-order sections (as described in section 11.6) at the cost of only a slight increase in roundoff noise [68]. In this case, $\mathbf{A}$ and thus $\tilde{\mathbf{A}} = \mathbf{A}^L$ are block diagonal matrices consisting of 2×2 submatrices along the main diagonal. The optimal block length is then

$$L_{opt} = 2\sqrt{N} \tag{12.4.15}$$

and the corresponding number of multiplications per output sample is about $2N + 2\sqrt{N} + 0.5$. Distributed arithmetic may also be employed to implement the vector and matrix multiplications (inner products) [68].

The reason for the reduced quantization effects in block implementations is that the poles of the multiple-input, multiple-output system $\tilde{\mathbf{H}}(z)$ (having $\underline{x}_k$ and $\underline{y}_k$ as input and output, respectively) are the original poles of $H(z)$ raised to the Lth power. Hence, the poles are moved toward the origin in the z plane and away from the unit circle. This fact is evident for the block state-space structures because the poles are the eigenvalues of $\tilde{\mathbf{A}} = \mathbf{A}^L$, and it has been proved in the general case, as well, by Barnes and Shinnaka [67]. They have also noted that if the original state-space structure is scaled, the corresponding block state-space structure will be, too, because the new state vectors are just samples of the original ones.

Quantization of the coefficients of $(\tilde{\mathbf{A}}, \tilde{\mathbf{B}}, \tilde{\mathbf{C}}, \tilde{\mathbf{D}})$ implies that the filter is no longer time invariant, but rather (slightly) periodically time varying. Such systems have been investigated by Meyer and Burrus [69, 70]. The roundoff noise is also no longer stationary, but has periodically varying variance with period L. Assuming rounding after summation (e.g., distributed arithmetic), the noise variance for the block structure is never greater than for the original structure and decreases during each period [67, 68]. The original variance is effectively spread out over the period, and hence the average variance for the block structure is $1/L$th the original variance. Limit cycles are also reduced because the poles are moved away from the unit circle; and if the block length is made sufficiently long such that $\|\tilde{\mathbf{A}}\| < 0.5$, limit cycles are eliminated altogether [67].

PROBLEMS

12.1 Add the two's-complement numbers 010011 (+19) and 010111 (+23) and observe that the six-bit sum is negative as the result of overflow. Then add 101010 (−22) to the sum and observe that the correct total sum (+20) is obtained, even though each partial sum overflowed.

12.2 When a digital filter is implemented in parallel form with serial arithmetic and multiplexing, a serial accumulator is required at the output to sum each set of L multiplexed section outputs. Draw a serial accumulator circuit.

12.3 A single second-order section is to be multiplexed to realize an eighth-order filter in each of 25 channels using 16-bit data and 10-bit coefficients with serial arithmetic. The sampling rate is 8 kHz. What clock rate is required for the digital circuits?

12.4 The sign of a two's-complement number may be changed (complemented) by leaving the least-significant "1" and all following "0's" unchanged and inverting the remaining (more significant) bits. For example, to complement the number 10101000, we leave 1000 unchanged and invert 1010 to produce 01011000. Design a simple serial two's-complementing circuit (without including an adder).

12.5 Sketch a distributed-arithmetic implementation of the allpass filter

$$H(z) = \frac{z^{-1} - a}{1 - az^{-1}}.$$

Gives the contents of the ROM for $a = 0.875$.

12.6 We wish to compute the linear convolution $y(n) = x(n) * h(n)$, where $x(n)$ is of length 4096, $h(n)$ is of length 256, and we have a machine that can only compute 512-point DFT's (and multiply, add, etc.).

a. How many DFT's and how many adds are required using the overlap-add method?

b. How many DFT's are required using the overlap-save method?

c. What is the length of $y(n)$?

12.7 For a second-order filter with state matrices

$$\mathbf{A} = \begin{bmatrix} 0.7 & 0.6 \\ -0.6 & 0.7 \end{bmatrix} \qquad \underline{b} = \begin{bmatrix} 0.4 \\ 0.3 \end{bmatrix}$$

$$\underline{c}^t = [0.5, \quad -0.2] \qquad d = 0.8$$

a. Find the block state matrices $\tilde{\mathbf{A}}$, $\tilde{\mathbf{B}}$, $\tilde{\mathbf{C}}$, $\tilde{\mathbf{D}}$ for $L = 3$.

b. Find the block coefficient matrices $\mathbf{A}_0^{-1}$, $\mathbf{A}_1$, $\mathbf{B}_0$, and $\mathbf{B}_1$ for the implementation in (12.4.8) with $L = 3$.

c. Find and compare the eigenvalues of $\mathbf{A}$, $\tilde{\mathbf{A}}$, and $-(\mathbf{A}_0^{-1}\mathbf{A}_1)$.

d. Compare the number of multiplications per output sample in (a) and (b).

12.8 Estimate the number of multiplications per output sample for a third-order filter using the block implementation in (12.4.8) with $L = 32$. Compare this with canonical implementation. Why might the block implementation still be advantageous for a filter with a sampling rate of 8 MHz?

13. FILTER AND SYSTEM EXAMPLES

13.0 INTRODUCTION

The variety of filters and systems that can be designed and implemented using the techniques presented in this book is virtually limitless. There has already been widespread practical application of these systems in many different fields, and their application will continue to increase as the speed of digital circuits increases and the size and cost decreases. Certain basic systems or subsystems recur often in different applications, and their design is thus of general interest. Four examples of these are presented in this chapter.

13.1 INTERPOLATION AND DECIMATION

In various applications, it is necessary or desirable to change the effective sampling rate (ω_s) of a discrete-time signal. Such systems are called *multirate* systems. Clearly, this change could always be made by converting the signal from digital to analog form and resampling the analog signal at the desired new rate. This approach, however, is expensive and introduces many potential sources of noise and distortion. Instead, it is possible and usually preferable to perform the rate conversion entirely by means of digital signal processing. The process of increasing the sampling rate by an integer factor L is called *interpolation*; reduction of the sampling rate by an integer factor K is called *decimation*. These two operations may be combined to change the sampling rate by any rational factor L/K [71, 72].

In numerical analysis, interpolation of arbitrary functions is often done by fitting polynomials to some number of discrete values. However, in digital signal processing, the

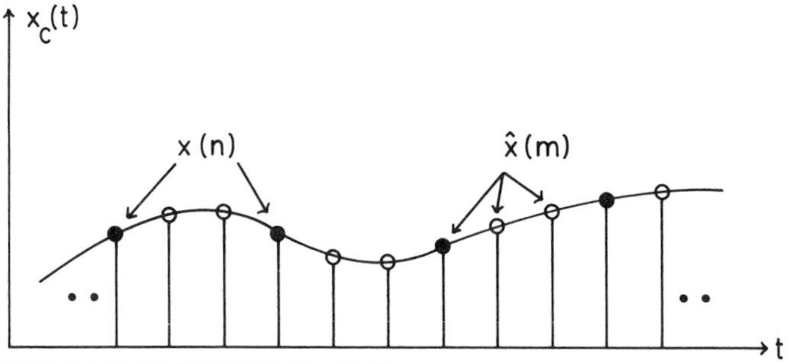

Figure 13.1. Time-domain illustration of interpolation for $L = 3$.

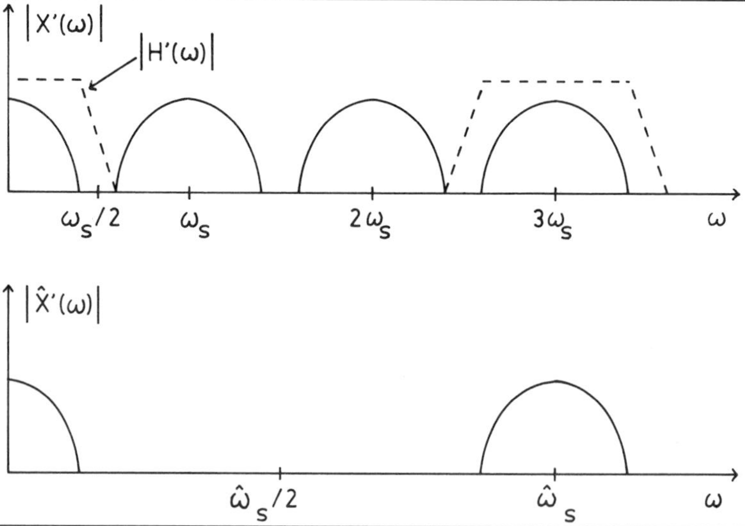

Figure 13.2. Frequency-domain illustration of interpolation for $L = 3$.

sampled signals must be bandlimited, and this additional constraint on the signal spectrum implies that lowpass filtering is the appropriate method of interpolation for these signals. The operation of interpolation is illustrated in the time domain in figure 13.1. The input sequence $x(n)$ with sampling rate ω_s is depicted by solid dots; the desired output sequence $\hat{x}(m)$ with sampling rate $\omega_s = L\omega_s$ (for $L = 3$) is indicated by both the dots and circles. The underlying continuous-time signal $x_c(t)$ is assumed to be bandlimited to frequencies less than $\omega_s/2$.

The operations required to implement interpolation become clear in the frequency domain, illustrated in figure 13.2, where the transforms $X'(\omega)$ and $\hat{X}'(\omega)$ are both depicted. To get from $X'(\omega)$ to $\hat{X}'(\omega)$, we must think of and process $X'(\omega)$ as if the sampling rate were $\hat{\omega}_s = 3\omega_s$, not ω_s. This requires the insertion of two zero samples between each pair of samples $x(n)$ and $x(n - 1)$, which clearly does not change the spectrum $X'(\omega)$. Then, as depicted, this modified signal can be lowpass filtered by $H'(\omega)$, operating at the new rate $\hat{\omega}_s$, to produce the interpolated signal $\hat{x}(m)$ with spectrum $\hat{X}'(\omega)$. Note that if the bandwidth of $x_c(t)$ is actually less than $\omega_s/2$, an ideal ("brickwall") lowpass filter is not

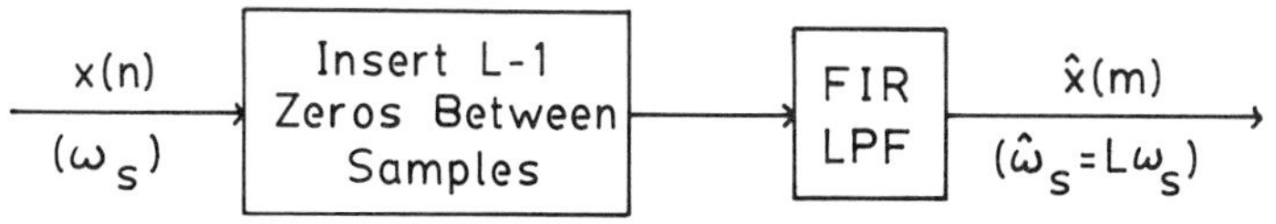

Figure 13.3. Interpolation system.

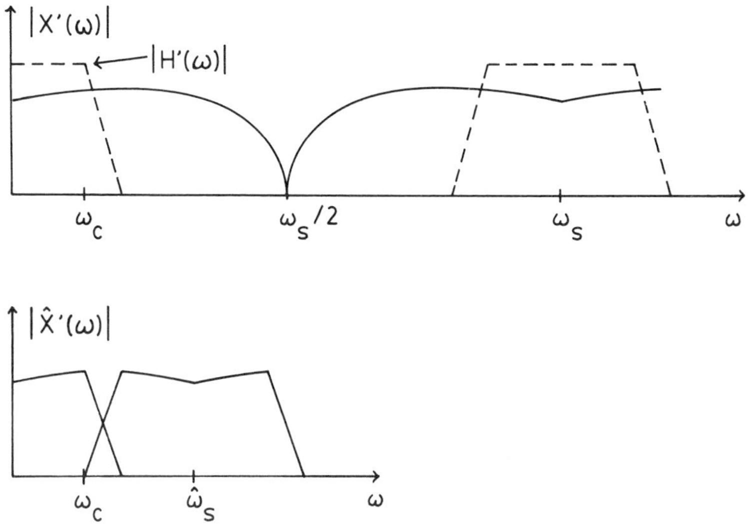

Figure 13.4. Frequency-domain illustration of decimation for $K = 3$.

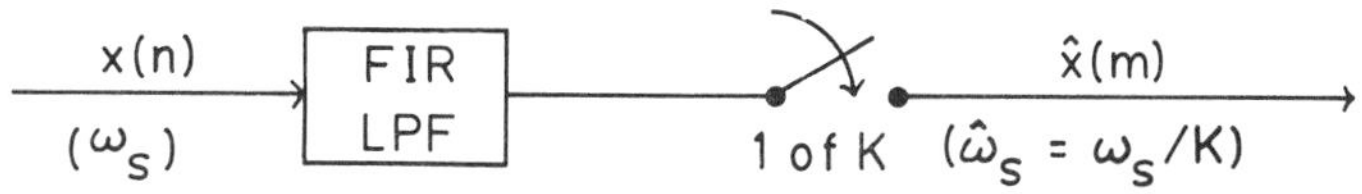

Figure 13.5. Decimation system.

required, even for ideal interpolation. The interpolation system is shown in figure 13.3. The reasons for choosing the FIR lowpass filter are discussed later.

Decimation implies a reduction of the sampling rate from ω_s to $\hat{\omega}_s = \omega_s/K$, and hence the signal bandwidth must be reduced accordingly by a lowpass filter. This is illustrated in figure 13.4 (for $K = 3$). The lowpass cutoff (ω_c) is somewhat less than $\omega_s/2K$ so that the aliased components below ω_c can be made as small as desired. The specification of this filter is completely analogous to that of the prefilter that precedes A/D conversion (described in section 6.3). After filtering, one of every K samples is retained, and the remaining $K - 1$ samples are discarded, to produce $\hat{x}(m)$, as depicted in figure 13.5.

A sampling rate alteration by a noninteger, but rational, factor L/K can clearly be accomplished by cascading the operations of interpolation and decimation in that order (to

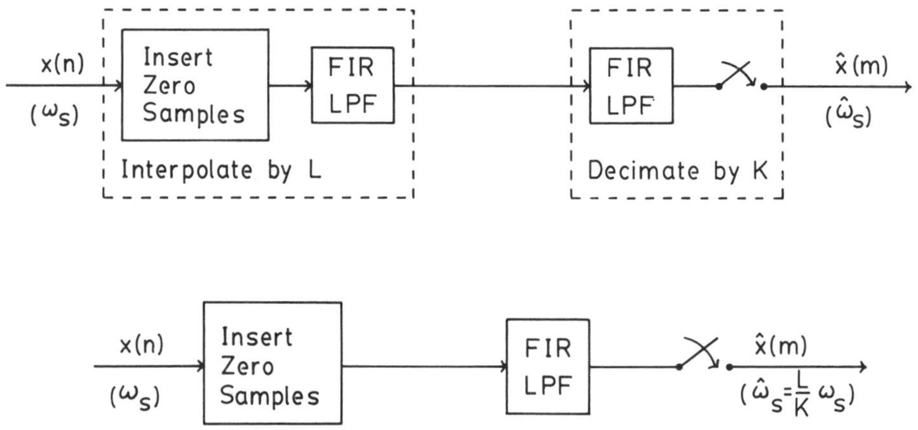

Figure 13.6. Combined interpolation/decimation for noninteger sampling-rate change, showing elimination of redundant FIR LPF.

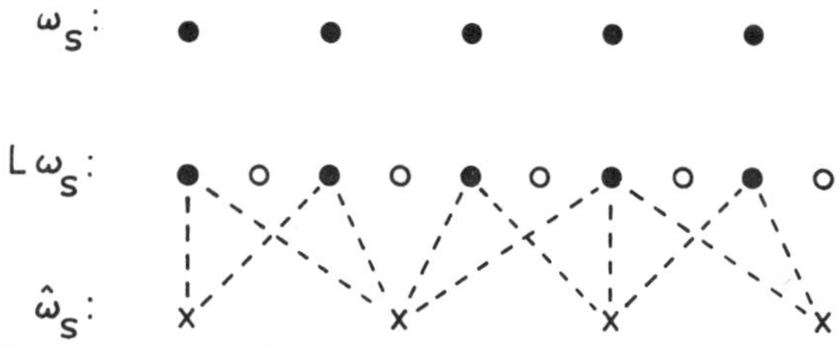

Figure 13.7. Time-domain illustration of sampling-rate change by factor $L/K = 2/3$ with sixth-order FIR LPF.

maintain the maximum possible bandwidth). This is shown in Figure 13.6, where it is noted that one of the lowpass filters (the one having the wider bandwidth) is redundant and can be eliminated.

It is usually advantageous to choose the lowpass filter to be a linear-phase FIR filter in direct form for the following reasons:

1. Linear phase implies no phase distortion.
2. In interpolation, $L - 1$ of the delayed input samples equal zero and thus do not require multiplication by filter coefficients.
3. In decimation, $K - 1$ of the output samples are to be discarded and thus need not be computed at all if the filter is nonrecursive.
4. The even symmetry of $h(n)$ for a linear-phase FIR filter implies a saving in multiplication of a factor of almost one-half.
5. For the common case of halfband filters, every other value of $h(n)$ equals zero, implying an additional saving of a factor of almost one-half in multiplications.

For a noninteger rate change by L/K both 2 and 3 above apply, but the bookkeeping is somewhat complicated. For example, figure 13.7 illustrates a rate change by 2/3 with an

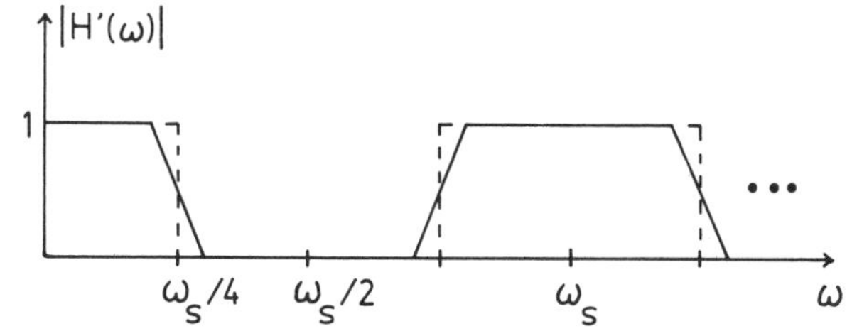

Figure 13.8. Magnitude response of halfband filter.

FIR lowpass filter of order $M = 6$. The input data are indicated by solid dots, and the zeros inserted between these samples are shown as circles. The output samples are indicated by x's, and coefficient multiplications by dotted lines. Note that the computation of each output sample requires only 3 or 4 multiplications, not $M + 1 = 7$, because of the zero samples. Hence, the multiplication rate is only $(M + 1)\hat{\omega}_s/L = 3.5\hat{\omega}_s$ or $(M + 1)\omega_s/K = 2.33\omega_s$. The even symmetry of $h(n)$ implies an additional reduction here by a factor of 4/7 in the multiplication rate to only $2\hat{\omega}_s$.

For a sampling-rate change by a factor of $L = 2$ or $K = 1/2$, the appropriate lowpass filter is the *halfband filter* depicted in figure 13.8. Note that the periodic frequency response of this filter is an approximation to a square wave in the frequency domain; and hence the corresponding (noncausal) Fourier series $h(n)$ equals zero for n even (except $n = 0$) and falls off approximately as $1/n$ for n odd. If the filter is designed to preserve the odd symmetry of the passband and stopband (with the passband ripple $\pm\delta_1$ equal to the stopband $\pm\delta_2$, etc.), the coefficients $h(n)$ for n even will remain zero, and the multiplication rate is reduced by a factor of almost one-half. The odd symmetry of $H'(\omega)$ about $\omega_s/4$ can be preserved using any of the FIR design techniques in chapter 9.

For very narrowband filters ($L \gg 1$ or $K \gg 1$), additional savings are possible by effecting the rate change in stages, not all at once. This is depicted in figure 13.9 for interpolation by a factor of $L = 4$. Instead of implementing the top filter $H'(\omega)$ to change the rate to $\hat{\hat{\omega}}_s = 4\omega_s$, we first implement the filter $H_1'(\omega)$ to double the rate to $\hat{\omega}_s = 2\omega_s$. This rate is then doubled again to $\hat{\hat{\omega}}_s = 2\hat{\omega}_s$ using the filter $H_2'(\omega)$. The advantage of this scheme is that the specifications on $H_1'(\omega)$ and $H_2'(\omega)$ are much less severe than on the original filter $H'(\omega)$ because their normalized transition bandwidths are wider, especially for $H_2'(\omega)$ which must operate at the higher rate $\hat{\hat{\omega}}_s$. Therefore, the combined orders of $H_1(z)$ and $H_2(z)$ will be much less than the order of $H(z)$—consequently, the overall multiplication rate will be much less. In this case, $H_1(z)$ and $H_2(z)$ are both halfband filters, as well, providing an additional saving of almost one-half in the multiplication rate.

The decimation case is similar, as illustrated in figure 13.10 for $K = 4$. The first filter $H_1'(\omega)$ reduces the bandwidth sufficiently to allow a rate reduction to $\hat{\omega}_s = \omega_s/2$. Then $H_2'(\omega)$, operating at $\hat{\omega}_s$, reduces the bandwidth further for a rate reduction to $\hat{\hat{\omega}}_s = \hat{\omega}_s/2$. The resulting output spectrum $Y'(\omega)$ is shown for the case of a white input spectrum. Again, both $H_1'(\omega)$ and $H_2'(\omega)$ are halfband filters.

This multiple-stage scheme can be useful for implementing narrowband filters even when the input and output sampling rates are the same. In that case, we decimate first by

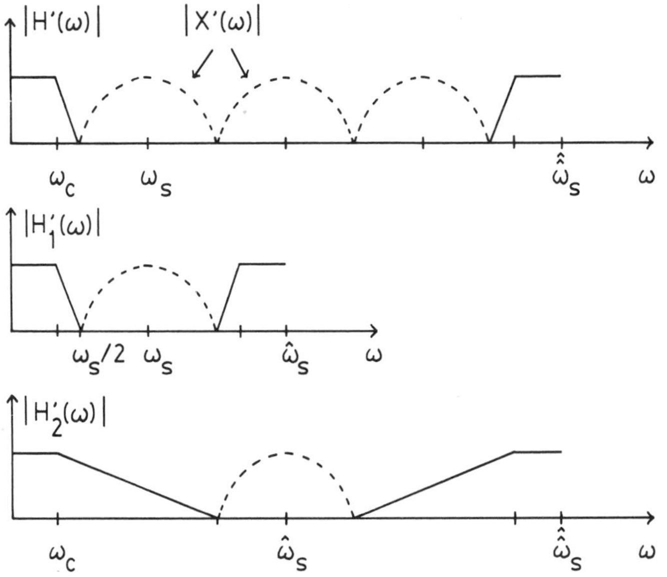

Figure 13.9. Two-stage implementation of interpolation by $L = 4$.

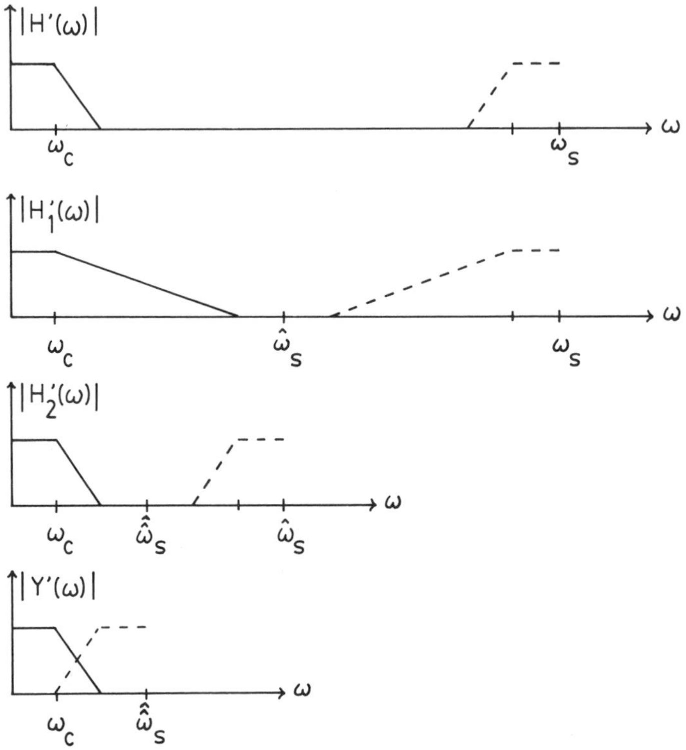

Figure 13.10. Two-stage implementation of decimation by $K = 4$.

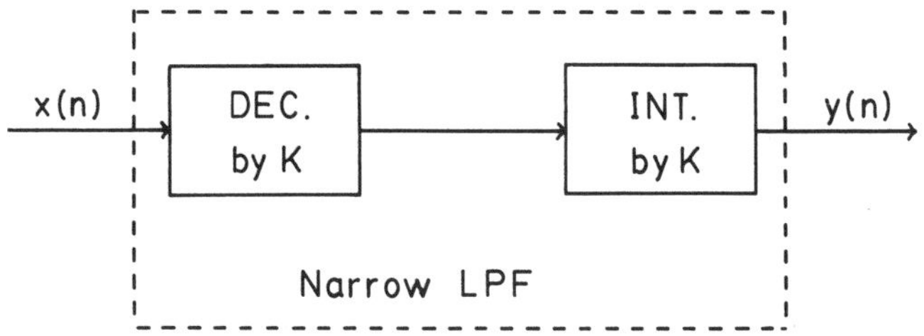

Figure 13.11. Implementation of narrowband LPF by multistage decimation and interpolation.

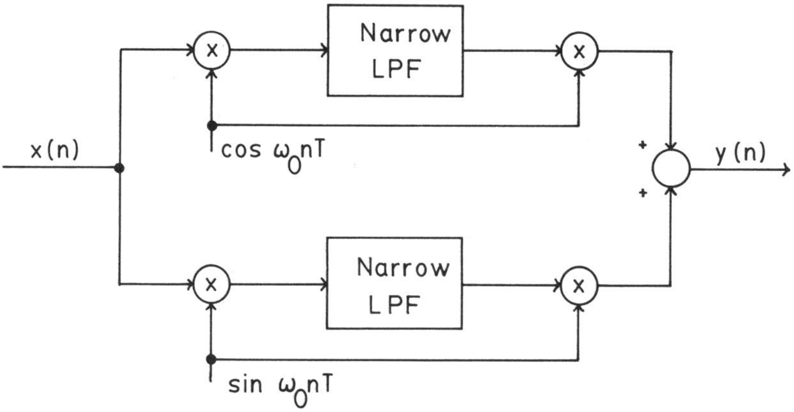

Figure 13.12. Implementation of narrowband BPF via two LPF's with frequency translation.

K and then interpolate by K, as depicted in figure 13.11. If K is a highly composite number so that the decimation and interpolation can be performed in many stages, the overall multiplication rate will be much less than that for the single filter $H(z)$. This scheme can also be used to implement a narrow bandpass filter centered at ω_0 by modulating the input signal by $e^{-j\omega_0 nT}$ to translate the center of the band of interest from ω_0 to dc. The real and imaginary parts of the complex-valued signal are then lowpass filtered in multiple stages using decimation and interpolation to reduce the computation rate. Finally, the signal is translated back to ω_0 (or to any other desired center frequency ω_0') by modulating by $e^{j\omega_0 nT}$, and the real part of this complex-valued signal is the desired narrowband output. The real-valued operations required to implement this quadrature (complex-valued) system are shown in figure 13.12.

13.2 HILBERT TRANSFORMATION

There are many applications in communications where it is necessary to delay all the (real) frequency components comprising a signal by 90°. One common application is the generation of the complex-valued *analytic signal* $\tilde{x}(n)$ having only the positive-frequency components of a real-valued signal $x(n)$. Letting $\hat{x}(n)$ be the *Hilbert transform* of $x(n)$ with all frequency components delayed by 90°, the analytic signal is given by

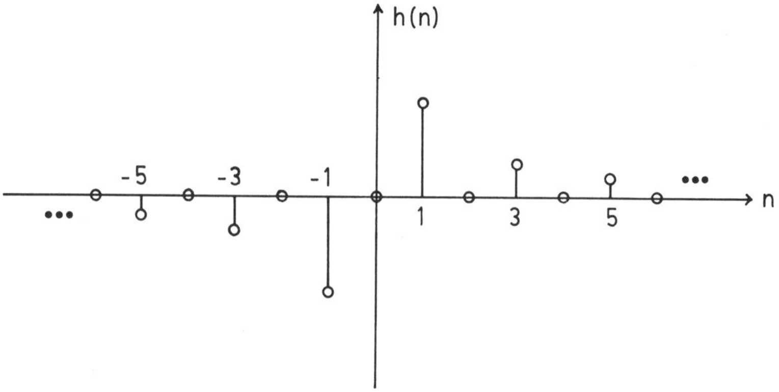

Figure 13.13. Ideal impulse response of Hilbert transformer.

$$\tilde{x}(n) = x(n) + j\hat{x}(n). \tag{13.2.1}$$

To see this, consider $x(n) = \cos\omega_0 nT$ with Hilbert transform $\hat{x}(n) = \sin\omega_0 nT$. Then,

$$\tilde{x}(n) = \cos\omega_0 nT + j\sin\omega_0 nT = e^{j\omega_0 nT}, \tag{13.2.2}$$

which indeed is only the positive-frequency component of $x(n)$. Since this holds for any component of $\tilde{x}(n)$, it holds for $\tilde{x}(n)$, in general, to the extent that we can realize the Hilbert transform $\hat{x}(n)$.

The ideal Hilbert transformation corresponds to a filter with frequency response

$$\begin{aligned} H'(\omega) &= \begin{cases} e^{-j\pi/2}, & 0 < \omega < \pi/T \\ e^{j\pi/2}, & -\pi/T < \omega < 0 \end{cases} \\ &= \begin{cases} -j, & 0 < \omega < \pi/T \\ +j, & -\pi/T < \omega < 0. \end{cases} \end{aligned} \tag{13.2.3}$$

As noted in section 5.3, the 90°-phase portion of this response can be realized exactly by a noncausal FIR filter with odd symmetry in its impulse response $h(n)$. The only approximation occurs in the magnitude response, which ideally equals unity except at $\omega = 0$ and π/T. That is, letting $H'(\omega) = jR'(\omega)$, we design the coefficients $h(n)$ to approximate

$$R'(\omega) = \begin{cases} -1, & 0 < \omega < \pi/T \\ +1, & -\pi/T < \omega < 0. \end{cases} \tag{13.2.4}$$

Note that, like the halfband filter in the preceding section, $R'(\omega)$ is a square wave in the frequency domain. In this case, however, the square wave is odd, and the corresponding Fourier series $r(n)$ is thus purely imaginary, equaling zero for n even and falling off as $1/n$.

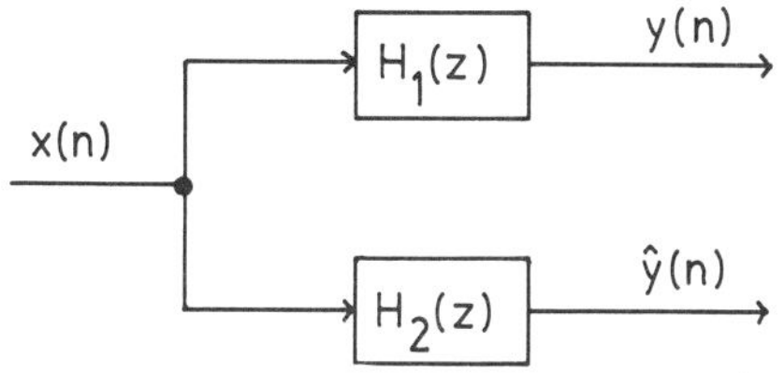

Figure 13.14. IIR phase-splitting network.

The resulting odd-symmetry impulse response $h(n) = jr(n)$ is thus real and is illustrated in figure 13.13.

Any of the design techniques in chapter 9 can be used to obtain the actual FIR design. An interesting and useful relationship exists between the Hilbert transform filter and the halfband lowpass filter by which, given an FIR halfband design $h_B(n)$, one can readily transform it into a Hilbert transform design $\hat{h}(n)$, or vice versa [73]. Without deriving it here, we note that the halfband and Hilbert designs are related by

$$h_B(n) = 0.5[\delta(n) + \hat{h}(n)(-1)^{(n-1)/2}]. \tag{13.2.5}$$

The passband and stopband ripple of the halfband filter is thus one-half the (passband) ripple of the corresponding Hilbert transformer.

It is also possible to approximate an analytical signal using IIR filters. In this case, however, we do not attempt to produce the Hilbert transform $\hat{x}(n)$; rather, two allpass filters $H_1(z)$ and $H_2(z)$ are employed to generate a Hilbert transform pair $y(n)$ and $\hat{y}(n)$, as depicted in figure 13.14. A great deal of work has been done on the design of such analog *phase-splitting* networks. This work is directly applicable to the digital case, as well, via the bilinear transformation [74]. Note that, in contrast to FIR Hilbert transformers, the desired unity-magnitude response is achieved exactly by the allpass filters, but the 90°-phase shift can only be approximated. A clever and efficient multirate scheme has also been proposed by Rader [75].

13.3 DIGITAL OSCILLATORS AND SYNTHESIZERS

There are many applications where a digital oscillator or frequency synthesizer is required to generate a discrete-time sinusoid of programmable frequency ω_0. Two basic approaches to this design problem are the recursive digital oscillator and the ROM table-lookup synthesizer. These units are briefly described in this section. The digital oscillator provides a simple example of some of the quantization and hardware issues discussed in chapters 11 and 12.

A second-order recursive digital filter with poles on the unit circle is "marginally stable"; and, with nonzero initial conditions, it ideally produces a sinusoidal output. The frequency ω_0 of the sinusoid is determined by the angle $\omega_0 T$ of the unit-circle poles. For certain applications, such as the narrow bandpass filter described in section 13.1, we require a quadrature oscillator with two sinusoidal outputs of the same frequency and amplitude, but with phases differing by 90°. A single second-order recursive structure can produce both outputs.

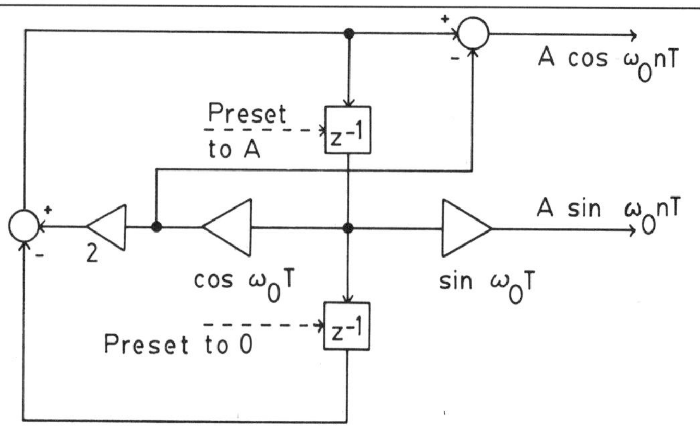

Figure 13.15. Recursive quadrature oscillator.

Consider two causal impulse responses of the form

$$\begin{aligned} h_1(n) &= A\cos\omega_0 nTu(n) \\ h_2(n) &= A\sin\omega_0 nTu(n), \end{aligned} \tag{13.3.1}$$

where $u(n)$ is the unit step. The corresponding system functions (without the gain A) are

$$\begin{aligned} H_1(z) &= \frac{1 - (\cos\omega_0 T)z^{-1}}{1 - 2(\cos\omega_0 T)z^{-1} + z^{-2}} \\ H_2(z) &= \frac{(\sin\omega_0 T)z^{-1}}{1 - 2(\cos\omega_0 T)z^{-1} + z^{-2}}. \end{aligned} \tag{13.3.2}$$

A two-output recursive structure having these system functions is shown in figure 13.15. Instead of an impulse input $A\delta(n)$ to start the oscillation, the z^{-1} registers can be present at $n = 1$ to initial conditions of A and 0, respectively, resulting in the desired quadrature sinusoidal outputs $h_1(n)$ and $h_2(n)$ for $n > 1$.

Implementation of this oscillator with finite wordlengths produces several quantization effects. Quantization of the coefficient $(\cos\omega_0 T)$ causes the actual output frequency to differ from the ideal frequency ω_0 by a small amount; whereas quantization of the coefficient $(\sin\omega_0 T)$ causes the amplitude of $h_2(n)$ to differ slightly from A. Finally, rounding of the output products from the $(\cos\omega_0 T)$ multiplier produces roundoff noise in the recursive loop, which cause the amplitude and phase of the output signals to wander slowly over time. This last effect is potentially the most severe because it is cumulative, and it necessitates periodic resetting of the initial conditions in the loop to reinitialize the proper amplitude and phase.

Assume, for example, that we are implementing the modulators and demodulators in a frequency-division-multiplexed (FDM) telephony system with 60 single-sideband voice channels separated by 4 kHz from 40 kHz to 280 kHz. Therefore, the frequencies to be generated are all multiples of 4 kHz and thus have an integer number of cycles in 0.25 msec.

Let the sampling rate be 640 kHz. Hence, the initial states (A and 0) in the z^{-1} registers should repeat at least every 0.25 msec, or 160 samples. Since these states will have slight errors due to rounding, we reset them again to A and 0, respectively, to keep the errors from accumulating further. That is, we reset the registers when $n = 1, 161, 321, \ldots$ (or $n = 0, 160, 320, \ldots$, or any other convenient sequence). The amplitude variation over 160 samples is negligible, and the frequency is now synchronized to a multiple of 4 kHz, eliminating that as a concern. This leaves only the phase errors over 160 samples due to quantization of the coefficient ($\cos \omega_0 T$) to consider.

Frequencies from 40 to 280 kHz are to be generated. At 40 kHz, the maximum possible frequency error (before resetting) due to rounding of the ($\cos \omega_0 T$) coefficient is ± 2 Hz if 16-bit coefficients are chosen, and thus the maximum phase error after 160 samples (10 cycles) is $\pm 0.18°$. At 280 kHz, the maximum possible frequency error is only ± 0.004 Hz, which is clearly negligible. For 9-bit amplitude accuracy (plus sign), we need about 9-bit phase accuracy in each 90° interval, i.e., $90°/512 \approx 0.18°$. Therefore, 16-bit coefficient accuracy is just right (actually conservative) at 40 kHz.

Having chosen 16-bit coefficients ($b + 1 = 16$), we do likewise for the data ($k + 1 = 16$). For 60-dB SNR at the outputs, we need only about 10 bits per data sample (including sign). However, at 40 kHz, the signal in the loop has an amplitude of about 2.1 A, and hence we need more than one bit of "headroom" above the output amplitude A to accommodate this loop signal without overflow. The 4 to 5 additional bits ensure that the cumulative rounding errors are negligible over 160 samples. Note that saturation arithmetic is not required because of the resetting.

As discussed in section 12.1, the fact that $b = k$ means that each multiplier represents a delay of $2(k + 1)$ bits in a bit-serial implementation. Hence, this recursive oscillator cannot be realized without multiplexing it to implement at least two channels. The required clock rate in such an implementation is thus $2 \times 16 \times 640$ kHz, or 20.48 MHz.

The second approach to accurate sinusoidal synthesis is the *digital frequency synthesizer* [76, 77]. This approach is simple in concept and often in implementation. The values of $\sin \theta$ are stored in a read-only memory (ROM), and the phase θ is computed as a linearly increasing function of time. Let the basic frequency increment be ω_1, i.e., the generated frequency is $\omega_k = k\omega_1$ for integer *frequency number k*. The output sinusoid is then

$$
\begin{aligned}
y(n) &= \sin (n\omega_k T + \Delta) \\
&= \sin (nk\omega_1 T + \Delta) \\
&= \sin (m\phi_1 + \Delta), \qquad (13.3.3)
\end{aligned}
$$

where $m = (nk)$ is the *phase number*, $\phi_1 = \omega_1 T$ is the basic phase increment, and Δ is a possible phase offset. The phase number m is generated by adding the *frequency number k* repeatedly to the contents of an accumulator at the clock rate $f_s = 1/T$, as depicted in figure 13.16. The accumulator operates modulo-K, where $K = \omega_s/\omega_1$, and thus there are K accumulator states. For example, for the quadrature oscillator examined earlier, the specifications of $\omega_1 = 2\pi(4\text{ kHz})$ and $f_s = 640$ kHz imply that $K = 160$. Frequencies up to the Nyquist frequency $\omega_{K/2} = \omega_s/2$ can be generated without aliasing.

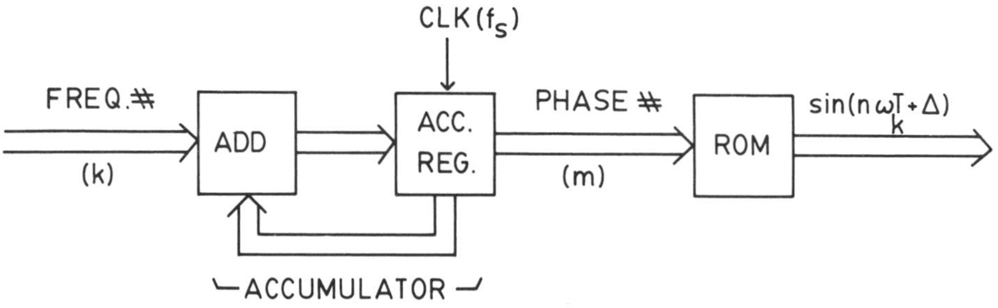

Figure 13.16. Digital frequency synthesizer using ROM table lookup.

The values sin $(m\phi_1 + \Delta)$ are stored in the ROM and are simply read out as a function of the address m. Note that if m contains more bits than the phase (amplitude) accuracy required, the least-significant bits can be deleted from the address (but not from the accumulator itself). If a full 360° is stored in the ROM, no other circuitry is required. However, the quadrant symmetry of the sinusoid indicates that only one quadrant (0–90°) need be stored, and the other quadrants can be generated from that one. This implies more circuitry to modify the address as a function of the two most-significant bits from the accumulator. With slight additional modification, the synthesizer can also be multiplexed to generate quadrature (sine and cosine) outputs. It is to simplify this quadrant-modification circuitry that a nonzero phase offset Δ may be useful.

13.4 SPEECH SYNTHESIS

A problem of widespread interest and application in digital signal processing is the synthesis of speech signals from discrete-time models of the human vocal tract [78, 79]. Major application areas include voice output from computers and instrumentation, and data reduction for digital voice communications. Speaking toys have even been produced using this technology. The heart of the speech synthesizer is a (slowly) time-varying digital filter which models the response of the mouth, throat, etc. to excitation from the vocal cords and turbulance. In this section, we give a brief overview of digital speech synthesis.

Speech sounds can be divided into two general classes: *voiced* and *unvoiced*. Voiced speech is produced as the response of the vocal tract to periodic excitation from the vocal cords. Therefore, voiced speech includes the majority of speech sounds or *phonemes*, such as the vowels, nasals (/m/, /n/, and /ng/), and semivowels (/l/, /w/, /y/, and /r/). Voiced transients include the stops /b/, /d/, and /g/. Unvoiced sounds, on the other hand, are created as the result of turbulence produced by a constriction somewhere along the vocal tract. These include the *fricatives* /s/, /sh/, /f/, and /θ/ and the unvoiced stops /p/, /t/, and /k/.

Figure 13.17 shows a short segment of voiced speech. Note that this segment is essentially periodic and that the signal in each period decays like the impulse response of a linear filter. A model that suggests itself is that voiced speech is the response of a linear filter to an impulse-train input. The period (P_0) of voiced speech for male speakers ranges from about 5 to 20 msec., and the fundamental frequency or *pitch* (f_0) thus varies from 50 to 200 Hz. The pitch of female speakers is about twice that of males. The zero-padded DFT

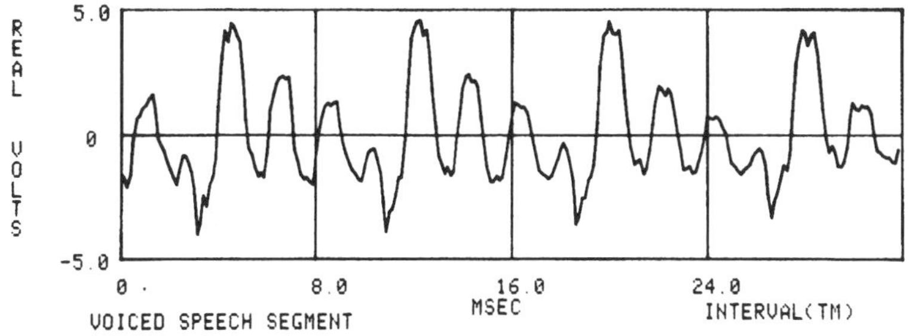

Figure 13.17. 32-msec segment of voiced speech.

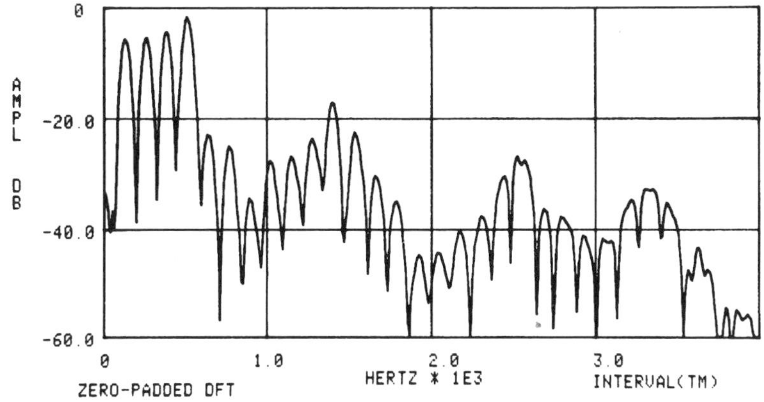

Figure 13.18. Spectrum (DFT) of voiced-speech segment.

of this voiced speech segment is shown in figure 13.18. Note the pitch harmonics at multiples of f_0 and the envelope of these pitch harmonics. By our model of voiced speech, this envelope should be the magnitude response of the vocal tract filter.

To extract the filter model $H(z)$, we apply an appropriate modeling method from chapter 10. Acoustical analysis of the vocal tract indicates that, except for the nasals, an all-pole or autoregressive (AR) model is appropriate. AR modeling is also the most straightforward computationally, as we saw in chapter 10. Hence, for many applications, an all-pole filter model is chosen although some degradation of nasal sounds does result. In other applications, a few nasal zeros are added to improve the speech quality. If the modeling is done *pitch synchronously*, i.e., from single pitch periods, the covariance method in (10.5.14) is the appropriate technique. Otherwise, either the covariance or autocorrelation method can be utilized. (The tradeoff is that the covariance method implies no windowing; while the autocorrelation method guarantees a stable filter model.)

The magnitude response of a 12th-order AR model of the data in figure 13.17, obtained by the covariance method, is shown in figure 13.19, along with the DFT from figure 13.18. Note the resonant peaks or *formants* in the response, designated by $F1$, $F2$, $F3$, and $F4$. The formants are key characterizing parameters of the speech model since they correspond to pole locations in $H(z)$. Their center frequencies vary greatly (but slowly) with time, having

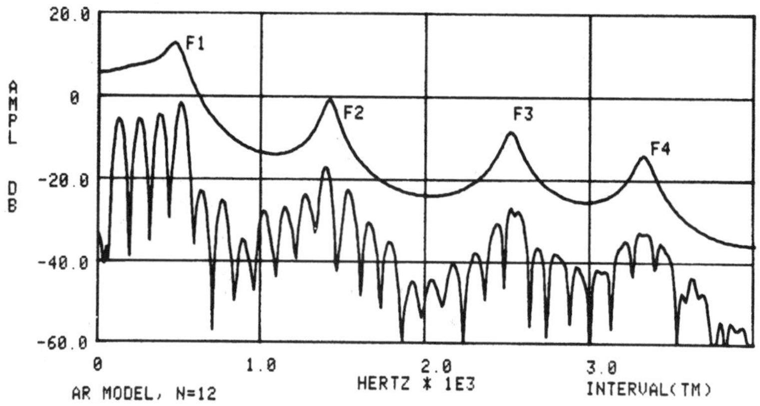

Figure 13.19. 12th-order AR model of voiced-speech segment showing formants, with DFT.

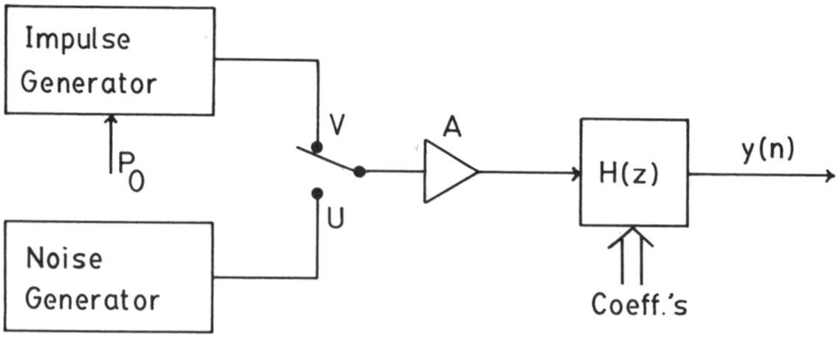

Figure 13.20. Basic digital speech synthesizer.

nominal values of 500, 1500, 2500, and 3500 Hz, respectively. The formant bandwidths also vary with time, but these variations are less dramatic and less significant to the intelligibility and quality of the speech. The first three formants (6 poles) are sufficient characterizing parameters for intelligibility, and the fourth adds significantly to the quality. A few additional poles are needed to model the shape of the glottal pulses from the vocal cords and the radiation from the lips. For many applications, therefore, a 10th- to 12th-order AR model suffices.

A basic digital speech synthesizer is shown in figure 13.20. The pulse generator produces an impulse train with time-varying period P_0 to simulate the periodic excitation from the vocal cords for voiced speech. The noise generator, on the other hand, produces white noise to simulate the unvoiced excitation from turbulence. A uniform probability density is adequate for the noise since later filtering will tend to make it Gaussian by the Central Limit Theorem. The selection of the appropriate excitation source is made via the voiced/unvoiced (V/U) switch, and the amplitude is controlled by the time-varying gain A. The filter $H(z)$ models the vocal tract response to the excitation, including the glottal and radiation characteristics, and thus provides the desired spectral shape of the speech signal. The filter order does not need to be as high for unvoiced speech as for voiced speech; but it is there anyway, so it is used.

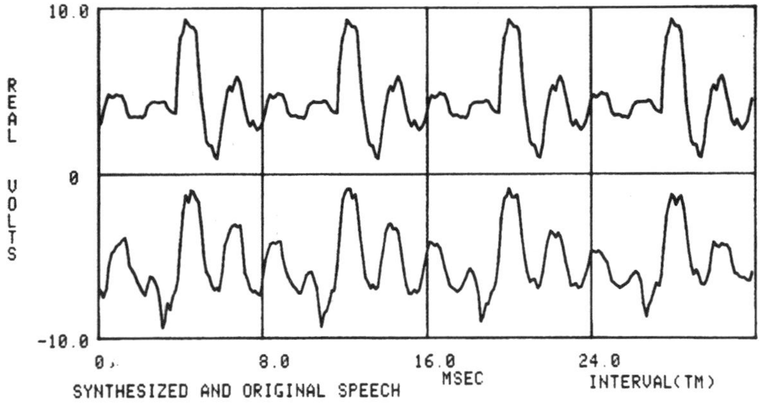

Figure 13.21. Synthesized speech segment (top) from 12th-order AR model, and original segment (bottom).

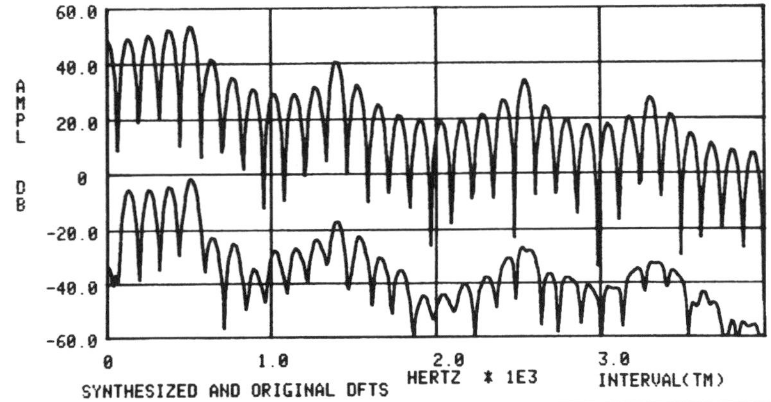

Figure 13.22. DFT's of synthesized (top) and original (bottom) speech segments in figure 13.21.

It has been determined experimentally that the parameters of the speech synthesizer need to be updated only every 10–20 msec, which is comparable to the pitch period. For voiced speech, the updating should be done "pitch synchronously" at the beginning of the pitch period so that transients from the changes in the filter coefficients are masked by the large (desired) transient from the impulse input.

The structure of the digital filter is usually chosen to be the recursive lattice or the cascade form—the lattice is the most common because of its low coefficient sensitivity and its direct relationship to the Levinson-Durbin algorithm. Typical quantizations of the synthesizer parameters are as follows [78]: 6 bits for the pitch period, 1 V/U bit, 5 bits (distributed logarithmically) for the gain A, and 5–6 bits for each reflection coefficient (or transformation thereof), for a total of about 72 bits per frame. At typical frame rates of 100, 67, and 33 per second, this implies bit rates of 7200, 4800, and 2400 bps, respectively.

As an example of these techniques, the voiced speech signal in figure 13.17 was resynthesized using the 12th-order model in figure 13.19 and a pitch period of 7.9 msec. The resulting synthetic speech signal is shown in figure 13.21, and its DFT in figure 13.22, along with the original speech signal and its DFT for comparison. The difference in the time-domain signals is due to the fact that the synthesized speech is minimum phase; while the original speech is not. The ear is not sensitive to such phase changes.

PROBLEMS

13.1 We wish to interpolate a signal $x(n)$ by a factor of $L = 4$, attenuating spectral images by 40 dB. The bandwidth of the signal is one-third the original sampling rate ($\omega_c = \omega_s/3$).

a. If the FIR lowpass filter is designed by frequency sampling, what order (M) is required?

b. If the filter is designed by Kaiser windowing, what order is required?

c. If the interpolation is done in two stages of $L = 2$ each, what order FIR filters are required in (a) and (b)?

13.2 We wish to reduce the sampling rate for a discrete-time signal $x(n)$ to 2/3 the original rate ω_s. The band of interest is the lower half of the original Nyquist band (i.e., $\omega_c = \omega_s/4$). Sketch the appropriate system, and design the FIR lowpass filter by frequency sampling for an aliasing suppression of 40 dB.

13.3 Sketch the impulse response $h_B(n)$ for an FIR halfband filter designed by Hamming windowing. Sketch the corresponding magnitude response $|H_B'(\omega)|$.

13.4 Sketch the impulse response $\hat{h}(n)$ for an FIR Hilbert transformer designed by Hamming windowing. Sketch the corresponding magnitude response $|\hat{H}'(\omega)|$.

13.5 Show an implementation of the recursive quadrature oscillator incorporating the recursive lattice structure. Briefly discuss the effects of coefficient quantization on the output.

13.6 Show an implementation of the recursive quadrature oscillator incorporating the normal-form structure. (An extra delay in both $h_1(n)$ and $h_2(n)$ is acceptable.) Briefly discuss the effects of coefficient quantization on the output.

13.7 The shape of the glottal pulses from the vocal cords can be approximated by the impulse response of a second-order filter with a double pole at $z = a$.

a. Sketch the glottal-pulse model $h(n)$.

b. Sketch the corresponding magnitude response $|H'(\omega)|$ in dB versus $\log \omega$ for a bandwidth of 250 Hz.

c. The effect of lip radiation can be modeled by a single zero at $z = 1$. Repeat (b) with the inclusion of this extra zero.

d. Draw a filter implementation of the combined glottal-pulse and lip-radiation models.

References

1. L. R. Rabiner and C. M. Rader, eds., *Digital Signal Processing*, New York: IEEE Press, 1972.
2. A. V. Oppenheim and R. W. Schafer, *Digital Signal Processing*, Englewood Cliffs, NJ: Prentice-Hall, 1975.
3. L. B. Jackson, J. F. Kaiser, and H. S. McDonald, "An Approach to the Implementation of Digital Filters," *IEEE Trans. Audio Electro.*, vol. AU-16, Sept. 1968, pp. 413–421; also in [1].
4. L. B. Jackson, "Digital Phase Equalizer," U.S. Patent 3,537,015, Oct. 1970.
5. B. Gold and K. L. Jordan, "A Note on Digital Filter Synthesis," *Proc. IEEE*, vol. 56, Oct. 1968, pp. 1717–1718; also in [1].
6. F. C. Marshall and G. C. Temes, "Binary Windows for the Discrete Fourier Transform," *Proc. IEEE*, vol. 63, Sept. 1975, pp. 1370–1371.
7. F. J. Harris, "On the Use of Windows for Harmonic Analysis with the Discrete Fourier Transform," *Proc. IEEE*, vol. 66, Jan. 1978, pp. 51–83.
8. P. D. Welsh, "The Use of Fast Fourier Transform for the Estimation of Power Spectra: A Method Based on Time Averaging over Short, Modified Periodograms," *IEEE Trans. Audio Electro.*, vol. AU-15, June 1967, pp. 70–73; also in [1].
9. A. I. Zverev, *Handbook of Filter Synthesis*, New York: Wiley, 1967.
10. A. G. Constantinides, "Spectral Transformations for Digital Filters," *Proc. Inst. Elec. Eng.*, vol. 117, Aug. 1970, pp. 1585–1590; also in [1].
11. J. F. Kaiser, "Nonrecursive Digital Filter Design Using the I_0-sinh Window Function," *Proc. Int. Symp. Cir. Systems*, April 1974, pp. 20–23; also in [12].
12. Digital Signal Processing Committee, ed., *Digital Signal Processing II*, New York: IEEE Press, 1975.
13. L. R. Rabiner, B. Gold, and C. A. McGonegal, "An Approach to the Approximation Problem for Nonrecursive Digital Filters," *IEEE Trans. Audio Electro.*, vol. AU-18, June 1970, pp. 83–106; also in [1].
14. J. H. McClellan, T. W. Parks, and L. R. Rabiner, "A Computer Program for Designing Optimum FIR Linear Phase Digital Filters," *IEEE Trans. Audio Electro.*, vol. AU-21, Dec. 1973, pp. 506–526; also in [12].
15. ———, "FIR Linear Phase Filter Design Program," in [16].
16. Digital Signal Processing Committee, *Programs for Digital Signal Processing*, New York: IEEE Press, 1979.
17. T. W. Parks and J. H. McClellan, "Chebyshev Approximation for Nonrecursive Digital Filters with Linear Phase," *IEEE Trans. Circuit Theory*, vol. CT-19, March 1972, pp. 189–194.
18. H. S. Hersey, J. T. Lewis, and D. W. Tufts, "Interactive Minimax Design of Linear-Phase Nonrecursive

Digital Filters Subject to Upper and Lower Function Constraints," *IEEE Trans. Audio Electro.*, vol. AU-20, June 1972, pp. 171–173.

19. B. Noble and J. W. Daniel, *Applied Linear Algebra*, Englewood Cliffs, NJ: Prentice-Hall, 1977 and 1969.
20. A. Papoulis, *Probability, Random Variables, and Stochastic Processes*, New York: McGraw-Hill, 1984.
21. J. Makhoul, "Linear Prediction: A Tutorial Review," *Proc. IEEE*, vol. 63, April 1975, pp. 561–580; also in [12].
22. J. P. Burg, "Maximum Entropy Spectral Analysis," Ph.D. Thesis, Stanford Univ., 1977.
23. K. Steiglitz, "On the Simultaneous Estimation of Poles and Zeros in Speech Analysis," *IEEE Trans. Acous. Speech Sig. Proc.*, vol. ASSP-25, June 1977, pp. 229–234.
24. K. Steiglitz and L. E. McBride, "A Technique for the Identification of Linear Systems," *IEEE Trans. Auto. Control*, vol. AC-10, Oct. 1965, pp. 461–464.
25. J. Durbin, "Efficient Estimation of Parameters in Moving-Average Models," *Biometrika*, vol. 46, 1959, pp. 306–316.
26. N. H. Judell, "Maximum Likelihood Parameter Estimation for Signals with Rational Spectra," M.S. Thesis, Univ. R.I., 1983.
27. C. S. Burrus and T. W. Parks, "Time Domain Design of Recursive Digital Filters," *IEEE Trans. Audio Electro.*, vol. AU-18, June 1970, pp. 137–141; also in [1].
28. A. A. Beex and L. L. Scharf, "Recursive Digital Filter Design via Covariance Sequence Approximation," *IEEE Trans. Acous. Speech Sig. Proc.*, vol. ASSP-29, Feb. 1981, pp. 51–57.
29. J. A. Cadzow, "High Performance Spectral Estimation—A New ARMA Method," *IEEE Trans. Acous. Speech Sig. Proc.*, vol. ASSP-28, Oct. 1980, pp. 524–529.
30. C. Gueguen and M. Mathieu, "Contribution des Zéros à la Modélisation du Signal de Parole," GALF, Nancy, France, vol. 2, 1976, pp. 101–108.
31. L. B. Jackson, "Simple, Effective MA and ARMA Techniques," *Proc. IEEE Int. Conf. Acous. Speech Sig. Proc.*, April 1983, pp. 1426–1429.
32. A. G. Evans and R. Fischl, "Optimal Least Squares Time-Domain Synthesis of Recursive Digital Filters," *IEEE Trans. Audio Electro.*, vol. AU-21, Feb. 1973, pp. 61–65.
33. R. Kumaresan, L. L. Scharf, and A. K. Shaw, "An Algorithm for Pole-Zero Modeling and Spectrum Analysis," submitted for publication, 1985.
34. A. H. Gray, Jr. and J. D. Markel, "Digital Lattice and Ladder Filter Synthesis," *IEEE Trans. Audio Electro.*, vol. AU-21, Dec. 1973, pp. 491–500; also in [12].
35. S. M. Kay and S. L. Marple, Jr., "Spectrum Analysis—A Modern Perspective," *Proc. IEEE*, vol. 69, Nov. 1981, pp. 1380–1419.
36. D. G. Childers, ed., *Modern Spectrum Analysis*, New York: IEEE Press, 1978.
37. S. L. Marple, Jr., personal communication.
38. R. Kumaresan, B. P. Agrawal, and M. R. Sridhar, "An Optimum Whitening Filter for Spectrum Analysis," *Proc. ASSP Spectrum Est. Workshop II*, Nov. 1983, pp. 245–246.
39. L. B. Jackson, "On the Interaction of Roundoff Noise and Dynamic Range in Digital Filters," *Bell Sys. Tech. J.*, vol. 49, Feb. 1970, pp. 159–184; also in [1].
40. ———, "Roundoff-Noise Analysis for Fixed-Point Digital Filters Realized in Cascade or Parallel Form," *IEEE Trans. Audio Electro.*, vol. AU-18, June 1970, pp. 107–122; also in [1].
41. ———, "Roundoff Noise Bounds Derived from Coefficient Sensitivities in Digital Filters," *IEEE Trans. Cir. Systems*, vol. CAS-23, Aug. 1976, pp. 481–485.
42. ———, "An Analysis of Limit Cycles due to Multiplication Rounding in Recursive Digital (Sub)Filters," *Proc. 7th Allerton Conf. Cir. Sys. Th.*, 1969, pp. 69–78; also in [1].
43. V. B. Lawrence and K. V. Mina, "A New and Interesting Class of Limit Cycles in Recursive Digital Filters," *Proc. IEEE Int. Symp. Cir. Systems*, April 1977, pp. 191–194.
44. J. F. Kaiser and I. W. Sandberg, "A Bound on Limit Cycles in Fixed-Point Implementation of Digital Filters," *IEEE Trans. Audio Electro.*, vol. AU-20, June 1972, pp. 110–112; also in [12].
45. J. L. Long and T. N. Trick, "An Absolute Bound on Limit Cycles Due to Roundoff Errors in Digital Filters," *IEEE Trans. Audio Electro.*, vol. AU-21, Feb. 1973, pp. 27–30; also in [12].
46. T. A. C. M. Claasen, W. F. G. Mecklenbrauker, and J. B. H. Peek, "Second-Order Digital Filter with Only One Magnitude-Truncation Quantiser and Having Practically No Limit Cycles," *Elec. Letters*, vol. 9, Nov. 1973; also in [12].
47. ———, "Some Remarks on the Classification of Limit Cycles in Digital Filters," *Philips Res. Rep.*, vol. 28, Aug. 1973, pp. 297–305; also in [12].
48. ———, "Frequency Domain Criteria for the Absence of Zero-Input Limit Cycles in Nonlinear Discrete-Time Systems, with Application to Digital Filters," *IEEE Trans. Cir. Systems*, vol. CAS-22, Mar. 1975, pp. 232–239; also in [12].

49. P. M. Ebert, J. E. Mazo, and M. G. Taylor, "Overflow Oscillations in Digital Filters," *Bell Sys. Tech. J.*, vol. 48, Nov. 1969, pp. 2999–3020; also in [1].
50. D. Mitra, "Summary of Some Results on Large Amplitude, Self-Sustaining Oscillations in High Order Digital Filter Sections Using Saturation Arithmetic," *Proc. IEEE Int. Symp. Cir. Systems*, April 1977, pp. 195–198.
51. C. T. Mullis and R. A. Roberts, "Synthesis of Minimum Roundoff Noise Fixed Point Digital Filters," *IEEE Trans. Cir. Systems*, vol. CAS-23, Sept. 1976, pp. 551–562.
52. S. Y. Hwang, "Roundoff Noise in State-Space Digital Filtering: A General Analysis," *IEEE Trans. Acous. Speech Sig. Proc.*, vol. ASSP-24, June 1976, pp. 256–262.
53. L. B. Jackson, A. G. Lindgren, and Y. Kim, "Optimal Synthesis of Second-Order State-Space Structures for Digital Filters," *IEEE Trans. Cir. Systems*, vol. CAS-26, Mar. 1979, pp. 149–153.
54. C. W. Barnes and A. T. Fam, "Minimum Norm Recursive Digital Filters that Are Free of Overflow Limit Cycles," *IEEE Trans. Cir. Systems*, vol. CAS-24, Oct. 1977, pp. 569–574.
55. W. L. Mills, C. T. Mullis, and R. A. Roberts, "Digital Filter Realizations without Overflow Oscillation," *IEEE Trans. Acous. Speech Sig. Proc.*, vol. ASSP-26, Aug. 1978, pp. 334–338.
56. L. B. Jackson, "Limit Cycles in State-Space Structures for Digital Filters," *IEEE Trans. Cir. Systems*, vol. CAS-26, Jan. 1979, pp. 67–68.
57. R. F. Lyons, "A Bit-Serial VLSI Architecture Methodology for Signal Processing," in *VLSI 81: Very Large Scale Integration*, J. P. Gray, ed., London: Academic, 1981.
58. R. F. Lyons, "Two's Complement Pipelined Multiplier," *IEEE Trans. Comm.*, vol. COM-24, April 1976, pp. 418–425.
59. L. B. Jackson and H. S. McDonald, "Digital Filters" (Multiplexing), U.S. Patent 3,522,546, Aug. 1970.
60. P. R. Cappello and K. Steiglitz, "Completely-Pipelined Architectures for Digital Signal Processing," *IEEE Trans. Acous. Speech Sig. Proc.*, vol. ASSP-31, Aug. 1983, pp. 1015–1023.
61. H. T. Kung, "Why Systolic Architectures?" Dept. Comp. Sci., Carnegie-Mellon Univ., CMU-CS-81-148, Nov. 1981.
62. A. Peled and B. Liu, "A New Hardware Realization of Digital Filters," *IEEE Trans. Acous. Speech Sig. Proc.*, vol. ASSP-22, Dec. 1974, pp. 456–462; also in [12].
63. A. Crosisier, D. J. Esteban, M. E. Levilion, and V. Rizo, "Digital Filter for PCM Encoded Signals," U.S. Patent 3,777,130, Dec. 1973.
64. T. G. Stockham, Jr., "High-Speed Convolution and Correlation," *AFIPS Conf. Proc.*, vol. 28, 1966, pp. 229–233; also in [1].
65. C. S. Burrus, "Block Realization of Digital Filters," *IEEE Trans. Audio Electro.*, vol. AU-20, Oct. 1972, pp. 230–235.
66. S. K. Mitra and R. Gnanesekaran, "Block Implementation of Recursive Digital Filters—New Structures and Properties," *IEEE Trans. Cir. Systems*, vol. CAS-25, April 1978, pp. 200–207.
67. C. W. Barnes and S. Shinnaka, "Finite Word Effects in Block-State Realizations of Fixed Point Digital Filters," *IEEE Trans. Cir. Systems*, vol. CAS-27, May 1980, pp. 345–349.
68. J. Zeman and A. G. Lindgren, "Fast Digital Filters With Low Round-Off Noise," *IEEE Trans. Cir. Systems*, vol. CAS-28, July 1981, pp. 716–723.
69. R. A. Meyer and C. S. Burrus, "A Unified Analysis of Multirate and Periodically Time Varying Digital Filters," *IEEE Trans. Cir. Systems*, vol. CAS-22, Mar. 1975, pp. 162–168.
70. ———, "Design and Implementation of Multirate Digital Filters," *IEEE Trans. Acous. Speech Sig. Proc.*, vol. ASSP-24, Feb. 1976, pp. 55–58.
71. R. W. Schafer and L. R. Rabiner, "A Digital Signal Processing Approach to Interpolation," *Proc. IEEE*, vol. 61, June 1973, pp. 692–702: also in [12].
72. R. E. Crochiere and L. R. Rabiner, *Multi-Rate Digital Signal Processing*, Englewood Cliffs, NJ: Prentice-Hall, 1983.
73. L. B. Jackson, "On the Relationship between Digital Hilbert Transformers and Certain Low-Pass Filters," *IEEE Trans. Acous. Speech Sig. Proc.*, vol. ASSP-23, Aug. 1975, pp. 381–383.
74. B. Gold, A. V. Oppenheim, and C. M. Rader, "Theory and Implementation of the Discrete Hilbert Transform," *Proc. Symp. Comp. Proc. Comm.*, 1970, pp. 235–250; also in [1].
75. C. M. Rader, "A Simple Method for Sampling In-Phase and Quadrature Components," *IEEE Trans. Aero. Elec. Sys.*, vol. AES-20, Nov. 1984, pp. 821–824.
76. J. Tierney, C. M. Rader, and B. Gold, "A Digital Frequency Synthesizer," *IEEE Trans. Audio Electro.*, vol. AU-19, Mar. 1971, pp. 48–58; also in [1].
77. L. B. Jackson, "Digital Frequency Synthesizer," U.S. Patent 3,735,269, May 1973.
78. L. R. Rabiner and R. W. Schafer, *Digital Processing of Speech Signals*, Englewood Cliffs, NJ: Prentice-Hall, 1978.

79. J. D. Markel and A. H. Gray, Jr., *Linear Prediction of Speech*, New York: Springer-Verlag, 1976.
80. E. J. Hannan, "The Estimation of Mixed Moving Average Autoregressive Systems," *Biometrika*, vol. 56, 1969, pp. 579–593.

INDEX